FUNCTIONAL DAIRY INGREDIENTS AND NUTRACEUTICALS

Physicochemical, Technological, and Therapeutic Aspects

Innovations in Agricultural and Biological Engineering

FUNCTIONAL DAIRY INGREDIENTS AND NUTRACEUTICALS

Physicochemical, Technological, and Therapeutic Aspects

Edited by

Megh R. Goyal, PhD
N. Veena, PhD
Santosh K. Mishra, PhD

APPLE ACADEMIC PRESS

First edition published 2023

Apple Academic Press Inc.
1265 Goldenrod Circle, NE,
Palm Bay, FL 32905 USA

760 Laurentian Drive, Unit 19,
Burlington, ON L7N 0A4, CANADA

CRC Press
6000 Broken Sound Parkway NW,
Suite 300, Boca Raton, FL 33487-2742 USA

4 Park Square, Milton Park,
Abingdon, Oxon, OX14 4RN UK

Library and Archives Canada Cataloguing in Publication

Title: Functional dairy ingredients and nutraceuticals : physicochemical, technological, and therapeutic aspects / edited by Megh R. Goyal, PhD, N. Veena, PhD, Santosh K. Mishra, PhD.
Names: Goyal, Megh R., editor. | Veena, N. (Nagaraj) editor. | Mishra, Santosh K., editor.
Series: Innovations in agricultural and biological engineering.
Description: First edition. | Series statement: Innovations in agricultural and biological engineering | Includes bibliographical references and index.
Identifiers: Canadiana (print) 20220222630 | Canadiana (ebook) 20220222649 | ISBN 9781774639917 (hardcover) | ISBN 9781774639924 (softcover) | ISBN 9781003277309 (ebook)
Subjects: LCSH: Dairy products. | LCSH: Functional foods. | LCSH: Dietary supplements.
Classification: LCC SF250.5 .F86 2023 | DDC 637—dc23

Library of Congress Cataloging-in-Publication Data

CIP data on file with US Library of Congress

ISBN: 978-1-77463-991-7 (hbk)
ISBN: 978-1-77463-992-4 (pbk)
ISBN: 978-1-00327-730-9 (ebk)

ABOUT THE BOOK SERIES: INNOVATIONS IN AGRICULTURAL AND BIOLOGICAL ENGINEERING

Under this book series, Apple Academic Press Inc. is publishing book volumes over a span of 8–10 years in the specialty areas defined by the American Society of Agricultural and Biological Engineers (www.asabe.org). Apple Academic Press Inc. aims to be a principal source of books in agricultural and biological engineering. We welcome book proposals from readers in areas of their expertise.

The mission of this series is to provide knowledge and techniques for agricultural and biological engineers (ABEs). The book series offers high-quality reference and academic content on agricultural and biological engineering (ABE) that is accessible to academicians, researchers, scientists, university faculty and university-level students, and professionals around the world.

Agricultural and biological engineers ensure that the world has the necessities of life, including safe and plentiful food, clean air and water, renewable fuel and energy, safe working conditions, and a healthy environment by employing knowledge and expertise of the sciences, both pure and applied, and engineering principles. Biological engineering applies engineering practices to problems and opportunities presented by living things and the natural environment in agriculture.

ABE embraces a variety of the following specialty areas (www.asabe.org): aquaculture engineering, biological engineering, energy, farm machinery and power engineering, food, and process engineering, forest engineering, information, and electrical technologies, soil, and water conservation engineering, natural resources engineering, nursery, and greenhouse engineering, safety, and health, and structures and environment.

For this book series, we welcome chapters on the following specialty areas (but not limited to):

1. Academia to industry to end-user loop in agricultural engineering.
2. Agricultural mechanization.
3. Aquaculture engineering.
4. Biological engineering in agriculture.

5. Biotechnology applications in agricultural engineering.
6. Energy source engineering.
7. Farm to fork technologies in agriculture.
8. Food and bioprocess engineering.
9. Forest engineering.
10. GPS and remote sensing potential in agricultural engineering.
11. Hill land agriculture.
12. Human factors in engineering.
13. Impact of global warming and climatic change on agriculture economy.
14. Information and electrical technologies.
15. Irrigation and drainage engineering.
16. Micro-irrigation engineering.
17. Milk Engineering.
18. Nanotechnology applications in agricultural engineering.
19. Natural resources engineering.
20. Nursery and greenhouse engineering.
21. Potential of phytochemicals from agricultural and wild plants for human health.
22. Power systems and machinery design.
23. Robot engineering and drones in agriculture.
24. Rural electrification.
25. Sanitary engineering.
26. Simulation and computer modeling.
27. Smart engineering applications in agriculture.
28. Soil and water engineering.
29. Structures and environment engineering.
30. Waste management and recycling.
31. Any other focus areas.

Books in the Innovations in Agricultural & Biological Engineering Series

- Advances in Food Process Engineering: Novel Processing, Preservation and Decontamination of Foods
- Advances in Green and Sustainable Nanomaterials: Applications in Energy, Biomedicine, Agriculture, and Environmental Science
- Biological and Chemical Hazards in Food and Food Products: Prevention, Practices, and Management

- Bioremediation and Phytoremediation Technologies in Sustainable Soil Management
  - Volume 1: Fundamental Aspects and Contaminated Sites
  - Volume 2: Microbial Approaches and Recent Trends
- The Chemistry of Milk and Milk Products: Physicochemical Properties, Therapeutic Characteristics, and Processing Methods
- Dairy Engineering: Advanced Technologies and Their Applications
- Developing Technologies in Food Science: Status, Applications, and Challenges
- Emerging Technologies in Agricultural Engineering
- Engineering Interventions in Agricultural Processing
- Engineering Interventions in Foods and Plants
- Engineering Practices for Agricultural Production and Water Conservation: An Interdisciplinary Approach
- Engineering Practices for Management of Soil Salinity: Agricultural, Physiological, and Adaptive Approaches
- Engineering Practices for Milk Products: Dairyceuticals, Novel Technologies, and Quality
- Enzyme Inactivation in Food Processing: Technologies, Materials, and Applications
- Field Practices for Wastewater Use in Agriculture: Future Trends and Use of Biological Systems
- Flood Assessment: Modeling and Parameterization
- Food Engineering: Emerging Issues, Modeling, and Applications
- Food Process Engineering: Emerging Trends in Research and Their Applications
- Food Processing and Preservation Technology: Advances, Methods, and Applications
- Food Technology: Applied Research and Production Techniques
- Functional Dairy Ingredients and Nutraceuticals: Physicochemical, Technological, and Therapeutic Aspects
- Handbook of Research on Food Processing and Preservation Technologies:
  - Volume 1: Nonthermal and Innovative Food Processing Methods
  - Volume 2: Nonthermal Food Preservation and Novel Processing Strategies
  - Volume 3: Computer-Aided Food Processing and Quality Evaluation Techniques

  - Volume 4: Design and Development of Specific Foods, Packaging Systems, and Food Safety
  - Volume 5: Emerging Techniques for Food Processing, Quality, and Safety Assurance
- Modeling Methods and Practices in Soil and Water Engineering
- Nanotechnology Horizons in Food Process Engineering, 3-volume set:
  - Volume 1: Food Preservation, Food Packaging and Sustainable Agriculture
  - Volume 2: Scope, Biomaterials, and Human Health
  - Volume 3: Trends, Nanomaterials, and Food Delivery
- Nanotechnology and Nanomaterial Applications in Food, Health, and Biomedical Sciences
- Nanotechnology Applications in Agricultural and Bioprocess Engineering: Farm to Table
- Nanotechnology Applications in Dairy Science: Packaging, Processing, and Preservation
- Novel and Alternate Methods in Food Processing: Biotechnological, Physicochemical, and Mathematical Approaches
- Novel Dairy Processing Technologies: Techniques, Management, and Energy Conservation
- Novel Processing Methods for Plant-Based Health Foods: Extraction, Encapsulation and Health Benefits of Bioactive Compounds
- Novel Strategies to Improve Shelf-Life and Quality of Foods: Quality, Safety, and Health Aspects
- Processing of Fruits and Vegetables: From Farm to Fork
- Processing Technologies for Milk and Milk Products: Methods, Applications, and Energy Usage
- Quality Control in Fruit and Vegetable Processing: Methods and Strategies
- Scientific and Technical Terms in Bioengineering and Biological Engineering
- Soil and Water Engineering: Principles and Applications of Modeling
- Soil Salinity Management in Agriculture: Technological Advances and Applications
- State-of-the-Art Technologies in Food Science: Human Health, Emerging Issues and Specialty Topics

- Sustainable Biological Systems for Agriculture: Emerging Issues in Nanotechnology, Biofertilizers, Wastewater, and Farm Machines
- Sustainable Nanomaterials for Biomedical Engineering: Impacts, Challenges, and Future Prospects
- Sustainable Nanomaterials for Biosystems Engineering: Trends in Renewable Energy, Environment, and Agriculture
- Technological Interventions in Dairy Science: Innovative Approaches in Processing, Preservation, and Analysis of Milk Products
- Technological Interventions in Management of Irrigated Agriculture
- Technological Interventions in the Processing of Fruits and Vegetables
- Technological Processes for Marine Foods, from Water to Fork: Bioactive Compounds, Industrial Applications, and Genomics

OTHER BOOKS ON AGRICULTURAL AND BIOLOGICAL ENGINEERING FROM APPLE ACADEMIC PRESS, INC.

Management of Drip/Trickle or Micro Irrigation
Megh R. Goyal, PhD, PE, Senior Editor-in-Chief

Evapotranspiration: Principles and Applications for Water Management
Megh R. Goyal, PhD, PE, and Eric W. Harmsen, Editors

Book Series: Research Advances in Sustainable Micro Irrigation
Senior Editor-in-Chief: Megh R. Goyal, PhD, PE

Volume 1: Sustainable Micro Irrigation: Principles and Practices
Volume 2: Sustainable Practices in Surface and Subsurface Micro Irrigation
Volume 3: Sustainable Micro Irrigation Management for Trees and Vines
Volume 4: Management, Performance, and Applications of Micro Irrigation Systems
Volume 5: Applications of Furrow and Micro Irrigation in Arid and Semi-Arid Regions
Volume 6: Best Management Practices for Drip Irrigated Crops
Volume 7: Closed Circuit Micro Irrigation Design: Theory and Applications
Volume 8: Wastewater Management for Irrigation: Principles and Practices
Volume 9: Water and Fertigation Management in Micro Irrigation
Volume 10: Innovation in Micro Irrigation Technology

Book Series: Innovations and Challenges in Micro Irrigation
Senior Editor-in-Chief: Megh R. Goyal, PhD, PE

Volume 1: Management of Drip/Trickle or Micro Irrigation
Volume 2: Sustainable Micro Irrigation Design Systems for Agricultural Crops
Volume 3: Principles and Management of Clogging in Micro Irrigation
Volume 4: Performance Evaluation of Micro Irrigation Management

Volume 5: Potential Use of Solar Energy and Emerging Technologies in Micro Irrigation
Volume 6: Micro Irrigation Management: Technological Advances and Their Applications
Volume 7: Micro Irrigation Engineering for Horticultural Crops
Volume 8: Micro Irrigation Scheduling and Practices
Volume 9: Engineering Interventions in Sustainable Trickle Irrigation
Volume 10: Management Strategies for Water Use Efficiency and Micro Irrigated Crops
Volume 11: Fertigation Technologies in Micro Irrigation

ABOUT THE SENIOR-EDITOR-IN-CHIEF

Megh R. Goyal, PhD, PE
Retired Professor in Agricultural and Biomedical Engineering, University of Puerto Rico, Mayaguez Campus; Senior Acquisitions Editor, Biomedical Engineering and Agricultural Science, Apple Academic Press, Inc..

Megh R. Goyal, PhD, PE, is, currently a retired professor of agricultural and biomedical engineering from the General Engineering Department at the College of Engineering at the University of Puerto Rico–Mayaguez Campus (UPRM); and Senior Acquisitions Editor and Senior Technical Editor-in-Chief for Agricultural and Biomedical Engineering for Apple Academic Press Inc.

During his long career, he has worked as a Soil Conservation Inspector; Research Assistant at Haryana Agricultural University and Ohio State University; Research Agricultural Engineer/Professor at the Department of Agricultural Engineering of UPRM; and Professor of Agricultural and Biomedical Engineering in the General Engineering Department of UPRM. He spent a one-year sabbatical leave in 2002–2003 at the Biomedical Engineering Department of Florida International University, Miami, USA.

Dr. Goyal was the first agricultural engineer to receive the professional license in agricultural engineering from the College of Engineers and Surveyors of Puerto Rico. In 2005, he was proclaimed the "Father of Irrigation Engineering in Puerto Rico for the Twentieth Century" by the American Society of Agricultural and Biological Engineers, Puerto Rico Section, for his pioneering work on micro irrigation, evapotranspiration, agroclimatology, and soil and water engineering.

During his professional career of 52 years, he has received many awards, including Scientist of the Year, Membership Grand Prize for the American Society of Agricultural Engineers Campaign, Felix Castro Rodriguez Academic Excellence Award, Man of Drip Irrigation by the Mayor of Municipalities of Mayaguez/Caguas/Ponce and Senate/Secretary of Agriculture of ELA, Puerto Rico, and many others. He has been recognized as one of the experts "who rendered meritorious service for the development of [the] irrigation sector in India" by the Water Technology Centre of Tamil Nadu Agricultural University in Coimbatore, India, and ASABE who bestowed on him the 2018 Netafim Microirrigation Award. VDGOOD Professional Association of India awarded Lifetime Achievement Award at 12th Annual Meeting on Engineering, Science and Medicine that was held on 20-21 of November of 2020 in Visakhapatnam, India.

Dr. Goyal has authored more than 200 journal articles and edited more than 100 books. AAP has published many of his books, including *Management of Drip/Trickle or Micro Irrigation; Evapotranspiration: Principles and Applications for Water Management;* ten-volume set on *Research Advances in Sustainable Micro Irrigation*. Dr. Goyal has also developed several book series with AAP, including Innovations in Agricultural & Biological Engineering, Innovations and Challenges in Micro Irrigation; and Innovations in Plant Science for Better Health: From Soil to Fork.

Dr. Goyal received his BSc degree in Engineering from Punjab Agricultural University, Ludhiana, India, and his MSc and PhD degrees from the Ohio State University, Columbus, Ohio, USA. He also earned a Master of Divinity degree from the Puerto Rico Evangelical Seminary, Hato Rey, Puerto Rico, USA.

Readers may contact him at goyalmegh@gmail.com.

ABOUT THE EDITORS

N. Veena, PhD
Assistant Professor, Department of Dairy Chemistry, College of Dairy Science and Technology, Guru Angad Dev Veterinary and Animal Sciences University, Ludhiana, Punjab, India

N. Veena, PhD, is an Assistant Professor in the Department of Dairy Chemistry, College of Dairy Science and Technology, Guru Angad Dev Veterinary and Animal Sciences University, Ludhiana, Punjab, India. She received her BTech degree in Dairy Technology from Karnataka Veterinary, Animal, and Fisheries Sciences University, Bidar, Karnataka, India, and her MTech and PhD degrees in Dairy Chemistry from ICAR-National Dairy Research Institute, Karnal, Haryana, India. Her area of expertise includes fortification of milk, development, and characterization of functional dairy foods, and nanoencapsulation of nutraceuticals and their delivery. She is the recipient of junior research and institutional fellowships during her master's and doctoral programs from the ICAR-National Dairy Research Institute, Karnal, Haryana, India. She has been involved in different institutional and externally funded research projects such as Milkfed (Punjab), ICAR, DST, and RKVY as PI and Co-PI. She has published many research papers and review articles in various national and international peer-reviewed journals. Also published several book chapters, popular articles, chapters in the compendium, and manuals. She also developed the e-learning content for post-graduate dairy chemistry courses under the NAHEP project. She has presented her research work at various national and international seminars and conferences and received several awards for best oral and poster presentations.

Readers may contact her at: veena.ndri@gmail.com.

Santosh Kumar Mishra, PhD
Assistant Professor, Department of Dairy Microbiology, College of Dairy Science and Technology, Guru Angad Dev Veterinary and Animal Sciences University, Ludhiana, Punjab, India

Santosh Kumar Mishra, PhD, is working as an Assistant Professor in the Department of Dairy Microbiology, College of Dairy Science and Technology, Guru Angad Dev Veterinary and Animal Sciences University, Ludhiana, Punjab, India. He received his BTech degree in Dairy Technology from Maharashtra Animal and Fisheries Sciences University, Nagpur, India; his MSc and PhD degrees from National Dairy Research Institute, Karnal, Haryana. Now he is working presently in the area of functional foods and dairy products incorporating live probiotics, the technology of functional lactic cultures for fermented and non-fermented dairy products. He has also served the dairy industry as Quality Assurance Executive in Mother Dairy, New Delhi. In India, he has handled externally funded projects granted by DST, MoFPI, and UGC as PI or Co-PI. He received several awards for best papers and posters presentations. He is the recipient of Junior and Senior Research fellowships during his Master and Doctoral programs in the National Dairy Research Institute, Karnal, Haryana. He was a university gold medalist during his graduation program and also got the best thesis award by his university after his PhD research. Recently he has got an Award of Honor in the International Conference sponsored by Partap College of Education, Ludhiana, in association with International Professionals Development Association, UK. He is a member of various scientific societies: Life member of SASNET-Fermented Foods, Anand; member of Indian Dairy Associations, New Delhi. He has published several research, review, and popular articles in national and international journals. Also published five international edited books, several book chapters, and teaching reviews in various training programs. He has recently completed the young scientist project by DST, SEED Department, Govt. of India, New Delhi on Isolation and characterization of novel oxalate degrading lactic acid bacteria for potential probiotic management of kidney stone.

Readers may contact him at: skmishra84@gmail.com.

CONTENTS

CONTRIBUTORS

Kunal K. Ahuja
Assistant Professor, College of Dairy Science, Kamdhenu University, Gandhinagar – 365601, Gujarat, India, Mobile: +91-8140210100, E-mail:kunalkumarahuja@gmail.com

Sumit Arora
Principal Scientist, Dairy Chemistry Division, ICAR-National Dairy Research Institute, Karnal – 132001, Haryana, India, Mobile: +91-9896054444, E-mail: sumitak123@gmail.com

Venus Bansal
Assistant professor, College of Dairy Science and Technology, Guru Angad Dev Veterinary and Animal Sciences University, Ludhiana – 141004, Punjab, India, Mobile: +91-9478476400, E-mail: venus3b3@gmail.com

Surendra N. Battula
Principal Scientist, ICAR-National Dairy Research Institute, Southern Regional Station, Bengaluru – 560030, Karnataka, India, Mobile: +91-9449028628, E-mail: bsn_ndri@yahoo.com

Sudheendra V. K. Chaganti
Assistant Professor, College of Dairy Science, Kamdhenu University, Gandhinagar – 365601, Gujarat India, Mobile: +91-8238162268, E-mail: sudheendracvk@yahoo.co.in

Archana Chandran
Assistant Professor, Department of Dairy Microbiology, College of Dairy Science and Technology, Pookode, Wayanad – 673576, Kerala, India, Mobile: +91-9744975460, E-mail: archanac@kvasu.ac.in

Anindita Debnath
Assistant Professor, Faculty of Dairy Technology, West Bengal University of Animal and Fishery Sciences, Mohanpur, Nadia – 741252, West Bengal, India, Mobile: +91-9163384369, E-mail: aninditadebnath2009@gmail.com

Kamal Gandhi
Scientist, Dairy Chemistry Division, ICAR-National Dairy Research Institute, Karnal – 132001, Haryana, India, Mobile: +91-9729134444, E-mail: kamalgandhindri@gmail.com

Priyae B. Gautam
PhD Research Scholar, Dairy Chemistry Division, ICAR-National Dairy Research Institute, Karnal – 132001, Haryana, India, Mobile: +91-7982738721, E-mail: priyae.gautam92@gmail.com

Apurba Giri
Assistant Professor and Head, Department of Nutrition, Mugberia Gangadhar Mahavidyalaya, Bhupatinagar, Purba Medinipur – 721425, West Bengal, India, Mobile: +91-9564289290, E-mail: apurbandri@gmail.com

Megh R. Goyal
Retired Faculty in Agricultural and Biomedical Engineering from College of Engineering at University of Puerto Rico-Mayaguez Campus; and Senior Technical Editor-in-Chief in Agricultural and Biomedical Engineering for Apple Academic Press Inc.; PO Box 86, Rincon-PR – 00677-0086, USA, Mobile: 787-536-0039, E-mail: goyalmegh@gmail.com

Ashvin S. Hariyani
Assistant Professor, College of Dairy Science, Kamdhenu University, Gandhinagar – 365601, Gujarat, India, Mobile: +91-7016255048, E-mail: ashvin.hariyani@gmail.com

Tanmay Hazra
Assistant Professor, College of Dairy Science, Kamdhenu University, Gandhinagar – 365601, Gujarat, India, Mobile: +91-9541053968, E-mail: tanmayhazra08@gmail.com

Payal Karmakar
PhD Research Scholar, Dairy Chemistry Division, ICAR-National Dairy Research Institute, Karnal – 132001, Haryana, India, Mobile: +91-9051093929, E-mail: payal.106.karmakar@gmail.com

Ronit Mandal
PhD Research Scholar, Faculty of Land and Food Systems, 2205 East Mall, University of British Columbia, Vancouver, BC, V6T 1Z4, Canada, Mobile: +1-2368639270, E-mail: ronit.mandal@ubc.ca

Bimlesh Mann
Principal Scientist, Dairy Chemistry Division, ICAR-National Dairy Research Institute, Karnal – 132001, Haryana, India, Mobile: +91-9896245110, E-mail: bimleshmann@gmail.com

Santosh K. Mishra
Assistant Professor, Department of Dairy Microbiology, College of Dairy Science and Technology, Guru Angad Dev Veterinary and Animal Sciences University, Ludhiana – 141004, Punjab, India, Mobile: +91-9464995049, E-mail: skmishra84@gmail.com

Sonia Mor
PhD Research Scholar, ICAR-National Dairy Research Institute, Southern Regional Station, Bengaluru – 560030, Karnataka, India, Mobile: +91-8053160162, E-mail: soniamor286@gmail.com

Brunda S. Murthy
PhD Research Scholar, ICAR-National Dairy Research Institute, Southern Regional Station, Bengaluru – 560030, Karnataka, India, Mobile: +91-8722418042, E-mail: brunda4592@gmail.com

Laxmana N. Naik
Scientist, ICAR-National Dairy Research Institute, Southern Regional Station, Bengaluru – 560030, Karnataka, India, Mobile: +91-9996123653, E-mail: laxmandcnaik@gmail.com

Srushty Patil
PhD Research Scholar, ICAR-National Dairy Research Institute, Karnal – 132001, Haryana, India, Mobile: +91-9896494415, E-mail: srushtypatil5@gmail.com

Subhash Prasad
Assistant Professor, College of Dairy Science, Kamdhenu University, Gandhinagar – 365601, Gujarat, India, Mobile: +91-8140720676, E-mail: prasadsubhash91@yahoo.com

M. P. Rahila
Assistant Professor, Department of Dairy Chemistry, Kerala Veterinary and Animal Sciences University, Wayanad – 673576, Kerala, India, Mobile: +91-9400587443, E-mail: rahilasona@gmail.com

Vimal M. Ramani
Principal, College of Dairy Science, Kamdhenu University, Gandhinagar – 365601, Gujarat, India, Mobile: +91-9879527043, E-mail: vimalmramani@gmail.com

Priyanka S. Rao
Scientist, Dairy Chemistry Division, ICAR-National Dairy Research Institute, Karnal – 132001, Haryana, India, Mobile: +91-9541285384, E-mail: priyanka.rao@icar.gov.in

R. L. Rathish
Assistant Professor, Department of Veterinary Epidemiology and Preventive Medicine, College of Veterinary and Animal Sciences, Pookode, Wayanad – 673576, Kerala, India, Mobile: +91-9387387023, E-mail: rathish@kvasu.ac.in

Monalisa Roy
Assistant Professor, Department of Food Processing, Mugberia Gangadhar Mahavidyalaya, Bhupatinagar, Purba Medinipur – 721425, West Bengal, India, Mobile: +91-8116303175, E-mail: monalisaroy1997@gmail.com

Heena Sharma
Scientist, Dairy Technology Division, ICAR-National Dairy Research Institute, Karnal – 132001, Haryana, India, Mobile: +91-8930515996, E-mail: heena.sharma@icar.gov.in

Neelima Sharma
Post-Doc Scholar, Dairy Chemistry Division, ICAR-National Dairy Research Institute, Karnal – 132001, India, Mobile: +91-9654885303, E-mail: sharmaneelima123@gmail.com

Rajan Sharma
Principal Scientist, Dairy Chemistry Division, ICAR-National Dairy Research Institute, Karnal – 132001, Haryana, India, Mobile: +91-9416120181, E-mail: rajansharma21@gmail.com

Vivek Sharma
Principal Scientist, Dairy Chemistry Division, ICAR-National Dairy Research Institute, Karnal – 132001, Haryana, India, Mobile: +91-9416651314, E-mail: vivek.sharma@icar.gov.in

Rohit G. Sindhav
Assistant Professor, College of Dairy Science, Kamdhenu University, Gandhinagar – 365601, Gujarat, India, Mobile: +91-9016767674, E-mail: rohitsindhav7@gmail.com

Richa Singh
Scientist, Dairy Chemistry Division, ICAR-National Dairy Research Institute, Karnal – 132001, Haryana, India, Mobile: +91-9466963930, E-mail: richa.singh@icar.gov.in

Akashkumar Solanki
Assistant Professor, College of Dairy Science, Kamdhenu University, Gandhinagar – 365601, Gujarat, India, Mobile: +91-9265219592, E-mail: akashkumarks@gmail.com

Ankitkumar J. Thesiya
Assistant Professor, College of Dairy Science, Kamdhenu University, Gandhinagar – 365601, Gujarat, India, Mobile: +91-9558247047, E-mail: ankitthesiya@gmail.com

N. Veena
Assistant Professor, Department of Dairy Chemistry, College of Dairy Science and Technology, Guru Angad Dev Veterinary and Animal Sciences University, Ludhiana – 141004, Punjab, India, Mobile: +91-9855886831, E-mail: veena.ndri@gmail.com

ABBREVIATIONS AND SYMBOLS

AA	amino acid
ACAT	acyl-CoA cholesterol acyltransferase
ACE	angiotensin-converting enzyme
ACO	acyl-CoA oxidase
ACP	amorphous calcium phosphate
Ala (A)	alanine
ALA	alpha-linolenic acid
AMP	antimicrobial peptides
AO	antioxidant
AOAC	Association of Official Analytical Chemists
Arg (R)	arginine
Asn (N)	asparagine
Asp (D)	aspartic acid
BAP	benzo(a)pyrene
BCM	β-casomorphin
BHT	butylated hydroxytoluene
BMO	bovine milk oligosaccharide
BOC	tert-butyloxycarbonyl
BP	bioactive peptide
BSA	bovine serum albumin
Ca	calcium
$CaCl_2$	calcium chloride
CETB	cetyltrimethylammonium bromide
CETP	cholesteryl ester transfer protein
CHD	coronary heart disease
CLA	conjugated linoleic acid
CN	casein
CNS	central nervous system
COX	cyclooxygenase
CPP	caseino phosphopeptide
CPT	carnitine palmitoyl transferase
Cr	chromium
Cu	copper
Cys (C)	cysteine

CZE	capillary zone electrophoresis
DHA	docosahexaenoic acid
dL	deciliter
DM	Diabetes mellitus
DNA	deoxyribonucleic acid
DPP-IV	dipeptidyl peptidase-IV
EFSA	European Food Safety Authority
ELISA	enzyme-linked immunosorbent assay
EPA	eicosapentaenoic acid
ESI	electrospray ionization
ESI-MS	electrospray ionization-mass spectrometry
FAB-MS	fast atom bombardment-mass spectrometry
Fe	iron
$FeCl_3$	ferric chloride
$FePO_4$	ferric phosphate
$FeSO_4$	ferrous sulfate
FMOC	9-fluorenylmethyloxycarbonyl
FOSHU	foods for specified health use
FPLC	fast protein liquid chromatography
FPP	farnesyl pyrophosphate
FR	free radical
FSSAI	Food Safety and Standards Authority of India
FSSR	Food Safety and Standards Regulations
Fuc	fucose
G+ve	gram-positive
GAG	glycosaminoglycan
Gal	galactose
GalNAc	N-acetylgalactosamine
GALT	gut-associated lymphoid tissue
GC	gas chromatography
GH	growth hormone
GI	gastrointestinal
GIP	glucose-dependent insulinotropic polypeptide
GLC	gas-liquid chromatography
Glc	glucose
GlcNAc	N-acetylglucosamine
Gln (Q)	glutamine
GLP-1	glucagon-like peptide-1
Glu (E)	glutamic acid

Gly (G)	glycine
GOS	galactooligosaccharides
GRAS	generally recognized as safe
G –ve	gram-negative
HAT	histone acetyltransferase
HCT	heat coagulation time
HDL	high-density lipoprotein
HDL-C	high-density lipoprotein-cholesterol
HE	hepatic encephalopathy
HIC	hydrophobic interaction chromatography
His (H)	histidine
HMG-CoA	3-hydroxy 3-methylglutaryl-CoA
HMO	human milk oligosaccharide
HPLC	high-performance liquid chromatography
HS	heparan sulfate
HTST	high-temperature short-time
IAC	immunoaffinity chromatography
IBD	inflammatory bowel disease
IC	inhibitory concentration
IDF	International Dairy Federation
IDL	intermediate-density lipoprotein
IEC	ion exchange chromatography
IFN-γ	interferon gamma
Ig	immunoglobulin
IgA	immunoglobulin A
IgE	immunoglobulin E
IGF	insulin-like growth factor
IGFBP	insulin-like growth factor binding protein
IGF-I	insulin-like growth factor-I
IgG	immunoglobulin G
IgM	immunoglobulin M
IL	interleukin
Ile (I)	isoleucine
IMAC	immobilized metal affinity chromatography
IPA	International Probiotic Association
IPP	isopentyl pyrophosphate
ISAPP	International Scientific Association for Probiotics and Prebiotics
ISO	International Organization for Standardization

IU	international unit
JNK	c-Jun N-terminal kinase
kDa	kilo Dalton
LA	linoleic acid
LA	Lobry de Bruyn-Alberda van Ekenstein
LAB	lactic acid bacteria
LC-MS	liquid chromatography-mass spectrometry
LDL	low-density lipoprotein
Leu (L)	leucine
LF	lactoferrin
LLI	lactose to lactulose isomerization
LPL	lipoprotein lipase
LPO	lactoperoxidase
LPS	lipopolysaccharide
LTB4	leukotriene B4
Lys (K)	lysine
MALDI	matrix-assisted laser desorption/ionization
MALDI-MS	matrix-assisted laser desorption/ionization-mass spectrometry
MAPK	mitogen-activated protein kinase
MCAC	metal chelate affinity chromatography
MDA	malondialdehyde
Met (M)	methionine
Mg	magnesium
mL	milliliter
mM	millimolar
Mn	manganese
MPa	megapascal
MS	mass spectrometry
MW	molecular weight
Na	sodium
NANA	N-acetylneuraminic acid
NeuGc	N-glycolylneuraminic acid
NF	nanofiltration
NIPI	National Iron Plus Initiative
NK	natural killer
OS	oligosaccharide
PAGE	polyacrylamide gel electrophoresis
PBMC	peripheral bone marrow cells

PGE_2	prostaglandin E_2
Phe (F)	phenylalanine
PIC	proinflammatory cytokines
PP	peroxisome proliferator
PPAR	peroxisome proliferator-activated receptor
Pro (P)	proline
PUFA	polyunsaturated fatty acid
QPS	qualified presumption of safety
QSAR	quantitative structure-activity relationship
RCT	rennet coagulation time
RDA	recommended dietary allowance
RI	refractive index
RM	reverse micelle
RNA	ribonucleic acid
ROS	reactive oxygen species
RP-HPLC	reversed phase-high performance liquid chromatography
RSA	radical scavenging activity
RSD	relative standard deviation
SCFA	short-chain fatty acid
SCID	severe combined Immunodeficient
SDS-PAGE	sodium dodecyl sulfate-polyacrylamide gel electrophoresis
Se	selenium
SEC	size exclusion chromatography
Ser (S)	serine
SMB	simulated moving bed
SREBP	sterol regulatory element-binding proteins
TBT	technical barriers to trade
TGF	transforming growth factor
Thr (T)	threonine
TLC	thin layer chromatography
TNF-α	tumor necrosis factor-alpha
T_{reg}	regulatory T cells
Trp (W)	tryptophan
Tyr (Y)	tyrosine
UF	ultrafiltration
UHT	ultra-high temperature
USA	United States of America
UV	ultraviolet
Val (V)	valine

VEGF	vascular endothelial growth factor
VLDL	very-low-density lipoprotein
WHO	World Health Organization
WP	whey protein
WPI	whey protein isolate
ZDF	Zucker diabetic fatty
Zn	zinc
μg	microgram
μL	microliter
μM	micromolar

PREFACE

Functional dairy ingredients and nutraceuticals are applied broadly to the constituents that offer specific health benefits beyond basic nutrition. Functional foods are natural foods in which a functional component has been added to provide beneficial health effects, or a component has been removed to reduce the adverse health effects. There has been increasing demand for the health-enhancing, biologically active compounds of foods as consumers have become increasingly aware of the important link between diet and health and the significant role that diet plays in preventing or treating various diseases or improving in physiological functions of the body.

Milk and milk products are naturally 'functional' because of the presence of various biologically active substances such as protein, peptides, oligosaccharides, phospholipids, sphingolipids, conjugated linoleic acid, lactoferrin, fat globule membrane proteins, calcium, etc., which exhibit specific health-promoting properties such as antioxidant, antimicrobial, anticancer, anti-hypertensive, anti-diabetic, hypo-cholesterolemic, immuno-modulatory, and mineral binding properties. The functional attributes of dairy foods are further enhanced by the addition of functional ingredients displayed to possess beneficial health effects such as probiotics, prebiotics, omega-3 fatty acids, phytosterols, fiber, vitamins, minerals, and others.

This book presents a comprehensive summary of research into the functional dairy ingredients and nutraceutical components. The book consists of three main parts. In Part I-Physicochemical and technological aspects of milk-derived components: the concept of production, extraction, and purification, functional, and technological applications of various functional dairy ingredients such as lactulose, casein, and whey protein-derived bioactive peptides have been discussed. In Part II-Therapeutic aspects of dairy ingredients: details of physiological and health effects of colostrum, oligosaccharides, conjugated linoleic acid, and lactoferrin have been highlighted. In Part III-Enhancing the functionality of dairy foods: functional attributes of the dairy products are further augmented by the addition of nutraceuticals such as probiotics, vitamins, and minerals or removal of cholesterol has been discussed, which demonstrated to have positive physiological effects in human beings.

We introduce this book volume under the book series Innovations in Agricultural and Biological Engineering. This book volume is a treasure house of information and excellent reference material for researchers, scientists, students, growers, traders, processors, industries, and others on physicochemical, technological, and therapeutic aspects of various nutraceuticals and functional dairy ingredients and their application in food and dairy industry.

This book has several potential users. It can be a reference book for those students who are taking a college- or university-level functional dairy foods and nutraceuticals course for the first time. Our objectives were to compile information that dairy and food science students are expected to be familiar with as part of their college or university programs before they seek career positions. This book will especially benefit academicians, health professionals, research, and development in functional foods and food product development.

This book has surpassed our vision and expectations due to the contribution by all cooperating authors to this book volume, who have been most valuable in this compilation. Their names are mentioned in each chapter and in the list of contributors. We are grateful and thankful to all of them for their expertise, commitment, and dedication. We hope that this book will be a useful source of information on various functional dairy ingredients and the design of functional dairy foods and their health benefits to consumers.

We request that readers offer their constructive suggestions that may help to improve the next edition.

Also, we would like to thank our families, who have taught us the importance of working hard, having clear goals, and standing for what we believe is right. It is a lesson that guides us in everything we do. Last but not least, we wish to thank our spouses Subhadra, Mohan Kumar, and Anamika, for their understanding and patience throughout this project.

—*Editors*

PART I

Physicochemical and Technological Aspects of Milk-Derived Components

CHAPTER 1

LACTULOSE: PRODUCTION AND TECHNO-FUNCTIONAL APPLICATIONS

TANMAY HAZRA, ROHIT G. SINDHAV,
SUDHEENDRA V. K. CHAGANTI, and SUBHASH PRASAD

ABSTRACT

Lactulose with prebiotic activity is the isomerized form of lactose, having chemical name as 4-O-β-D-galactopyranosyl-D-fructofuranose and molecular weight of 342.3 g/mol. The sweetness for lactulose is greater than lactose but lesser than sucrose. Lactulose can be derived or prepared from lactose by chemical, electro-activation, or enzymatic methods. The versatile applications of lactulose are well acknowledged by dairy-, food-, and pharma industries. The digestive enzymes from human gut are unable to digest lactulose; therefore, this indigested lactulose can be utilized for the growth of human gut's *Bifidobacterium* species. *Lactobacillus* and *Bifidobacterium* species can degrade lactulose to different organic acids (such as: acetic, lactic, and formic acids) that have potent laxative properties. Scientific studies have shown that lactulose has a potent prebiotic effect by protecting the probiotic bacteria in dairy and food products. The effect of lactulose is well acknowledged for the treatment of constipation as well as controlling insulin, blood glucose level, and serum lipids, etc.

Functional Dairy Ingredients and Nutraceuticals: Physicochemical, Technological, and Therapeutic Aspects. Megh R. Goyal, N. Veena, and Santosh K. Mishra (Eds.)

1.1 INTRODUCTION

Milk, a biological fluid obtained by healthy milch animal's mammary glands, where protein as well as minerals present in partly soluble and partly in colloidal form; milk sugar (mainly lactose) present in soluble form; however, the fat present in emulsion form. Since from pre-civilization era, an integral relation between milk with human diet has been established. It is nectar for modern civilization, as milk not only provides basic nutrition but also provides physiological growth and development for newborn. It is a nutritive, scrumptious, and nature's perfect food that contains almost all macromolecules and bioactive components [8].

The term lactose derived from Latin word 'Lactis' -milk and 'Ose' -sugar. It is unique and principal milk carbohydrate that is composed of two monosaccharides *viz* glucose and galactose. Commercially, lactose is manufactured from whey, obtained from cheese, paneer or casein followed by purification. Edible lactose has wide applications in pharmaceutical as well as food industries. However, lactose derivatives, such as galacto-oligosaccharides (GOS), lactulose, lactobionic acids, lactitol, etc. (Figure 1.1 and Table 1.1), are widely acceptable in food and pharmaceutical industries for different health and techno-functional attributes. Global production of lactose derivatives [60] is presented in Table 1.2.

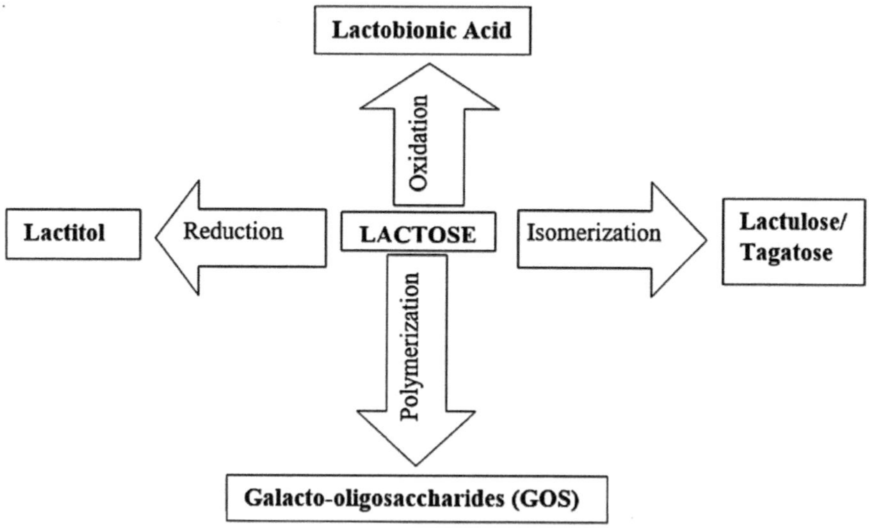

FIGURE 1.1 Different derivatives of lactose [69].

TABLE 1.1 Lactose Derivatives and Their Applications [3, 69]

Lactose Derivatives	Systematic Name	Applications
Lactitol	4-O-β-D-galactopyranosyl-D-glucitol	Bulking agent, prebiotic, sweetner
Lactosucrose	4(G)-beta-D-galactosylsucrose	Prebiotic
Lactobionic acid	4-O-β-galactopyranosyl-D-gluconate	Acidulant
Lactulose	4-O-β-D-galactopyranosyl-D-fructofuranose	Prebiotic, laxative effect

The commercial market for lactose derivatives is inflating very rapidly. Apart from pharma industries, the lactose derivatives are an alternative choice as a functional ingredient for food researchers and manufacturers. The health beneficial effects of lactulose, attracting food and pharma industries for designing novel foods and drugs by using lactulose as a major ingredient. Therefore, it is considered as one of the most valuable derivatives of lactose [51, 78].

This chapter explores the different techno-functional and health aspects of lactulose, which is a unique lactose derivative.

TABLE 1.2 Global Production of Lactose Derivatives [60]

Lactose Derivatives	Estimated Production (Tons/Year)
GOS	1,75,700
Lactitol	11,000–12,000
Lactobionic acid	1,000
Lactosucrose	4,000–4,600
Lactulose	50,000

1.2 LACTULOSE: A UNIQUE LACTOSE DERIVATIVE

Lactulose (Figure 1.2) is a disaccharide consisting of two sugar molecules, viz, fructose, and galactose and it is synthetic in nature. These sugars are attached together by β-1,4-glycosidic bond or linkage [51, 57]. Glucose molecule of lactose isomerizes to fructose (at 100°C, pH near about 6.7) to form Lactulose [15, 69]. Therefore, concentration of lactulose is considered as an indicator of fluid milk undergoing high heat treatment [51]. Extending heating temperatures and time has no effect on the concentration of lactulose due to the drop in pH and the lactulose degrades during heating [4, 51].

Lactulose can also be degraded by β-elimination and produces low molecular weight products like tagatose and saccharine [57]. Lactulose is hygroscopic in nature, which is soluble in water, methanol, ethanol, propanol, and isopropanol. Lactulose is more soluble in water than lactose [40].

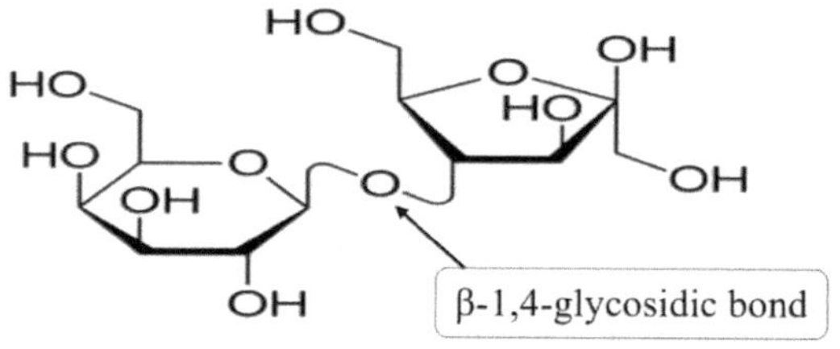

FIGURE 1.2 Structure of lactulose.

Due to hygroscopic in nature, lactulose has been sold as syrup for pharmaceutical purpose only [48]. It has been estimated that the production of lactulose for pharmaceutical purpose may be 12,000 tons/year [79] as shown in Table 1.2. Solvay Pharmaceuticals (Canada and Netherland) are the major lactulose manufacturers in the world and share almost 50% of the total lactulose production of the whole world [62]. Fresenius Kabi (Austria) is another world's top lactulose manufacturer [54]. Historical aspects of lactulose are presented in Table 1.3.

TABLE 1.3 Historical Aspects of Lactulose

Year	Historical Aspects of Lactulose	References
1930	First time: Lactose and lime were heated at 35°C for preparation of lactulose.	[43]
1950	*Bifidus* activities of lactulose were recognized	[61]
1960	Lactulose was marketed by Morinaga (Japan) in infant formula	[62]

1.3 DIFFERENT FORMS OF LACTULOSE

There are two forms of lactulose, i.e., anhydride, and trihydrate forms. The anhydride form is commercially available [69]. Trihydrate form of lactulose contains three moles of water [27] and its chemical formula is $C_{12}H_{22}O_{11}.3H_2O$ [69]. The β-Pyranose and α- or β-furanose forms of lactulose isomer have also been identified during production of lactulose [51].

Lactulose syrup is yellow in color and is odorless. The intensity of sweetness for lactulose is higher than the lactose but lesser than sucrose (almost 60%) [54]. The ability of lactulose to reduce the Fehling's solution on heating is characterized as a reducing nature of sugar [57]. Table 1.4 represents some of the physicochemical properties of lactulose [51, 69].

TABLE 1.4 Physicochemical Properties of Lactulose

Physicochemical Properties	Anhydride	Trihydrate
Molecular formula	$C_{12}H_{22}O_{11}$	$C_{12}H_{22}O_{11}3H_2O$
Molecular weight	342	396
Melting temperature (°C)	169	68
Heat of solution (KJ/ mol)	–4	34
Glass transition temperature (°C)	–47	–
Sweetness	1.5 times higher than lactose	1.5 times higher than lactose
Water solubility	76.4% w/w at 30C	76.4% w/w at 30°C
Specific rotation, α (at 20°C)	–51.5°	–

1.4 MANUFACTURING PROCESS OF LACTULOSE

Lactulose can be prepared from lactose by a chemical method [3]. During isomerization process, the glucose molecule present in lactose converts into fructose (ketose sugar) by replacing the molecule of glucose [57]. Three different methods have been adopted for the isomerization of lactose [51], such as: chemical, enzymatic, and electro-activation method.

1.4.1 CHEMICAL METHOD

The chemical method of lactulose synthesis (Table 1.5) involves isomerization of lactose under alkaline conditions in the presence of different catalysts, followed by separation and purification [54]. This method involves two different processes for production of lactulose.

TABLE 1.5 Chemical Methods for Lactulose Production from Lactose

Method	Description	References
Lobry de Bruyn-Alberda van Ekenstein (LA) method (transformation/ synthesis)	In alkaline medium, lactose is epimerized to form epilactose; finally, glucose present in lactose transferred to fructose.	[3, 57]
Maillard reaction	During heating of milk, lactose reacts with amines; and ammonia results in the formation of lactosyl-amine complex. In the subsequent steps, the Amadori rearrangement and hydrolysis leads to the formation of lactulose.	[51]

Commercially accepted method for lactulose synthesis is the LA method involving different catalysts and other alkaline reagents (such as: sodium hydroxide, magnesium oxide and tertiary amines, etc.), that are commonly used during the production of lactulose [3, 54, 82].

Production of lactulose is a complex process (Figure 1.3) [51]. However, different other sugars along with lactulose are formed during heating; although the heat treatment may partially degrade some lactulose too. Therefore, different catalysts have been introduced for production of lactulose. A wide range of catalysts including alkaline agents (synthetic and natural) and complex reagents have been used to make the production process faster, robust, and low energy-dependent [3, 51, 57].

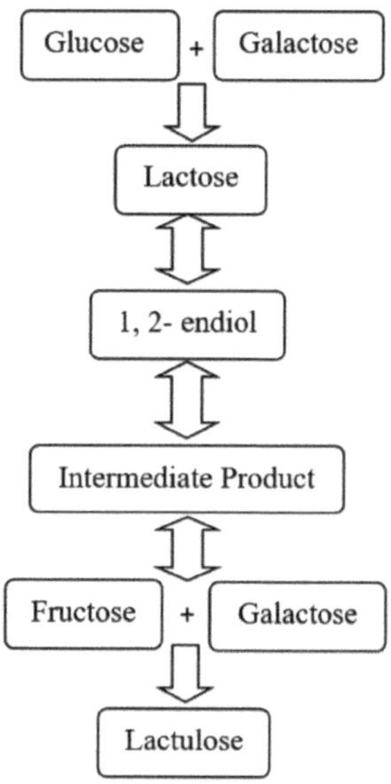

FIGURE 1.3 Flow diagram for the production of lactulose by LA method [51].

Ideal characteristics of catalysts that can be used during lactulose synthesis are: Low cost; Non-toxic; Safe for environment; Easily available; Easy to remove from the isomerized medium; Maximum isomerization but production of minimum by-products; and High repeatability.

1.4.1.1 USE OF ALKALINE CATALYSTS

1.4.1.1.1 Synthetic

Alkaline catalysts (Table 1.6; such as, sodium hydroxide, magnesium oxide, triethylamine, and sulfites) have been successfully used for the production of lactulose from lactose [57]; and the temperature should be between 70° to 100°C [51, 68] during the use of such catalyst. Along with traditional heating process, microwave heating process has also been used to accelerate the synthesis of lactulose [23, 51]. A study reported that isomerization time during lactulose production can be reduced by using microwave heating for 60 seconds at 600 W [50]. The interaction between ionic molecules with microwave radiation produces rapid heat; consequently, decreasing the lactose to lactulose isomerization (LLI) time [23].

TABLE 1.6 Catalysts Used in Production of Lactulose [54, 59]

Name of the Catalyst	LLI Temperature (°C)
Calcium carbonate	96
Calcium hydroxide	35
Eggshell	98
Phosphates	104
Sodium hydroxide	80

1.4.1.1.2 Natural

Lactulose can be produced using naturally available calcium carbonate-based substances, such as: egg shells and oyster shells [44, 68]. All these naturally available catalysts are generally used in the powder form. The main advantages of these catalysts are:

- Easy to remove from final product by the centrifugation or filtration;
- High yield of lactulose;
- Proper utilization of waste materials.

Sometimes complex agents (such as: borate and aluminate) that are able to make complex compounds with lactulose have been used as catalyst under alkaline conditions [33].

1.4.1.2 PURIFICATION OF LACTULOSE

Purification is a very essential step for production of commercial lactulose. Different by-products or other sugars including lactose, epilactose, etc., are formed during the synthesis of lactulose [69]. Purification steps are also essential for the recovery of catalyst that is used during the production of lactulose [78]. Tamura et al. [74] reported that chromatographic separation followed by crystallization and drying are the essential steps for commercial lactulose production. Activated carbon or borate-carbohydrate mixture was successfully used for the separation of lactulose from other sugars or by-products [34, 44]. Sodium aluminate catalyst was used to recover more than 98% catalysts by simple precipitation without using any other additional steps [78]. Disadvantages while using chemical catalysts are the production of numbers of unwanted secondary products and expensive separation and purification methods are required [51].

1.4.2 ENZYMATIC METHOD

These methods are environment friendly approaches. The β-galactosidase and glycosidase class enzymes are mainly used for the production of lactulose. However, low price and easily available β-galactosidase enzyme derived from microorganisms, animals or plants are mainly used for this purpose [55, 56]. Commercially, β-galactosidase enzyme hydrolyzes the lactose to yield galactose and glucose and further forming a complex called galactosyl-β-galactosidase complex. Thereafter lactulose is produced via trans-galactosylation process or pathway in the presence of fructose, which act as nucleophilic acceptor [57, 65, 77]. The yield of lactulose is always lower by this method as compared to chemical methods [18]. In this process, both free and bound (immobilized) form of enzymes has been used for production of lactulose.

1.4.2.1 FREE FORM OF ENZYMES

β-galactosidase enzymes obtained from different microbial sources (such as: *Lactobacillus acidophilus*, *Aspergillus oryzae*, *Aspergillus niger*, etc.), have been tried for the production of lactulose [19–21]. However, β-galactosidase enzyme's origin does play an essential role in the production as well as yield of lactulose.

The β-galactosidase and hyper thermo-stable β-glycosidase (from *Aspergillus oryzae* and *Pyrococcus furiosus*, respectively) have been used during the enzymatic production of lactulose. Lactulose production was reported to be higher in case of *Pyrococcus furiosus* as compared to *Aspergillus oryzae* [41, 54]. β-Galactosidase enzymes isolated from *B. circulans* was reported to produce lactulose with higher yield than β-galactosidase enzyme isolated from yeast, e.g., *Kluyveromyces lactis* [1].

Genetically modified *Escherichia coli* containing *Sulfolobus solfataricus* thermo-stable β-galactosidase is used in the production of lactulose [31]. Efficiency of lactulose synthesis using free enzymes mainly depends on factors, such as: sources of β-galactosidase, substrate concentration, type of substrate or raw materials, etc. (Figure 1.4) [51].

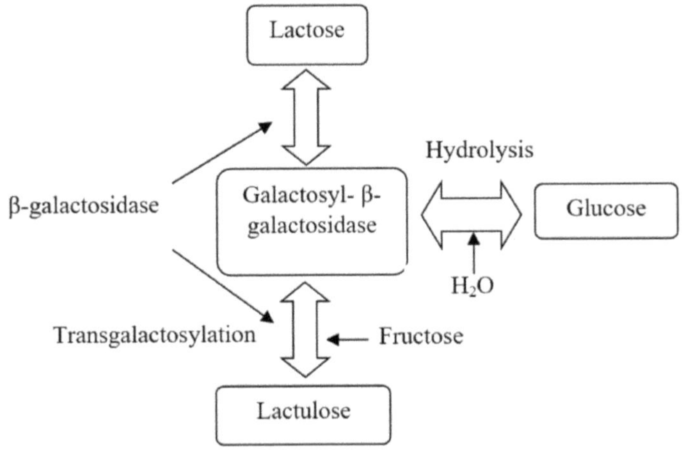

FIGURE 1.4 Schematic diagram for the production of lactulose by enzymatic method [51].

1.4.2.2 IMMOBILIZED ENZYMES

β-galactosidase enzyme can be immobilized using different approaches, such as: entrapment, adsorption, covalent binding and cross-linking [19]. Fe_3O_4-chitosan magnetic microspheres (Table 1.7) are used as a medium for immobilization of β-galactosidase that was successfully used for the production of lactulose, which has increased the yield of lactulose almost by 19–20%

than the batch method [24, 51]. A continuous enzymatic process employed β-glycosidase extracted from *Pyrococcus furiosus* that was used in free and immobilized forms for production of lactulose through transgalactosylation. β-Galactosidase immobilized on silica gel (bed) was proved to be effective for the production of high yield lactulose from whey [71, 72].

TABLE 1.7 Immobilized Enzyme System for the Production of Lactulose

Enzyme Support System	Type of Interaction	References
Eupergit	Covalent	[41]
Magnetic chitosan microspheres	Covalent	[24]
Silica	Covalent	[70–72]

1.4.3 ELECTRO-ACTIVATION

It is a clean and green technology, but the yield of lactulose is low [29, 51]. Self-generated alkaline state in the reaction medium is the basic requirement for this method [5, 29]. The reactor contains two electrodes where the electric field is applied [4]. Kareb et al. [29] used cheese whey and applied 400 mA/40 min electric field (10°C) for lactulose synthesis. Aider and Gimenez-Vidal [4] used the electro-activation technique for synthesis of lactulose at 23–24°C. However, commercial application for production of lactulose by this method is very rare.

1.5 HEALTH BENEFITS OF LACTULOSE

The possible mechanisms for the health effects of lactulose [53, 54] are summarized in this section (Figure 1.5). Certain bacteria (e.g., lactic acid bacteria and *Bifidobacterium* spp.) in the human colon can metabolize lactulose to carbon dioxide, lactic, acetic, and formic acid and short-chain fatty acids (SCFAs). This results in lowering of intestinal buffering conditions or pH, making a conducive environment for the growth of probiotics as well as stimulates immune system and vitamin synthesis. Simultaneously, these conditions suppress the growth of undesirable putrefactive bacteria that decrease the production of ammonia. Therefore, stools get softened and can easily pass out. Prebiotic effect of lactulose promotes the growth of probiotic bacteria that indirectly reduces the chances of many diseases including cancer.

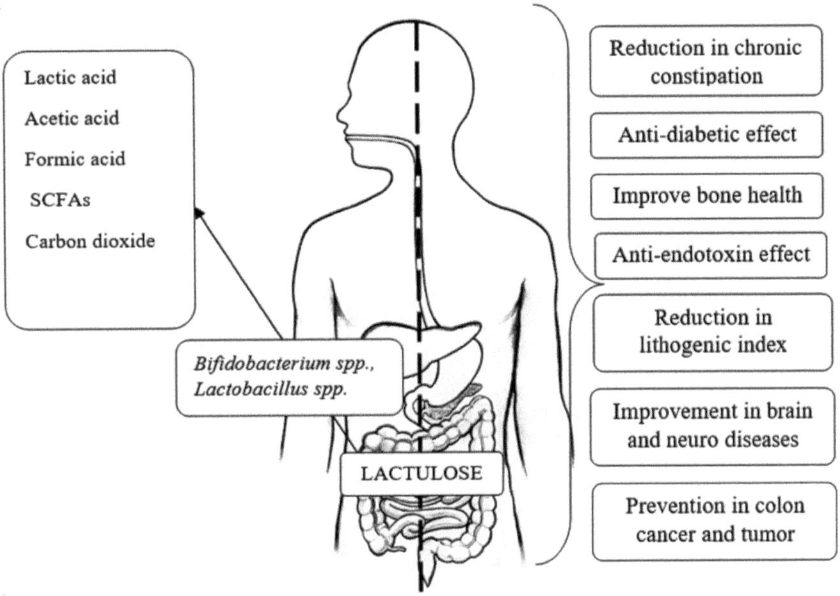

FIGURE 1.5 Health benefits of lactulose.

1.5.1 PREVENTION OF CONSTIPATION

The effective action of lactulose against chronic constipation (laxative action) irrespective of all age groups has been proved in many scientific investigations [51, 54]. Since long time the USA, Europe, and Japan pharmacopeia has accepted it as a drug for the treatment of constipation [57].

It gives a laxative effect @ 0.25 g/kg body weight for daily consumption [68]. The possible mechanisms for laxative action of lactulose can be attributed in the following ways [5, 67]:

- It increases the osmolarity of colon by increasing the production of SCFAs;
- pH level in colon is dropped and liquid level is found to be balanced;
- It also reduces the intestinal transit time.

Therefore, stools become softened and can be easily excreted. The common factors affecting the laxative action of lactulose are [68]: health condition of the individual, age, body weight, dietary habits and ingested dosage.

1.5.2 ANTI-DIABETIC EFFECT

Lactulose can control the blood glucose level. The anti-diabetic effect of lactulose has been demonstrated by controlling of pancreatic insulin production [5, 57].

1.5.3 IMPROVES BONE HEALTH

Scientific studies have confirmed that absorption of calcium and magnesium is increased during oral administration of lactulose [57], thereby improving the bone health. Moreover, lactulose also provides favorable conditions for the growth of probiotic bacteria in the human colon, resulting in the reduction of intestinal pH. Under these conditions, minerals become more soluble, and the permeability of the intestinal mucosa is increased leading to the increased absorption of minerals [37].

1.5.4 ANTI-ENDOTOXIN EFFECTS

Lactulose provides favorable conditions for the proliferation of health-promoting probiotics like *Lactobacilli* spp. by producing fatty acids with less carbon, which in turn inhibits the growth of coliforms in the colon [5, 57]. The changes occurring in the microbial environment in the colon, reduces the endotoxins production and their absorption in the colon [54]. Anti-endotoxin effect of lactulose has proved to be effective for the treatment of exocrine pancreatic dysfunction [39] and hypercholesterolemia [35].

1.5.5 SALMONELLA CARRIERS

Ait-Aissa and Aider [5] reported that 60 g/day dose of lactulose could decrease the colonic pH that promotes the growth of *Bifidus* bacteria including non-typhoid salmonella.

1.5.6 REDUCTION IN THE LITHOGENIC INDEX

Lithogenic action increases the chances of bile stone. Thornton and Heaton [75] reported that lactulose may reduce the lithogenic index; therefore, the chances of bile stone formation could be prevented.

1.5.7 IMPROVEMENT OF BRAIN AND NEURO-DISEASES

Hepatic encephalopathy (HE) is a common disease that directly affects the function or structure of human brain. Prasad et al. [63] found the positive effect of lactulose for the patients suffering from HE. Some researchers have claimed that lactulose decreased the methamphetamine (METH) (elevates neurotoxicity) by decreasing the oxidative stress in the body [80].

1.5.8 COLON CANCER AND TUMOR PREVENTION

Biochemical changes occurring in the colon due to the microflora present will increase the chances of colon cancer. It is reported that lactulose can decrease the occurrence of colon cancer [36, 54]. Verma and Shukla [76] reported the preventive effect of lactulose against DNA damage in colon mucosa of rat model. Probiotic bacteria can help to metabolize lactulose and produce SCFAs, which can reduce the growth of cancer cells. Lactulose influences Bifidobacterial growth, which indirectly suppresses the cancer cell growth or increase in anti-tumorous effect by increasing the immunogenic activity [5].

Lactulose can be used in pre-operation treatment by preventing the complications that may arise during surgery for obstructive jaundice [54]. Lactulose therapy has also proved effective against urinary and respiratory infections [54, 69]. The application of lactulose is not only restricted to therapeutic drugs but it is also used as a potent agent for diagnosis purpose. The diagnosis of colon disorder is performed breath hydrogen test after oral administration of a single dose of lactulose [54]. This test also helps in diagnosing the bacterial overgrowth in the small bowel [52].

1.6 POTENTIAL APPLICATIONS OF LACTULOSE IN FOOD INDUSTRIES

Lactulose's multi-functional applications in the food industry is well acclaimed [51, 69]. The versatile functions including prebiotic or bifidogenic activities or other functional attributes attract the attention of food researchers since a long time. Detailed applications of lactulose are elaborated in this section.

1.6.1 BIFIDOGENIC AGENT

Introduction of lactulose in the infant's diet increased the *Bifidus* flora in the gut [61]. High heat and acid stability properties of lactulose promote its applications in a wide range of dairy products, infant formulas and fruit juices [51]. Nagendra et al. [49] reported that the infant formula supplemented with 0.5% lactulose can stimulate the survivability and activity of *Bifidobacteria* in children. Moreover, a laxative effect of lactulose promotes its applications in infant formulas. The concentration of lactulose in infant formula varies from 29 to 108 mg/L depending on its composition as well as other formulation parameters [2, 57].

1.6.2 PROTECTION OF PROBIOTIC CULTURE

Kontula et al. [32] reported that survivability of probiotic yogurt culture in colon was 10–60% more due to the presence of lactulose. Saarela et al. [64] reported that lactulose had protecting effect in cold storage and they elaborated the possible cause that lactulose may provide extra energy for the probiotic culture during storage of fermented food. Also, in the presence of lactulose, probiotics cultures like *Lb. rhamnosus* and *B. bifidum* in yogurt could survive even for 5 weeks at 4°C [73]. The positive effect of lactulose as prebiotic was also acknowledged by Hernandez-Hernandez et al. [22], who concluded that probiotic culture specially *Lactobacillus* strains had higher survival rate in the gastrointestinal tract. Lactulose could improve harsh acid and bile stress in the gut for the probiotic culture. The applications of lactulose as prebiotic are not only restricted for the production of yogurt, but can also be used in other fermented dairy products like kefir [47].

Coelho et al. [10] reported the beneficial effects of lactulose as a prebiotic in dry fermented probiotic sausage. They reported that addition of lactulose at 2% may increase the survivability of probiotic strain *Lb. paracasei* without any negative effect on our body and texture of the final product. The positive attributes of lactulose as prebiotic in animal feed was highlighted by Ait-Aissa and Aider [5], who reported that apart from reducing the amount of antibiotic supplementation, lactulose helps to increase the number of *Bifidobacteria* and *Lactobacilli* in the animal gut.

1.6.3 IMPROVES THE SENSORY QUALITY

Flavor enhancing attributes of lactulose in different foods were well-established [69]. For the improvement of browning features, lactulose is also used in bakery products, including cakes and cookies [51, 67].

1.6.4 SUCROSE REPLACER

Sweetness of lactulose is higher than lactose; and it is almost 60% as sweet as sucrose. Therefore, it is used as an alternative to sucrose in different confectionaries [54, 69]. Higher water solubility and viscosity close to sucrose can enhance its application as a sugar replacement in different dairy products [10, 51]. However, replacement of lactulose to sugar can directly affect the free moisture or viscosity of the final product [12]. The energy value contributed by lactulose is 2 kcal/g; therefore, lactulose is used for the formulation of different low calorie dairy or food products [51, 69].

1.6.5 ENCAPSULATING MATERIAL

Lactulose can be used as a delivery system for bioactive components and health beneficial micro-organisms targeted for the colon. Its sweet taste could be very effective in delivering bitter tasted bioactive components [51].

1.7 ANALYTICAL TECHNIQUES FOR THE DETERMINATION OF LACTULOSE

Wide range of different analytical techniques (Table 1.8) are being extensively used in the determination of lactulose, such as: chromatography-based methods (e.g., HPLC, GLC, and TLC), spectrophotometric based methods, enzymatic based or other analytical techniques (e.g., electrophoresis, flow analysis methods) [17, 51, 65]. International Dairy Federation (IDF) [25] recommended that UHT milk or in bottle sterilized milk could be differentiated on the basis of lactulose concentration. The formation of lactulose alone or in combinations with other markers has been used to evaluate the thermal history or thermal damage of the milk or dairy products. The concentration of lactulose present in different heat-treated milk is given in descending order:

in container sterilization > indirect UHT > direct UHT > high-temperature short-time (HTST) pasteurization [51].

TABLE 1.8 Different Analytical Approaches for Estimation of Lactulose

Analytical Technique	References
GC	[45]
HPLC	[39]
LC-MS	[28]
Nuclear magnetic resonance	[26]
Spectrophotometer	[30, 81]
TLC	[14, 40]

Parrish et al. [59] used HPLC technique for the separating different carbohydrates like lactulose, lactose, etc., and they have been used a common carbohydrate column for this purpose. Types of HPLC column and pretreatment of samples plays a significant role for effective determination of lactulose by HPLC method. HPLC analysis using refractive index (RI) detector has been suggested by IDF for the quantification of lactulose in milk or dairy products [11]. Kim et al. [31] used the HPLC with RI detector and cartridge column for effective separation of lactulose from other sugars. Gonzaga et al. [17] used same detector but column was cation-exchange column for determination of lactulose. Apart from using RI detector; other detectors for HPLC have been used for effective determination of lactulose. Schmidt et al. [65] applied HPLC along with evaporative light scattering detector (ELSD) for separation and quantification of lactulose with 98% accuracy and precision.

Other chromatographic methods like TLC and GC have been applied for the determination or separation of lactulose. In TLC based approach, where acetonitrile and water (50:20) was used as mobile phase and silica gel-boric acid was used as stationary phase for estimation of lactulose even in very low concentration in dairy products [40]. Flick et al. [14] used the silica gel 60 plates and a propanol-borate solvent system in TLC for the determination of lactulose in urine.

In rapid flow system, immobilized β-galactosidase enzyme in a reactor which rapidly hydrolyzes lactulose to fructose and galactose was a very efficient and rapid method for the determining lactulose [46]. A more sensitive method for the determination of lactulose in milk was developed by Amine et al. [6] which involves the chromogenic reaction (Seliwanoffs) followed by spectrophotometric method. Various enzyme-based kits are also available commercially for determining the content of lactulose in milk. Ayala et al. [7] applied highly efficient front-face fluorescence spectroscopy and tryptophan as fluorescence marker for quantifying lactulose in milk. Few studies suggested that Capillary zone electrophoresis (CZE) with UV detector was also used in detecting lactulose present in heated milk [11]. This method was reported to be very efficient and could be detected lactulose with high precision; hence, this analytical protocol could be used as an alternative for detection and quantification of lactulose in milk.

1.8 SAFETY ASPECTS

Health benefits of lactulose have been established for more than fifty years; although there are some adverse health effects like gas or intestinal cramps [5]. European Food Safety Authority [13] claimed that consumption of 10 g of lactulose (in a single serving) could inhibit the growth of pathogenic microorganisms in the human intestine and can reduce intestinal transit time. WHO reported that daily dose of lactulose is 6.7 g is safe and effective against constipation [57]. More than 100 countries have registered lactulose as therapeutic and medicinal drug [5]. Foods for Specified Health Use (FOSHU) systems in Japan have also acknowledged the applications of lactulose as functional food with proper labeling [57].

1.9 SUMMARY

Lactulose is the unique and most valuable derivative of lactose. Chemical, enzymatic, and electro-activation methods can be used for the synthesis of lactulose. Commercially, chemical method is well acknowledged all over the world. Different analytical methods including TLC, GLC, and HPLC have been applied for the identification of lactulose in different food products. The therapeutic applications of lactulose against different diseases like chronic constipation, reducing blood ammonia levels, therapy of portal-systemic encephalopathy, IBD, colon carcinogenesis, etc., have been well acclaimed throughout the world since a long time. The unique physicochemical properties of lactulose attract the food and pharma industries for its applications in designing different novel food or medicines. The prebiotic effect of lactulose helps to stimulate health beneficial bacteria that help to combat against wide spectrum of diseases including cancer. Moreover, application of lactulose as a low-calorie sweetener is also attracting food researchers for developing anti-diabetic foods. The possible application of lactulose as nano or micro-carrier for delivering bioactive components in food or drugs also opens a new horizon of targeted therapy technology.

KEYWORDS

- **health benefits**
- **lactose**
- **lactulose**
- **laxative properties**
- **prebiotic**

REFERENCES

1. Aburto, C., Guerrero, C., Vera, C., & Illanes, A., (2020). Improvement in the yield and selectivity of lactulose synthesis with *Bacillus circulans* β-galactosidase. *LWT-Food Science and Technology, 118*(9). Online article ID: 108746.
2. Adriana, S. P., Gabrieala, B. N., Laura, S. M., & Maria, S. V., (2002). Available lysine, protein digestibility and lactulose in commercial infant formulas. *International Dairy Journal, 13*(2, 3), 95–99.
3. Aider, M., & De Halleux, D., (2007). Isomerization of lactose and lactulose production: Review. *Trends in Food Science and Technology, 18*(7), 356–364.
4. Aider, M., & Gimenez-Vidal, M., (2012). Lactulose synthesis by electro-isomerization of lactose: Effect of lactose concentration and electric current density. *Innovative Food Science and Emerging Technologies, 16*(1), 163–170.
5. Aït-Aissa, A., & Aïder, M., (2014). Lactulose: Production and use in functional food, medical, and pharmaceutical applications. Practical and critical review. *International Journal of Food Science & Technology, 49*(5), 1245–1253.
6. Amine, A., Moscone, D., & Palleschi, D., (2000). Rapid determination of lactulose in milk using seliwanoff's reaction. *Analytical Letters, 33*(1), 125–129.
7. Ayala, N., Zamora, A., Gonzalez, C., Saldo, J., & Castillo, M., (2017). Predicting lactulose concentration in heat-treated reconstituted skim milk powder using front-face fluorescence. *Food Control, 73*(1), 110–116.
8. Chavan, R. S., Sehrawat, R., Mishra, V., & Bhatt, S., (2016). Milk: Processing of milk. In: Caballero, B., Paul, M. F., & Fidel, T., (eds.), *Encyclopedia of Food and Health* (pp. 729–735). London, UK: Academic Press.
9. Claeys, W. L., Ludikhuyze, L. R., & Hendrickx, M. E., (2001). Formation kinetics of hydroxymethylfurfural, lactulose, and furosine in milk heated under isothermal and non-isothermal conditions. *Journal Dairy Research, 68*(2), 287–301.

10. Coelho, S. R., Lima, Í. A., & Martins, M. L., (2019). Application of *Lactobacillus paracasei* LPC02 and Lactulose as a potential symbiotic system in the manufacture of dry-fermented sausage. *LWT-Food Science and Technology, 102*(1), 254–259.
11. De Oliveira, N. L. N., Marques, R., Da Silva, P. H. F., & De Oliveira, M. A. L., (2018). Lactulose determination in UHT milk by CZE-UV with indirect detection. *Food Chemistry, 258*, 337–342.
12. Dorohovich, A., Dorohovich, V., & Kambulova, J., (2016). The study of the rheological properties of pectin gels with mono- and disaccharides. *Eureka: Life Sciences, 4*, 14–19.
13. EFSA, (2010). EFSA Panel on dietetic products, nutrition, and allergies (NDA). Scientific Opinion on the substantiation of health claims related to lactulose and decreasing potentially pathogenic gastro-intestinal microorganisms (ID 806) and reduction in intestinal transit time (ID 807) pursuant to Article 13(1) of Regulation (EC) No 1924/2006. *EFSA Journal, 8*(10), 26. Online article ID: 1806.
14. Flick, J. A., Schnaar, R. L., & Perman, J. A., (1987). Thin-layer chromatographic determination of urinary excretion of lactulose, simplified, and applied to cystic fibrosis patients. *Clinical Chemistry, 33*(7), 1211–1220.
15. Ganzle, M. G., Haase, G., & Jelen, P., (2008). Lactose: Crystallization, hydrolysis, and value added derivatives. *International Dairy Journal, 18*(7), 685–694.
16. Ganzle, M. G., (2012). Enzymatic synthesis of galacto-oligosaccharides and other lactose derivatives (hetero-oligosaccharides) from lactose. *International Dairy Journal, 22*(2), 116–122.
17. Gonzaga, N., Watanabe, L. S., Mareze, J., Madeira, T. B., Tamanini, R., Rios, E. A., Nixdorf, S. L., & Beloti, V., (2019). Green method using water for lactose and lactulose extraction and determination in milk by high-performance liquid chromatography (HPLC) with refractive index detection. *LWT-Food Science and Technology, 113*(7), 9. Online article ID: 108288.
18. Gosling, A., Stevens, G. W., Barber, A. R., Kentish, S. E., & Gras, S. L., (2010). Recent advances refining galactooligosaccharide production from lactose. *Food Chemistry, 121*(2), 307–318.
19. Guerrero, C., Vera, C., & Illanes, A., (2017). Fed-batch operation for the synthesis of lactulose with β-galactosidase of *Aspergillus oryzae*. *Bioresource Technology, 237*(1), 126–134.
20. Guerrero, C., Vera, C., Acevedo, F., & Illanes, A., (2015). Simultaneous synthesis of mixtures of lactulose and galactooligosaccharides and their selective fermentation. *Journal of Biotechnology, 209*, 31–40.
21. Hashem, A. M., Wafaa, S. A. E. A. I., Helmy, A., El-mohamady, Y., & Abou-romia, R., (2013). Factors affecting the production of lactulose by *Lactobacillus acidophilus* NRRL 4495 β-galactosidase and its biological activity. *Malaysian Journal of Microbiology, 9*(1), 1–6.
22. Hernandez-Hernandez, O., Muthaiyan, A., Moreno, F. J., Montilla, A., Sanz, M. L., & Ricke, S. C., (2012). Effect of prebiotic carbohydrates on the growth and tolerance of *lactobacillus*. *Food Microbiology, 30*(2), 355–361.
23. Hoai, N. T., Sasaki, A., Sasaki, M., Kaga, H., Kakuchi, T., & Satoh, T., (2011). Selective synthesis of 1,6-anhydro-β-D-mannopyranose and mannofuranose using microwave assisted heating. *Carbohydrate Research, 346*(13), 1747–1751.

24. Hua, X., Yang, R., Zhang, W., Fei, Y., Jim, Z., & Jiang, B. O., (2010). Dual-enzymatic synthesis of lactulose in organic-aqueous two-phase media. *Food Research International, 43*(3), 716–722.
25. IDF, (1992). Influence of technology on the quality of heat-treated milk and fluid milk products. *Bulletin of the International Dairy Federation* (p. 31). B-Doc.222; Brussels.
26. Jayalakshmi, K., Ghoshal, U. C., Kumar, S., Misra, A., Roy, R., & Khetrapal, C. L., (2009). Assessment of small intestinal permeability using 1H-NMR spectroscopy. *Journal of Gastrointestinal and Liver Diseases, 18*(1), 27–32.
27. Jeffrey, G. A., Huang, D. B., Pfeffer, P. E., Dudley, R. L., Hicks, K. B., & Nitsch, E., (1992). Crystal structure and NMR analysis of lactulose trihydrate. *Carbohydrate Research, 226*(1), 29–42.
28. Jia, S., & Dong, X., (2016). Recent advances in analytical techniques for the determination of lactulose. *International Journal of Chemical Studies, 4*(6), 51–53.
29. Kareb, O., Champagne, C. P., & Aider, M., (2016). Contribution to the production of lactulose-rich whey by in situ electro-isomerization of lactose and effect on whey proteins after electro-activation as confirmed by matrix-assisted laser desorption/ionization-time of-flight-mass spectrometry and sodium dodecyl sulfate-polyacrylamide gel electrophoresis. *Journal of Dairy Science, 99*(4), 2552–2570.
30. Khan, M. A., Iqbal, Z., Jan, M. R., Shah, J., Ahmad, W., & Haq, Z. U., (2006). A Spectrophotometric method for quantitative determination of lactulose in pharmaceutical preparations. *Journal Analytical Chemistry, 61*(1), 32–36.
31. Kim, Y. S., Park, C. S., & Oh, D. K., (2006). Lactulose production from lactose and fructose by a thermostable β-galactosidase from *Sulfolobus solfataricus*. *Enzyme and Microbial Technology, 39*(4), 903–908.
32. Kontula, P., Nollet, L., Saarela, M., Vilpponen-Salmela, T., Verstraete, W., Mattila-Sandholm, T., & Wright, A. V., (2002). The effect of lactulose on the survival of *Lactobacillus rhamnosus* in the simulator of the human intestinal microbial ecosystem (SHIME) and *in vivo*. *Microbial Ecology in Health and Disease, 14*(2), 90–96.
33. Kozempel, M., & Kurantz, M., (1994). The isomerization kinetics of lactose to lactulose in the presence of borate. *Journal of Chemical Technology and Biotechnology, 59*(1), 25–29.
34. Kozempel, M. F., Kurantz, M. J., Craig, J. C., & Hicks, K. B., (1995). Development of a continuous lactulose process: Separation and purification. *Biotechnology Progress, 11*(5), 592–595.
35. Liao, W., & Florin, C. H., (1995). Endotoxin, cytokines, and hyperlipidemia. *Scandinavian Journal of Gastroenterology, 28*(2), 97–103.
36. Macfarlane, G. T., Steed, H., & Macfarlane, S., (2008). Bacterial metabolism and health-related effects of galacto-oligosaccharides and other prebiotics. *Journal of Applied Microbiology, 104*(2), 305–344.
37. Macfarlane, S., Macfarlane, G. T., & Cummings, J. H., (2006). Prebiotics in the gastrointestinal tract. *Alimentary Pharmacology and Therapeutics, 24*(5), 701–714.
38. Mack, D. R., Flick, J. A., Durie, P. R., Rosentein, B. J., Ellis, L. E., & Perman, J. A., (1992). Correlation of intestinal lactulose permeability with exocrine pancreatic dysfunction. *Journal of Pediatrics, 120*(5), 696–701.
39. Manzi, P., & Pizzoferrato, L., (2013). HPLC determination of lactulose in heat-treated milk. *Food Bioprocess Technology, 6*(3), 851–857.

40. Martinez-Castro, I., & Olano, A., (1981). Ready detection of small amount of lactulose in dairy products by thin-layer chromatography. *Chromatographia, 14*(11), 621–622.
41. Mayer, J., Kranz, B., & Fischer, L., (2010). Continuous production of lactulose by immobilized thermostable β-glycosidase from *Pyrococcus furiosus*. *Journal of Biotechnology, 145*(4), 387–393.
42. Montanes, F., Olano, A., & Ibanez, E., (2007). Modeling solubilities of sugars in alcohols based on original experimental data. *Thermodynamics, 53*(9), 2411–2418.
43. Montgomery, E. M., & Hudson, C. S., (1930). Relations between rotatory power and structure in the sugar group XXVII. Synthesis of a new disaccharide ketose (lactulose) from lactose. *Journal of the American Chemical Society, 52*(5), 2101–2106.
44. Montilla, A., Del, C. M. D., Sanz, M. L., & Olano, A., (2005). Eggshell as catalyst of lactose isomerization to lactulose. *Food Chemistry, 90*(4), 883–890.
45. Montilla, A., Moreno, F. J., & Olano, A., (2005). A reliable gas capillary chromatographic determination of lactulose in dairy samples. *Chromatographia, 62*(5, 6), 311–314.
46. Moscone, D., Bernardo, R. A., Marconi, E., Amine, A., & Palleschi, G., (1999). Rapid determination of lactulose in milk by microdialysis and biosensors. *Analyst, 124*(3), 325–329.
47. Nacheva, I., Loginovska, K., & Valchkov, A., (2018). Acidification rate and survival of microorganisms in kefir from goat's milk, enriched with lactulose. *Khranitelnovkusova Promishlenost, 1*, 29–33.
48. Nagasawa, T., Kawashima, T., Suzuki, K., & Kawase, K., (1973). *Process for Preparing a Lactulose Powder* (p. 8). Washington-DC: United States Patent No. 3,716,408.
49. Nagendra, R., Viswanatha, S., Kumar, S. A., Murthy, B. K., & Rao, S. V., (1995). Effect of feeding milk formula containing lactulose to infants on fecal bifidobacterial flora. *Nutrition Research, 15*(1), 15–24.
50. Nooshkam, M., Madadlou, A., (2016). Microwave-assisted isomerization of lactose to lactulose and Maillard conjugation of lactulose and lactose with whey proteins and peptides. *Food Chemistry, 200*, 1–9.
51. Nooshkam, M., Babazadeh, A., & Jooyandeh, H., (2018). Lactulose: Properties, techno-functional food applications, and food-grade delivery system. *Trends in Food Science & Technology, 80*(1), 23–34.
52. Nucera, G., Gabrielli, M., Lupascu, A., Lauritano, E. C., Santoliquido, A., Cremonini, F., Cammarota, G., et al., (2005). Abnormal breath tests to lactose, fructose, and sorbitol in irritable bowel syndrome may be explained by small intestinal bacterial overgrowth. *Aliment Pharmacology and Therapeutics, 21*(11), 1391–1395.
53. Olano, A., & Corzo, N., (2009). Lactulose as a food ingredient. *Journal of the Science of Food and Agriculture, 89*(12), 1987–1990.
54. Panesar, P. S., & Kumari, S., (2011). Lactulose: Production, purification, and potential applications. *Biotechnology Advances, 29*(6), 940–948.
55. Panesar, P. S., Kumari, S., & Panesar, R., (2010). Potential applications of immobilized β-galactosidase in food processing industries. *Enzyme Research, 2010*, 7. Online article ID: 473137.
56. Panesar, P. S., Panesar, R., Singh, R. S., Kennedy, J. F., & Kumar, H., (2006). Microbial production, immobilization, and applications of β-D-galactosidase. *Journal Chemical Technology and Biotechnology, 81*(4), 530–543.

57. Parekh, S. L., Smitha, B., Hati, S., & Aparnathi, K. D., (2016). Lactulose: Significance in milk and milk products. *International Journal of Current Microbiology and Applied Science, 5*(11), 721–732.
58. Parrish, F. W., Hicks, K., & Doner, L., (1980). Analysis of lactulose preparations by spectrophotometric and HPLC methods. *Journal of Dairy Science, 63*(11), 1809–1814.
59. Paseephol, T., Small, D. M., & Sherkat, F., (2008). Lactulose production from milk concentration permeate using calcium carbonate-based catalysts. *Food Chemistry, 111*(2), 283–290.
60. Paterson, A. H. J., & Kellam, S. J., (2009). Transformation of lactose for value-added ingredients. In: Corredig, M., (ed.), *Dairy-Derived Ingredients: Food and Nutraceutical Uses* (pp. 625–643). Boca Raton: CRC Press.
61. Petuely, V. F., (1957). The bifidus factor. *Deutsche Medizinische Wochenschrift (German Medical Weekly), 82*(46), 1957–1960.
62. Playne, M. J., & Crittenden, R. G., (2009). Galacto-oligosaccharides and other products derived from lactose. In: Fox, P. F., & McSweeney, P. L. H., (eds.), *Advanced Dairy Chemistry. Lactose, Water, Salts, and Minor Constituents* (Vol. 3, pp. 121–202). New York: Springer.
63. Prasad, S., Dhiman, R. K., Duseja, A., Chawla, Y. K., Sharma, A., & Agarwal, R., (2007). Lactulose improves cognitive functions and health-related quality of life in patients with cirrhosis who have minimal hepatic encephalopathy. *Hepatology, 45*(3), 549–559.
64. Saarela, M., Hallamaa, K., Mattila-Sandholm, T., & Matto, J., (2003). The effect of lactose derivatives lactulose, lactitol, and lactobionic acid on the functional and technological properties of potentially probiotic *Lactobacillus* strains. *International Dairy Journal, 13*(4), 291–302.
65. Schmidt, C. M., Nedele, A. K., & Hinrichs, J., (2020). Enzymatic generation of lactulose in sweet and acid whey: Feasibility study for the scale-up towards robust processing. *Food and Bioproducts Processing, 119*, 329–336.
66. Schmidt, C. M., Zürn, T., Thienel, K. J. F., & Hinrichs, J., (2017). Development, optimization, and validation of an HPLC-ELSD method for the analysis of enzymatically generated lactulose and saccharide by-products. *Food Chemistry, 215*, 347–353.
67. Schumann, C., (2002). Medical, nutritional, and technological properties of lactulose: An update. *European Journal of Nutrition, 41*(1), 17–25.
68. Schuster-Wolff-Bühring, R., Fischer, L., & Hinrichs, J., (2010). Production and physiological action of the disaccharide lactulose. *International Dairy Journal, 20*(11), 731–741.
69. Seki, N., & Saito, H., (2012). Lactose as a source for lactulose and other functional lactose derivatives. *International Dairy Journal, 22*(2), 110–115.
70. Song, Y. S., Shin, H. Y., Lee, J. Y., Park, C., & Kim, S. W., (2012). β-galactosidase immobilized microreactor fabricated using a novel technique for enzyme immobilization and its application for continuous synthesis of lactulose. *Food Chemistry, 133*(3), 611–617.
71. Song, Y. S., Lee, H. U., Park, C., & Kim, S. W., (2013). Batch and continuous synthesis of lactulose from whey lactose by immobilized β-galactosidase. *Food Chemistry, 136*(2), 689–694.
72. Song, Y. S., Suh, Y. J., Park, C., & Kim, S. W., (2013). Improvement of lactulose synthesis through optimization of reaction conditions with immobilized β-galactosidase. *Korean Journal of Chemical Engineering, 30*(1), 160–165.

73. Tabatabaie, F., & Mortazavi, A., (2008). Influence of lactulose on the survival of probiotic strains in yoghurt. *World Applied Sciences Journal, 3*(1), 88–90.
74. Tamura, Y., Mizota, T., Shimamura, S., & Tomita, M., (1993). Lactulose and its application to the food and pharmaceutical industries (*Bifidus* factor). *International Dairy Federation Bulletin, 289*, 43–53.
75. Thornton, J. R., & Heaton, K. W., (1981). Do colonic bacteria contribute to cholesterol gallstone formation? Effects of lactulose on bile. *British Medical Journal (Clinical Research Edition), 282*(6269), 1018–1020.
76. Verma, A., & Shukla, G., (2013). Administration of prebiotic inulin suppresses 1,2 dimethylhydrazine dihydrochloride induced procarcinogenic biomarkers fecal enzymes and preneoplastic lesions in early colon carcinogenesis in Sprague Dawley rats. *Journal of Functional Foods, 5*(2), 991–996.
77. Wang, H., Yang, R., Hua, X., Zhao, W., & Zhang, W., (2013). Enzymatic production of lactulose and 1-lactulose: Current state and perspectives. *Applied Microbiology and Biotechnology, 97*(14), 6167–6180.
78. Wang, M., Gasmalla, M. A. A., Tessema, H. A., Hua, X., & Yang, R., (2017). Lactulose production from efficient isomerization of lactose catalyzed by recyclable sodium aluminate. *Food Chemistry, 233*, 151–158.
79. Westhoff, G. M., Kuster, B. F. M., Heslinga, M. C., Pluim, H., & Verhage, M., (2014). Lactose and derivatives. *Ullmann's Encyclopedia of Industrial Chemistry, 2014*, 1–9.
80. Xie, X. L., He, J. T., Wang, Z. T., Xio, H. Q., Zhou, W. T., Du, S. H., Xue, Y., & Wang, Q., (2018). Lactulose attenuates METH Induced neurotoxicity by alleviating the impaired autophagy, stabilizing the perturbed antioxidant system and suppressing apoptosis in rat striatum. *Toxicology Letters, 289*(1), 107–113.
81. Zhang, Z., Wang, H., Yang, R. J., & Jiang, X. Y., (2010). A novel spectrophotometric method for quantitative determination of lactulose in food industries. *International Journal of Food Science Technology, 45*(2), 258–264.
82. Zokaee, F., Kaghazchi, T., Zare, A., & Soleimani, M., (2002). Isomerization of lactose to lactulose-study and comparison of three catalytic systems. *Process Biochemistry, 37*(6), 629–663.

CHAPTER 2

CASEIN-DERIVED BIOACTIVE PEPTIDES: ISOLATION, CHARACTERIZATION, AND FUNCTIONAL ASPECTS

BRUNDA S. MURTHY, SONIA MOR, LAXMANA N. NAIK, N. VEENA, and SURENDRA N. BATTULA

ABSTRACT

Milk is a wholesome food containing proteins, fat, lactose, vitamins, and minerals. Total proteins in bovine milk consist of 80% of casein (CN) and 20% of whey or serum proteins. Apart from nutritional benefits, milk proteins have biological activities and serve as a good source of bioactive peptides (BPs). BPs are food-derived peptides with a specific amino acid sequence of about 3–20 residues that exert physiological benefits in the body besides their nutritional value. BPs are the constituents of primary amino acid (AA) sequences, which are not active within the protein sequence and are released during gastrointestinal (GI) digestion or food processing; these are known to regulate some specific physiological functions in the body. The bio-functionalities exhibited by bio-peptides include anticarcinogenic, antidiabetic, antihypertensive, antioxidative (AO), antimicrobial, antithrombotic, hypo-cholesterolmic, mineral binding, and opioid activities. Due to environmental and lifestyle modifications, the population is more prone to various diseases. While undoubtedly drugs play a significant role in the treatment of such diseases, the BPs from CN would also serve as a good natural

Functional Dairy Ingredients and Nutraceuticals: Physicochemical, Technological, and Therapeutic Aspects. Megh R. Goyal, N. Veena, PhD & Santosh K. Mishra (Eds.)

source either as preventative or therapeutic complements to the drugs. In the last few years, there has been an increased research in the field of BPs due to their health promoting factors. Presently, proteins of milk are considered as the best source of BPs. The synthesis of BPs with high concentration and specific bioactivity poses a great challenge, while using different proteolytic enzymes, their isolation, characterization, and utilization.

2.1 INTRODUCTION

Milk is a biological liquid secreted by female mammals as a source of nutrients and immunological protection for their young ones. It is a complex fluid consisting of excellent quality proteins, lipids, carbohydrates, minerals, vitamins, enzymes, and other essential components. Therefore, milk is recognized as 'nearly complete food' existing in nature for the growth and maintenance of health of human beings [157]. Apart from the conventional role of milk in nutritional value, it is also known to contain certain bioactive components. Bioactive components are defined as the compounds present in food, which exert physiological function in addition to their nutritive value. Milk contains several bioactive components, such as peptides, fatty acids, enzymes, oligosaccharides, immunoglobulins (Igs), and growth factors that exert a plethora of health benefits [58, 174].

Cow milk is known to consist of about 3.5% protein. The most important function of protein is to serve as a source of the amino acids (AAs) and nitrogen to the offspring and also provides the part of dietary proteins for the human beings to meet their nutrient requirements. Besides, they also exhibit important biological function; they help in the uptake of minerals and vitamins and also contain specific proteins, which display a protective role in the body. Milk proteins are composed of 80% of caseins (CNs) and 20% of whey proteins (WPs). The CNs are classified as α_{S1}-, α_{S2}-, β-, and κ-CN; while serum proteins consist of β-lactoglobulin (β-Lg), α-lactalbumin (α-La), bovine serum albumin (BSA) and Igs, and some minor proteins [243].

The primary biological function of CN is to supply AAs necessary for the growth and development of neonates. Research studies have highlighted diverse physiological roles of CN that include prevention of pathological calcification of mammary gland [79]. Milk proteins have a protein efficiency ratio (PER) of 3.1 compared to 2.9 for CN and 3.6 for WP. The WPs have higher biological value of 104 compared to 100 for egg protein and 77 for CN [7]. Many evidences suggest that milk proteins apart from providing energy and nutritional functions also act as physiological regulators [195, 220].

This chapter focuses on isolation, technological, and health aspects of bioactive peptides that are derived from casein.

2.2 BIOACTIVE PEPTIDES: GENERAL INFORMATION

Bio-peptides or bioactive peptides (BPs) are food-derived peptides with a definite amino acid sequence containing 3–20 residues, which are known to exert a positive impact on health, apart from their nutritional value. The beneficial health impact is attributed to the particular sequence and composition of AAs of that peptide [196]. BPs are the constituents of primary AA sequences, which are not active within the polypeptide chain, also called "cryptids" or hidden peptide but are liberated during GI digestion or food processing [224].

Generally, the BPs vary from di-, tri-, oligopeptides to higher molecular weight (MW) peptides [46, 75]. Once BPs are liberated, they can act as regulatory substrates with hormonal-like action. The effect shown by such BPs may be encouraging or depressing. Some activities exhibited of BPs include: antihypertensive, antioxidative, antimicrobial, antithrombotic, antidiabetic, and immunomodulatory activities.

The type of BPs released in the GI tract depends on the quantity and composition of milk and other foods consumed, pH, and activity of enzymes in the tract, which are in turn influenced by factors like parity, genetics, nutrition, and health condition of the human beings. Thus, healthier food is a basic requirement for accomplishing complete benefits of these bio-peptides [202].

2.2.1 MILK PROTEINS AS A SOURCE OF BIOACTIVE PEPTIDES

Physiological, genetic, and nutritional factors influence the protein concentration in milk [45]. They have been extensively studied for over two centuries [54] and are perhaps one of the best-characterized food protein systems. The biological role of milk peptides exhibiting multifunctional properties has been increasingly acknowledged. These can have beneficial effects on the physiology and metabolism of human beings by direct means or by enzymatic hydrolysis *in vivo* or *in vitro* [95, 96].

BPs from milk proteins have been claimed to enhance health factors by lowering the risk of disease or improve the physiological function by exerting antihypertensive, anti-oxidative, immune-stimulating, cyto-modulatory, antimicrobial activities. The bioactivity of the peptides is dormant or inactive, incomplete, or lacking in the native protein. Nevertheless, the biological effect could be accomplished by proteolysis to liberate the BPs from the protein. Milk of different mammalian species (including cow, buffalo, caprine, yak, camel, and ass) is being investigated largely for the production of BPs and to elucidate their influence on well-being and biological effects. Among all, bovine milk has been demonstrated to be the greatest source and has been extensively researched [164]. BPs usually consists of 2–20 AA residues and have MW less than 6,000 Daltons (Da) [192].

2.2.1.1 *CASEIN DERIVED BIOPEPTIDES*

CN is a phosphoprotein, obtained by the acidification of skim milk to a pH of 4.6 at 20°C. The electrophoretic technique and AA sequence homology evaluation revealed that CN is a heterogeneous protein consisting of αs_1-, αs_2-, β- and κ-CNs [233]. Each of the CN differs in the number and sequence of AAs, number of phosphate containing residues and calcium-binding properties. The α_{S1}-, α_{S2}-, β-CNs can bind Ca^{2+} and precipitate at relatively lower concentrations of Ca^{2+}; and thus, they are named as Ca-sensitive CNs. However, κ-CN is not sensitive to Ca^{2+} concentrations. This can stabilize up to ten times the weight of the Ca-sensitive CNs and contains single phosphoryl serine residue [5].

The bioactivity imparted from the CN-derived peptides depends on their specific AA sequence, which may comprise of minimum 3 AAs to a maximum of 20 AAs, with the possibility for some peptides to display multifunctional properties [139]. However, the larger MW peptides are known to exhibit technological properties and smaller MW peptides give bioactive effects [206]. The α_S- and β-CN has potential to release >20,000 BPs [241]. In 1950, the first BP was derived from CN that promoted Vitamin D independent bone calcification in rachitic infants [141]. The α_S-CN (α_{S1} and α_{S2}) is abundant in hydrophobic, aromatic, and negatively charged amino acids.

The proteins with hydrophobic AAs are known to be precursors for biopeptides with ACE inhibitory, AO, and antidiabetic properties. The presence of higher number of proline (Pro) residues in the case of α_{S1}- and β-CN, followed by absence of cysteine residue gives randomness, flexibility, and more susceptibility towards proteolysis [71, 214].

2.3 PRODUCTION OF BIOACTIVE PEPTIDES

BPs derived from milk become active only upon release from their parent protein. There are various ways by which the hidden BPs are released from their parent source. The mechanisms by which BPs can be produced are [138]:

- Combination of fermentation and hydrolysis;
- Enzymatic hydrolysis with digestive enzymes;
- Fermentation with proteolytic starter cultures;
- Hydrolysis by enzymes derived from proteolytic organisms.

The BPs liberated from food proteins are intermediate products that are extracted from complex peptidic hydrolysates, wherein the concentrations of these peptides are low. A typical protein related process consists of extraction, separation, and quantification, identification, data analysis and interpretation. There are different extraction and fractionation techniques based on physico-chemical and structural properties, such as: solubility, hydrophobicity, MW, isoelectric point, etc. Advanced isolation and fractionation methods for food proteins and bio-peptides are a foremost step of proteomic analysis [123].

The main principle involved in the production of peptides is the break-down of proteins to smaller peptides by the action of enzymes directly or by the enzymes released from the starter culture during fermentation. This is referred to hydrolysis of proteins. There are different types of hydrolysis: enzymatic and chemical hydrolysis. Chemical method, with the use of acid or alkali, is difficult to control and is known to reduce the nutritional value by destroying L-AAs and formation of toxic compounds [103]. In contrast, enzymatic hydrolysis is the best option because the destruction of AAs is reduced to a greater extent by avoiding extreme temperature and pH conditions as required in the case of chemical hydrolysis and thus the nutritional quality of hydrolysates is not largely affected [23].

Further, desired peptides could be produced due to the specificity of the enzyme. It has been used as an eco-friendly and safe method to extract peptides, and this process can improve nutritional and techno-functional properties and prevents severe damage to the proteins as formed in other methods [107]. No toxic chemicals are produced, but show some limitations, such as, requirement of lengthy-time period and highly-priced enzymes, which can be overcome by the use of immobilized enzymes [123].

Hydrolysis of proteins by enzymes can attribute to both the technological and bio-functional properties, thus resulting in hydrolysates [44, 136, 161, 207]. Controlled partial hydrolysis by proteinases will result in the production of partly degraded proteins that are significant for obtaining novel functional foods; if continued in excess, it can even destroy the activity of the peptides [196]. The choice of enzymes selected for hydrolysis needs to be assessed critically because it affects the composition of the hydrolysates [150].

2.3.1 HYDROLYSIS WITH DIGESTIVE ENZYMES

In humans, during digestion of food, two main processes occur simultaneously: (i) mechanical transformation to reduce the size of the food particles; and (ii) enzymatic transformations, where macromolecules are hydrolyzed into smaller molecules that can be absorbed into the bloodstream [64]. Initially, food proteins pass through the stomach before they get cleaved by pancreatic proteinases in the intestine, where food proteins stay for a long time, which depends on composition and extent of size reduction during chewing of food. In the stomach, the proteins undergo hydrolysis by hydrochloric acid and pepsin at acidic pH in the presence of gastric juice. Then, they are degraded by pancreatic proteases, trypsin, and chymotrypsin at neutral pH in the small intestine [190]. The un-hydrolyzed and partially hydrolyzed proteins that reach the large intestine are further specifically acted upon by the microbial enzymes [8]. GI digestion can also direct to the production of bioactive fractions from dormant or less active precursors [145].

Several researchers have tried to develop more or less complex *in vitro* static or dynamic, mono- or multi-phase model simulation systems to mimic GI digestion and to evaluate the stability and/or production of BPs [177, 187]. This simulation system follows sequential degradation with pepsin and pancreatic enzymes to mimic the GI conditions.

Many peptides are produced using GI enzymes like trypsin and pepsin. ACE inhibitory peptides and casein-phosphopeptides (CPPs) are usually released by trypsin [53]. In a study on ACE-inhibitory activity of fermented milk before and after simulated physiological digestion [72], it was observed that simulated physiological digestion promoted the production of BPs from the proteins present in the fermented milk.

Miquel et al. [146] studied the formation and identification of CPPs by *in vitro* simulation of physiological digestion. Picariello et al. [169] reported that peptides having more neutral and basic AAs were degraded faster than those with more acidic AAs. Contreras et al. [29] reported the resistance of CN-derived BPs, RYLGY, AYFYPEL, and YQKFPQY, with hypotensive activity, against the simulated GI digestion.

2.3.2 HYDROLYSIS BY ENZYMES: DERIVED FROM PROTEOLYTIC MICROORGANISMS

Proteinases (such as: alcalase, flavoenzyme, protease K and pronase) obtained from microorganisms (e.g., *Aspergillus*, *Bacillus*), plants (e.g., papain) and animal (e.g., pepsin, and trypsin) sources are used in the production of BPs. Commercially available food-grade endo- and exo-peptidase enzymes alone and in combination could potentially create a wide range of BPs, including bioactive sequences based on the specificity of proteolysis [150]. Various enzymes and their combinations (alcalase, pepsin, chymotrypsin, and thermolysin) and bacterial and fungal-derived enzymes sources are used for the production of BPs from dietary proteins [161].

CN hydrolyzed by the proteases associated with the cell wall of *Lb. helveticus* CP790 exerts hypotensive activity when tested on spontaneously hypertensive rats [241]. Various ACE inhibitory bio-peptides were extracted from the hydrolysate. Antihypertensive peptide was identified as the fragment β-CN (169–175), whose AA sequence was reported to be KVLPVPQ [117].

2.3.3 FERMENTATION WITH PROTEOLYTIC STARTER CULTURES

Majority of the starter cultures employed in the dairy industry is extremely proteolytic. BPs can be generated by these cultures used for the manufacture of fermented dairy products. The proteolytic activity of lactic acid bacteria (namely, *Lb. helveticus, L. lactis*, *Lb. delburkeii,* etc.), is well characterized. They contain cell wall bound proteases and various distinct intracellular peptidases, such as: endo-, amino-, di-, and tri-peptidases [56].

BPs generated during fermentation of milk are further acted upon by the GI enzymes and liberate other BPs. In addition, it has been revealed that the peptide sequence of milk proteins is extensively diverse after fermentation by starter culture, which suggests that the bacterial proteolysis can be an impending source for production of BPs [8].

2.3.4 COMBINATION OF FERMENTATION AND HYDROLYSIS

The hydrolysis is carried out using microorganisms in combination with enzymes either, digestive enzymes, or enzymes derived from plants or microbes. Ahtesh et al. [4] conducted an experiment to evaluate the effect of combination of enzyme (flavoenzyme) and microorganisms (*Lb. helveticus*) on the ACE inhibitory activity. Reconstituted skim milk (12% solids) and whey protein concentrate (4%) inoculated with *Lb. helveticus* were incubated at 37°C for 0, 4, 8, and 12 hours with and without flavoenzyme. The strains of *Lb. helveticus* were capable of increasing the ACE inhibition to more than 80% after eight hours of fermentation in combination with flavoenzyme in comparison to the ones incubated with *Lb. helveticus* but without the addition of enzyme.

2.4 SEPARATION AND ISOLATION OF BIOACTIVE PEPTIDES

Numerous techniques have been used for the separation and isolation of BPs in a laboratory as well as at industrial scale, but lack of suitable large-scale technology limits their commercial production. Selected techniques are discussed in this section.

2.4.1 CENTRIFUGATION

It can be used for separating crude mixtures into various layers based on MW, size, and shape of the components considering the coefficient of sedimentation. The peptides having a lower coefficient of sedimentation can be separated from the un-hydrolyzed molecules by centrifugation at higher speed [201].

2.4.2 PRECIPITATION

The commonly used precipitating agent is ammonium sulphate. The addition of high concentration of salts (e.g., sodium chloride) to a solution results in the increased interaction in the protein thereby results in the aggregation and precipitation, which is also called the salting-out method. The amount of salt required for precipitation of protein differs from one type to other, which allows selective protein separation. An alternative to salting-out technique, lower concentration of salts can be employed to enrich protein fractions previously precipitated [35].

2.4.3 MEMBRANE FILTRATION

Ultrafiltration (UF) is used for separation of peptides and for removal of proteins and other larger molecules mainly on the basis of their MW. Stepwise use of UF of different MW cut-off membranes has been helpful in extracting smaller bio-peptides from high MW proteins and enzymes.

Turgeon et al. [222] employed a two-step UF process and they were able to generate fractions that are rich in low MW peptides with molecular mass <2000 Da. Utilizing pepsin hydrolyzed α-La and pepsin and trypsin hydrolyzed β-Lg as source, Pihlanto [170] enriched opioid BPs (namely, alpha-, and beta-lactorphin, respectively) through the use of selective UF membranes (30 and 1 kDa) [172].

UF has been used successfully to concentrate ACE inhibitory [25, 78, 176] and AO peptides [186]. Ulukoet et al. [226] fractionated hydrolysates with ACE inhibitory activity using milk protein concentrate as substrate and neutrase as enzyme at its optimal conditions of temperature and pH using a 3-step USP enzymatic hydrolysis-membrane separation at lab and pilot scale. Membrane separation may enhance the AO activities of hydrolysates. He et al. [71] observed enhanced antioxidative activity of rapeseed protein hydrolysates in the majority of samples after fractionation.

Mendis et al. [142] reported that the low MW BPs have better antioxidant activities and the observation was similar to that reported by various workers [49, 87, 112, 149].

The nanofiltration (NF) technique has high selectivity in the isolation of fractions, which have a MW in the range of 300–1000 Da. Inorganic NF membranes, in particular, have considerably greater separative properties due to the amphiprotic behavior with varying pH. Butylina et al. [21] reported the fractionation and isolation of BPs in the sweet whey using UF (regenerated cellulose membrane) and NF (sulphonated polyether sulphone) membrane with cut-off limits of 10 and 1 kg/mol, respectively. Ting et al. [218] reported a NF treatment for the fractionation of whey protein derived BPs, which were selectively isolated on the basis of their charges and sizes.

Electro-membrane filtration is a potential alternative methodology to generate charged BPs. This merges traditional membrane filtration (UF or NF) with electrophoretic techniques. In comparison to pressure-driven membranes, a rise in selectivity for the separation of charged fragments can therefore be expected [49].

2.4.4 ELECTROPHORETIC TECHNIQUES

This technique separates the proteins on the basis of charge, mass/charge (m/z) ratio, size, or shape, etc. It can be applied as an analytical and preparative tool, where one dimension polyacrylamide gel electrophoresis (PAGE) is used as pre-fractionation step [47, 66]. Based on the nature of the sample and its complexity, the fractions that are separated can be analyzed directly by MS or can employ subsequent separation method in a 2nd dimension as a means to separate the proteins based on their MW. The electrophoretic bands can also be used to confirm the protein hydrolysis and the MW of the BPs formed [230]. Tricine sodium dodecyl sulfate (SDS) PAGE technique of milk proteins has been achieved using digital densitometry; and using this method the isolation, detection, and quantification of cow milk CN fractions and their hydrolysates was developed [162]. Hydrolysis of CN in cheese samples was analyzed using PAGE. Based on the electrophoretic analysis, it was observed that BPs were generated from the β- and α_{S1}-CN fractions due to the action of a trypsin-like enzyme, during ripening of cheese [3].

2.4.5 CHROMATOGRAPHIC TECHNIQUES

Liquid chromatographic (LC) methods are frequently applied in the protein pre-fractionation before comprehensive analysis. Based on charge, size, hydrophobic or specific property, the separation of various proteins can be accomplished. Among all the methods, ion-exchange chromatography (IEC) is the most common, where proteins are separated based on their charge. Acidic proteins are separated by anion-exchangers and the basic proteins are separated using cation-exchangers. Ruiz et al. [187] employed IEC with a cation-exchanger to fractionate ewes' milk CNs.

In size-exclusion chromatography (SEC) method, proteins are separated based on their MW. SEC in aqueous separation system and gel-permeation chromatography (GPC) in non-aqueous separation system are the permeation methods that are solely based on the molecular size. At laboratory scale, peptides have been separated by either SEC or reverse phase-high performance liquid chromatography (RP-HPLC) followed by characterization using MS [72, 160]. This has resulted in the generation of purified peptides with high potency [114]. In RP-HPLC, proteins are separated based on their hydrophobic property; the peptides are adsorbed on the stationary phase bearing hydrophobic residues and are eluted by raising the organic solvent concentration, acetonitrile being the most commonly used solvent. This technique is widely used in proteomics along with IEC and MS. Pre-fractionation with chromatographic techniques is also applied to study the post-translational changes, e.g., glycosylation, and phosphorylation. These modifications are studied using affinity chromatography [72], which employs extremely specific biological interactions, i.e., antigen-antibody, enzyme-substrate/inhibitor, etc.

2.5 CHARACTERIZATION OF BIOACTIVE PEPTIDES

There are various techniques for characterization of BPs. SDS-PAGE is commonly applied to estimate the MW and the purity of BPs [36]. Glycine-SDS-PAGE is applied to comparatively larger BPs. For shorter BPs, the resolution of glycine-SDS-PAGE gel is typically low. Hence, tricine-SDS-PAGE is commonly used to isolate and characterize the proteins smaller than 30 kDa. It gives a better resolution than the glycine-SDS-PAGE [194].

With suitable columns and conditions, HPLC may give valuable information for peptide characterization. The three major types of HPLC usually applied in BPs extraction are: distinctions in peptide size (SEC), net charge (IEC) or hydrophobicity (RP-HPLC) [62]. Size-exclusion-HPLC can be used for more precise evaluation of the peptide size, with suitable columns and conditions. IEC has been proven to be exceptionally helpful for BPs separations since column packing is capable of retaining both basic (+ve charge) and acidic (–ve charge) peptides. Semi-preparative/preparative RP-HPLC remains the more common method for BPs separation and fractionation. It has greater choice than other LC techniques in terms of speed and efficiency [120, 121].

The AA analyzer and sequencing of protein are generally applied to estimate the AA composition and their sequence in unidentified BPs. MS analysis based on Edman degradation method is most common in sequencing of BPs [27, 101, 109, 133, 148]. It is carried out by tryptic hydrolysis of sample accompanied by either matrix-assisted laser desorption/ionization (MALDI) or tandem mass spectrometric (MS/MS) analysis of each BPs in the hydrolysate.

In MS/MS-based protein identification, performed usually on an electron spray/quadrupole time-of-flight (Q-TOF) or MALDI/tandem time-of-flight, the MS/MS spectrum achieved from each BP (approximately 10–25 AA residues) having adequate information, which enables its native protein to be identified making use of database searching of the uninterrupted MS/MS spectrum. One of the useful programs is Mascot that is accessible to on the internet; hence, MS/MS is applied for identification of proteins [65] more precisely than the Electrospray ionization (ESI)-MS/MS [237] and MALDI-TOF-MS [119, 182, 229] methods. Because of the greater selectivity, speed, and sensitivity, MS-based techniques have been demonstrated to be highly potent and ideal for the analysis of phosphorylation of proteins when compared with the other biochemical methods [132].

Different types of cheese were evaluated for their ACE-inhibitory property and BPs with <1000 Da were estimated by HPLC-MS/MS and MS/MS. The κ-CN hydrolysate by RP-HPLC-MS/MS allowed the detection of BP sequences with impending anti-oxidative property [61]. The antimicrobial BPs in a pepsin digest of α_{S2}-CN were separated by IEC followed by fractionation using semi-preparative RP-HPLC; and the detection of these BPs was achieved by online and off-line ESI-MS/MS [113, 116].

2.6 FUNCTIONAL BENEFITS OF BIOACTIVE PEPTIDES

A number of scientific evidences have proposed that BPs control specific biological functions in our body. BPs liberated from CNs may influence the cardiovascular, nervous, immune, and GI systems (Figure 2.1) [68, 199, 205, 210]. Some peptides are also known to exhibit AO and mineral absorption properties [63].

2.6.1 ANTIOXIDANT ACTIVITY

During oxidative metabolism, which is important for the survival of living body cells, many free radicals (FRs) are produced as byproducts that result in oxidative changes. The FRs are chemical compounds having an unpaired electron in the outermost orbit. During stress, these FRs have the ability to attack macromolecules, resulting in lipid peroxidation, protein degradation and DNA damage leading to mutations. This is referred to as oxidative stress, during which a rise in production of FRs, a decline in synthesis of AOs, an interruption in cellular redox equilibrium and oxidation of cellular components. This also leads to commencement of diseases such as atherosclerosis, diabetes, rheumatoid arthritis and cancer [2, 67, 77]. Arresting the FRs produced in the biological systems and food products helps to defend from these grave disorders. In this regard, the AOs play an important role.

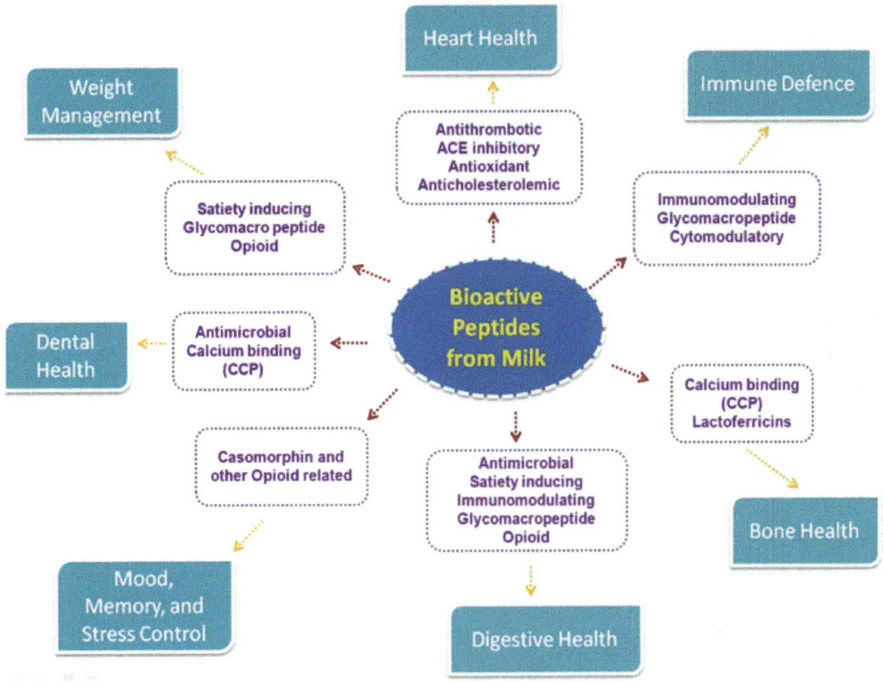

FIGURE 2.1 Role of milk derived BPs in human health.
Source: Reprinted with permission from Elsevier; from Journal of Functional Foods, Volume 1, Korhonen [98], milk-derived bioactive peptides; from science to applications, 177–187, @ Copyright 2020.

Several AOs are found in natural plants and some food proteins also possess AO capacity [34, 155]. Proteins have excellent potential as AO additive in foodstuffs, as they can hamper fat oxidation through several ways, such as: inhibition of reactive oxygen species, scavenging FRs, chelation of metal ions, reduces the formation of hydro-peroxides, and modification of the physical attributes of food systems. Both plant and animal proteins contain BPs with AO properties. Peptides from bovine milk proteins [73, 83, 168, 170] and yak milk CN [122] have also been used as natural source of AOs.

The mechanism of AO activity of a protein or a peptide depends on the physicochemical properties of the AAs. Preliminary studies suggest that the AO activity of peptides depends on its AA composition, sequence, structure, and hydrophobicity [26]. Studies on quantitative structure-activity relationship (QSAR) suggest that N- and C-terminal AAs of the biopeptide are major indicators of AO property [111]. During hydrolysis, when the peptide bonds are cleaved, the protein loses its original structure, and the number of free AAs are increased. These free AAs act as a source of protons or electrons that maintain high redox potential. AO activity is relying on the presence of specific AA residues and the position of these AAs within the peptide sequence [175].

The specificity of protease action influences the peptides that are formed. Hydrolysis with pepsin results in exposure of hydrophobic AAs as it has its specific cleavage site at hydrophobic and aromatic AAs. The exposure of these hydrophobic residues facilitates electron transfer from the peptides [246]. It is observed that majority of the AO bio-peptides identified so far have a hydrophobic group at the N-terminal. Net charge and molecular polarity of the C-terminus AAs are significant indicators of AO activity. The AA next to the C-terminal is demonstrated to be the most important in AO activity and if this AA has greater H-bonding, steric property and less hydrophobic nature, it increases the AO potential of the peptide. Hence, the position of acidic/basic AAs (Asp, Glu, His, Arg, Lys) or hydrophilic AAs (Ser, Thr) is crucial for the AO activities [26]. Of the potent CN-derived AO peptides, the 2nd AA nearby the C-terminal is classically Glu, Pro or Leu; and BPs obtained from β-Lg specifically have Glu, Arg or Asp in this position [42]. Photochemical oxidation with hydroxyl (OH) method was used to measure the oxidative reactivity of wide range of AAs, which was observed to be in the order [202]:

Cys > Trp/Tyr > Met > Phe > His > Ile > Leu > Pro

2.6.1.1 AO PEPTIDES FROM CASEIN

BPs derived from CN hydrolysis have been demonstrated to have AO property [24, 209, 234] as shown in Table 2.1. The AO hydrolysates can be produced from cow and buffalo milk CN using alcalase, trypsin, pepsin, and papain. The greatest degree of hydrolysis was obtained using alcalase followed by trypsin. Other than papain hydrolysates, all hydrolysates displayed defense against plasmid DNA nicking [203].

Phosphorylated serine groups are present in the polar domains of the different CNs (α_{S1}, α_{S2} and β) and their specific sequence serP-serP-serP-Glu-Glu is a potent chelator of cations by forming complex with iron (Fe), zinc (Zn) and calcium (Ca). CNs in native form; and their hydrolytic products in the solution serve as natural chelators to prevent fat autoxidation in food emulsions [92].

Suetsuna et al. [209] isolated antioxidant peptide from the casein hydrolysate obtained after digestion with pepsin and it was identified to have a sequence of Try-Phe-Try-Pro-Glu-Leu; this peptide has also been reported to have strong superoxide anion activity mainly at Glu-Leu residues.

TABLE 2.1 Anti-Oxidative Peptides Derived from Bovine Casein

Peptide Sequence	Fragment	Casein Type	References
APFPQVF	26–32	α_{S1}-CN	[217]
DKIHPF	47–52	β-CN	[217]
FALPQYLK	174–181	α_{S2}-CN	[114]
GLPQEVL	10–16	α_{S1}-CN	[217]
HIQKEDVPSER	95–104	α_{S1}-CN	[110]
IQY	28–30	κ-CN	[114]
KVLPVPE	169–175	β-CN	[188]
KVLPVPQK	169–176	β-CN	[184]
PVQPF	115–119	β-CN	[217]
PYVRYL	203–208	α_{S2}-CN	[114]
TVQVTSTAV	161–169	κ-CN	[188]
VAPFPQ	25–30	α_{S1}-CN	[217]
VKEAMAPK	98–105	β-CN	[184]
VPYPQ	178–182	β-CN	[184]
YVL	30–32	κ-CN	[114]

2.6.2 *ANTI-DIABETIC PEPTIDES*

To overcome Diabetes mellitus (DM), therapeutics in use include: insulin injections, inhalers, sensitizers, incretin mimetic, alpha-amylase, alpha-glucosidase inhibitors (Acarbose, Voglibose, and Miglitol) and dipeptidyl peptidase-IV (DPP-IV) inhibitors (Sitagliptin, Saxagliptin, Vildagliptin, and Alogliptin). Synthetic alpha-glucosidase inhibitors are linked with adverse effects, such as: stomach pain, gas production and diarrhea [213]. It was reported that milk proteins, mostly, BPs produced by them were also associated with the control of insulin production and postprandial glycemia in Type-2 diabetic and normal individuals [48].

Alpha-amylase is an important digestive enzyme secreted by salivary and pancreatic glands, is involved in the initial hydrolysis of starch and glycogen by breakdown of α (1–4) glycosidic bond and thus form oligosaccharides [191]. There is a reduction in postprandial hyperglycemia by inhibiting the alpha-amylase especially in diabetic situation [40, 106, 238]. Alpha-glucosidase found in the enterocytes of jejunum, hydrolyses the glycosidic bonds of oligosaccharides and leads to the formation of monosaccharides, which are absorbed by the blood leading to increased blood glucose levels. Thus, the inhibition of alpha-glucosidase results in delayed carbohydrate hydrolysis, reduced level of glucose absorption and resulting in reduced blood glucose and insulin levels [165] as shown in Figure 2.2.

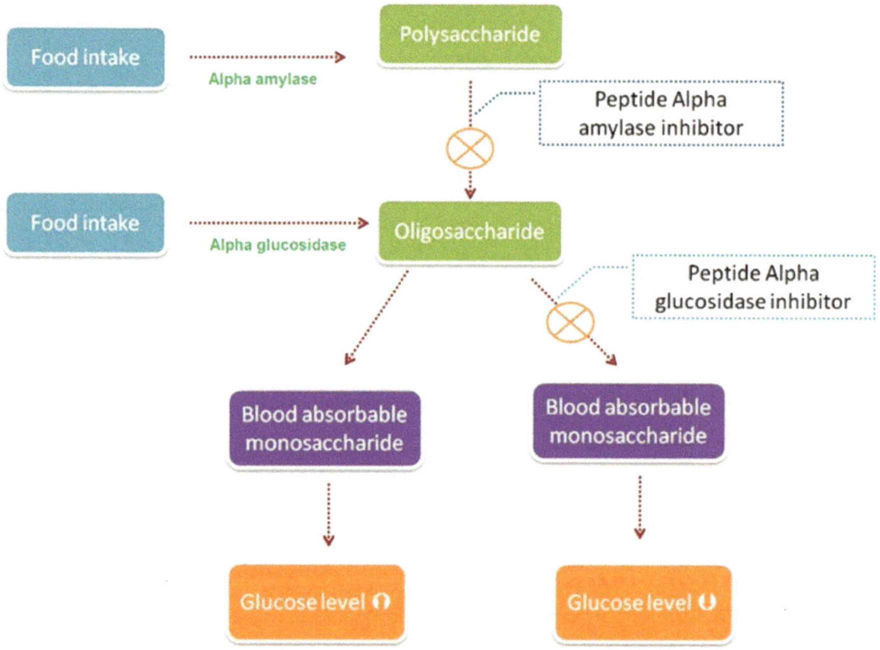

FIGURE 2.2 Action of alpha-amylase and alpha-glucosidase with and without the presence of peptide inhibitors.

Source: Reprinted from: Comprehensive Reviews from Food Science and Food Safety, Volume 17, Iwaniak et al. [84], Peptides derived from foods as supportive diet components in the prevention of metabolic syndrome, 63–81. Copyright (2020), with permission from John Wiley and Sons.

DPP-IV (a glycol protein and a plasmatic membrane enzyme) separates dipeptides (X-Pro and X-Ala) from the N-terminal side of the protein. It favors Pro, Ala, and Ser in the penultimate position, while it separates N-terminus dipeptides with Gly, Val, Trp, Lue, hydroxyl-, and dehydro proline. Incretins are gut hormones to accelerate a decline in blood glucose content. They are liberated after ingestion of food into the bloodstream and enhance the secretion of insulin by a plasma glucose-dependent system. The chief incretins are glucose dependent insulinotropic polypeptide (GIP) and the glucagon-like peptide-1 (GLP-1), which are deactivated by DPP-IV. Both the incretins contain Ala in the penultimate N-terminal position [175]. GLP-1, secreted in the gut, has a role in increased insulin secretion. It is degraded by DPP-IV within a few minutes, while 7–36 fragment of the GLP-1 amide is broken down to dipeptide and GLP-1 amide with 9–36 fragments [216]. It was reported that 95% of GLP-1 degradation by the action of DPP-IV and hence retardation of this enzyme in DM control has turned in to area of research interests. Both human and animal studies on DPP-IV indicated that inactivation of the enzyme increased the half-life of GLP-1, decreased blood glucose, and improved the defective glucose tolerance [81].

DPP-IV-inhibiting BPs were confirmed to inhibit the degradation of GIP, and other substances functioning as glucose homeostasis regulators. Moreover, DPP-IV inhibitors are more active orally than the incretin mimetics. Thus, DPP-IV inhibitors have a beneficial role in the treatment of type-2 DM [60].

2.6.2.1 ALPHA-AMYLASE AND ALPHA GLUCOSIDASE INHIBITORY PEPTIDES

The role of food proteins and peptides derived is increasingly being recognized in controlling diabetes. Many studies reveal the importance of natural food-derived peptides in combating type-2 DM. Few milk-derived peptides with DPP-IV and alpha-glucosidase inhibition property have also been reported [154, 225]. However, there is limited data available on the inhibitory aspects of alpha-amylase peptides from milk.

Janet et al. [86] found that the CN hydrolysate from raw sheep milk hydrolyzed with chymotrypsin showed better alpha-amylase inhibition of 43% after 3 hours of hydrolysis. Ramachandran and Shah [178] evaluated the alpha-glucosidase inhibitory potential of reconstituted skim milk containing Raftiline (an inulin) fermented with selective cultures of *S. thermophilus, Lb. delbrueckii, Lb. casei, Lb. acidophilus,* and *B. longum.* All strains, except *Lb. delbrueckii*, showed improvement in the alpha-glucosidase inhibitory activity in the reconstituted skim milk containing inulin.

Whey protein isolate (WPI), α-La, β-Lg, BSA, and lactoferrin were hydrolyzed using pepsin; and the hydrolysates were evaluated for their alpha-glucosidase inhibitory potentials. The BPs derived from the whey proteins had a beneficial effect on glycemic control and could serve as functional components for the treatment of type-2 DM. The alpha-glucosidase inhibitory activity of peptic hydrolysate of WP and β-Lg was 36 and 33% with IC_{50} values 4.5 and 3.5 mg/mL, respectively [102]. The alpha-glucosidase inhibitory activity of WP hydrolyzed using serine protease (an enzyme extracted from Asian pumpkin) was evaluated. It was found that the 3 kDa fractions appeared to have maximum inhibition with IC_{50} values < 2.0 mg/mL [97].

2.6.2.2 DPP-IV Inhibitory Peptides from Milk Proteins

Milk protein derived BPs (i.e., both from CN and WPs) showed *in vitro* DPP-IV inhibitory action and *in vivo* decline in plasma glucose levels. CN from cow milk is chosen to be the greatest source of DPP-IV inhibitory peptides as it contains a high content of proline residues, which are known to increase the resistance to hydrolysis by DPP-IV and thus release the potential BPs with DPP-IV inhibitory activity. Among all CN fractions, β-CN was most promising proteins to act as a source of DPP-IV inhibitors. DPP-IV inhibitory peptide VPYPQ from β-CN had the IC_{50} value of 41.45 μM [245].

2.6.3 ANTIHYPERTENSIVE PEPTIDES

Reduction in certain cardiovascular diseases could be achieved by proper treatment of hypertension (or high blood pressure) and thus can prevent myocardial infarction, stroke, and heart failure. By its virtue and prevalence, broad application of pharmacological treatments and dietary approaches are employed to treat the hypertension. Hypertension is biologically regulated by renin-angiotensin system (RAS) and kinin-nitric oxide system [198].

RAS includes the key enzyme, angiotensin-converting enzyme (ACE) (dipeptidyl carboxypeptidase, zinc-metalloprotease), a dipeptide hydrolase, which plays a significant role in the control of blood pressure. Renin acts on angiotensinogen to produce a deca-peptide angiotensin-I. ACE hydrolyses angiotensin-I to produce angiotensin-II. It is a powerful vasoconstrictor, and mainly acts on smooth muscle (in blood vessel) cells thereby elevating hypertension; additionally, it accelerates liberation of aldosterone from the adrenal cortex, which elevates the resorption of water and sodium, and excretion of potassium by the kidney; as a result, the overall effect is an increased blood pressure. Besides, ACE hydrolyzes the vaso-dilatator bradykinin, inhibiting its pressure minimizing effects [11, 125].

One of the major adverse effects of angiotensin-II is an increased generation of reactive oxygen compounds, since these compounds stimulate a powerful membrane oxidase and release superoxide anion and hydrogen peroxide [38].

Inactivation of ACE lowers blood pressure; and hence, it is a key clinical target for regulation of hypertension [15, 32, 85]. The initial natural inhibitors of ACE were stated to be the BPs derived from the snake venom [158]. Recently, a variety of ways for producing ACE inhibitory peptides were obtained, including chemical synthesis, recombinant DNA technology, extract from natural foods, enzymatic hydrolysis, and fermentation. Among all these methods, synthesis of these peptides from food proteins by enzymatic hydrolysis is known to be well-proven, and the peptides have several advantages, such as: low cost, safety, no side effects, easily absorbed and only acts on persons with hypertension [16, 240]. The consumption of food products containing ACE inhibitory BPs has appeared to considerably lower the hypertension of moderately hypertensive subjects. However, these peptides are needed at greater dosage compared to the antihypertensive drugs [108]. Till now numerous antihypertensive peptides have been stated from a huge list of dietary proteins, particularly: milk, fish, sweet potato, sesame protein, egg white and soy proteins [29, 69, 150, 152, 156, 199, 240].

MW of antihypertensive peptides is usually <3 kDa [74]. Potent sequences contain 2–12 AA residues and many BPs are well-known to have multifunctional properties [139, 153]. Some larger sequences up to 27 AAs also have ACE inhibitory activity [185, 189]. However, crystallographic studies suggested that the active site of ACE could not acclimatize larger peptides [153].

Various studies have reported general structural properties either by *in vitro* studies, QSAR or by synthetic peptides. They are short-chain (2–9 AAs) BPs containing hydrophobic AA residues (branched and aromatic); and occurrence of Pro, Lys, or Arg at the C-terminal and aliphatic AAs (Leu, Ile, and Val) at the N-terminal are necessary structural properties of ACE inhibitory BPs [52, 95, 128, 236]. Positively charged AAs (like Lys and Arg containing ε-amino and guanidine groups, respectively) contribute largely to ACE inhibition when located at the C-terminus of BPs, which is mainly due to the interaction with the anionic allosteric binding sites that are unlike the active site of the enzyme [139, 228]. The majority of the short ACE inhibitory peptides commonly contain a "Pro" at the C-terminal end. This AA is not degraded by the GI proteases and can pass from the intestine to blood as shorter peptides [241].

2.6.3.1 Casein Derived Antihypertensive Peptides

The antihypertensive peptide (i.e., YQKFPQYLQY) was isolated from CN fraction. CN hydrolyzed with trypsin and pepsin was separated using UF technique and Sephadex G-15 chromatography and the IC_{50} value of target peptide (i.e., YQK) was 11.1 μM. The peptide YQK retained its antihypertensive property not only in a wide range of temperature and pH environment but was also stable against the pepsin and trypsin [240]. The hydrolysates obtained from a controlled hydrolysis by trypsin were fractioned according to their MW by UF and RP-HPLC. The obtained fractions were evaluated with respect to their *in vitro* ACE inhibitory activity. The peptides identified were: FFVAPFPEVFGK, FALPQYLK, and ALNEINQFYQK [156].

The antihypertensive peptides derived from both CNs and WPs [28, 140] are known as casokinins and lactokinins, respectively. They are found in fermented products or hydrolysates of milk proteins obtained using specific enzymes for example pepsin, trypsin, and chymotrypsin [124]. The antihypertensive peptides isolated from CN are presented in Table 2.2.

TABLE 2.2 ACE Inhibitory Peptides Identified from Bovine Casein

Peptide Sequence	Fragment	Casein Type	References
DAYPSGAW	157–164	α_{S1}-CN	[171]
FFVAPFPEVFG	23–34	α_{S1}-CN	[68]
FFVAPFPEVFGK	38–49	α_{S1}-CN	[110]
HIQKEDVPSER	95–105	α_{S1}-CN	[110]
KKYNVPQL	102–109	α_{S1}-CN	[144]
KVLPVPG	169–175	β-CN	[117]
NMAINPSK	25–32	α_{S2}-CN	[221]
NMAINPSKENLCSTFCK	25–41	α_{S2}-CN	[221]
RYLGYLE	90–96	α_{S1}-CN	[68]
TPVVVPPFLQP	80–90	β-CN	[1]
VPP	84–86	β-CN	[151]
YKVPQL	104–109	α_{S1}-CN	[117]
YLGYLEQLLR	106–115	α_{S1}-CN	[110]

2.6.4 ANTI-MICROBIAL PEPTIDES (AMPS)

Milk proteins exhibit various physiological actions for the newborn and also show great protection against various infections. Milk components, which are responsible for such protection includes: antibacterial factors, anti-inflammatory and immune-modulatory compounds [115, 116]. Majorly, the antibacterial property of milk is generally attributed to Igs, lactoferrin, lactoperoxidase, lysozyme, and antimicrobial peptides (AMPs) [68].

The action of AMPs on sensitive bacteria relay on the AA composition, structural, and physicochemical characteristics of bacteria [12]. They have selective toxicity towards pathogenic bacteria without damaging the host cells. AMPs mainly act on the cell wall of the microbes, which is made up of lipoproteins.

These peptides are cationic and amphipathic in nature and largely have alpha-helical structures, while a few are hydrophobic BPs. The cationic component of BPs is accountable for the electrostatic interaction with anionic phospholipids, which leads in the pore formation in the cytoplasmic membrane; this causes ionic imbalance across the membrane resulting in the death of cell. In addition, these AMPs can self-assemble above the surface of the cell membrane and get incorporated into it [200].

2.6.4.1 CASEIN-DERIVED AMPS

Numerous AMPs, which are present within the primary structure of CNs, have been reported [12, 58]. Jones and Simms [89] discovered the first AMP, lactenin; and after that casecidins (a set of basic high MW polypeptides) were identified from CN hydrolysates obtained using chymosin at neutral pH [3, 4]. These BPs were apparently isolated from α_{S1}- and κ-CN and revealed *in vitro* antibacterial activity against *Lactobacillus* spp. and diverse pathogens including *S. aureus* [3, 4, 58, 116].

The first AMP identified in the primary structure of cow milk α_{S1}-CN was isracidin, which was obtained after hydrolysis using chymosin, and represented the fragment [f (1–23)] of the casein [76]. This peptide retarded the growth of both Gram-positive (G +ve) and Gram-negative (G –ve) bacteria [105]. Another cationic peptide [f (99–109)] of bovine α_{S1}-CN released by pepsin showed a wide range of activities against G +ve and G –ve organisms (such as: *B. subtilis, Lb. innocua, S. typhimurium, and S. enteritidis*) [130]. This peptide could be liberated in the stomach by the digestion of CN with pepsin and can provide defense against bacterial action in the GI tract [115].

Casocidin-I [f (150–188)], the AMP derived from α_{S2}-CN [247], is resistant to high temperature and acidic conditions, corresponding to peptide residues obtained from heat-treated (boiled) and acidified milk.

Casocidins were expected for therapeutic use, i.e., for treating contagious diseases as they were shown to have antibacterial property against *Staphylococcus spp., Sarcina spp., B. subtilis, D. pneumonia, and S. pyogenes* [28, 105]. Two potent antibacterial peptides obtained by pepsin digestion were derived from bovine α_{S2}-CN [f (164–179)]: a fragment of the formerly known as casocidin-I, and the second peptide at the C-terminal of the CN [f(183–207)] [179]. These fragments exhibited antimicrobial property against G +ve and G –ve organisms. The AMPs (e.g., f(181–207), f(175–207), and f(164–207)) from bovine CN hydrolysate obtained using chymosin were derived from the C-terminus of α_{S2}-CN, and these showed antibacterial activity against broad spectrum of G +ve and G –ve organisms [130]. Also, these peptides suppressed sensitive G +ve bacteria as efficiently as nisin and lactoferricin B. Thus, they can be utilized for food preservation purposes.

Kappacin is a non-glycosylated and phosphorylated bovine caseino macropeptide [f(106–169)], which showed growth inhibition against G +ve and G –ve bacteria [118]. It is also reported that kappacin can bind enterotoxins and suppress the attachment of virus and bacteria, such as: the binding of cariogenic organisms to oral surfaces [12, 33, 118]. Matin et al. [127] isolated κ-casecidin a pentapeptide, which was released by tryptic digestion of κ-CN. This BP, generated from bovine κ-CN [f (17–21)], hindered the growth of pathogens for instance *S. aureus, E. coli,* and *S. typhimurium.*

An easy and safest method was used by Elbarbary et al. [41] to separate CN hydrolysate on the basis of amphoteric character of the BPs. The AMPs, which showed the highest antimicrobial activity obtained from the basic fraction, were purified by applying a series of SEC and RP-HPLC. The three BPs (that were isolated from α_{S1}-, α_{S2}- and κ-CN fractions) showed the most potent antimicrobial activity.

2.6.5 MINERAL-BINDING PEPTIDES

Commonly consumed foods, for example, cereals and legumes, often have phytates [146], which chelate the minerals and make them unavailable. The adverse effects of metal salts and supplementation of multi-mineral in foods to overcome the issue of mineral deficiencies bring about a change in the physical and sensory attributes of foods [82]. Low bioavailability of multi-mineral supplements is also a hurdle to control the deficiency conditions [173]. Therefore, having natural mineral-enriched food in usual diets is a better option [92]. Considering general health and dietary choices, food derived supplements serve as the better option.

Several peptides from milk proteins with mineral-binding capabilities have been reported. The diverse BPs obtained from *in vitro* proteolysis are buried in a dormant state within the AA chain of the proteins. α_{S1}-, α_{S2}-, β- and k-CN are usually revealed as precursors of BPs containing binding sites such as phosphoserine and carboxylic groups, for various mineral elements [10, 19, 134, 196, 167]. Various BPs with definite sequences form complexes at intestinal pH by binding with elements like calcium, phosphorus, and many others in solution have been reported. BPs having a large proportion of –ve charges can competently bind divalent mineral elements (Mg, Mn, Cu, Fe, and Se) through the formation of soluble complexes. The large amount of –ve charges in these peptides can resist further breakdown.

2.6.5.1 CASEINO PHOSPHOPEPTIDES (CPPS) AND MINERAL BINDING PROPERTY

CPPs are strongly phosphorylated peptides obtained by enzymatic digestion of CNs. These peptides have been most commonly studied for their capacity to bind and solubilize minerals especially calcium (Ca). *In vitro* studies have been demonstrated that CPPs may avoid the Ca precipitation in the presence of phosphate at alkaline pH [55, 80]. The highly anionic phosphorylated sequences in CPPs have a significant role in mineral binding. Nevertheless, studies conducted in synthetic peptides as well as CPPs obtained from proteases bearing diverse substrate specificities revealed that AA residues next to hydrophilic regions also have the ability to bind the minerals [30]. Phosphorylated peptides bind the minerals (Ca, Zn, and Fe) so efficiently than peptides devoid of phosphorylated residues. Because CPPs can bind and dissolve minerals, they have been considered as functional ingredients in the dietary supplementation of minerals like Ca, Mg, and Fe.

The attraction of CPPs to mineral elements such as Fe, Zn, Cu is considered to be greater than for Ca. Accordingly, they have been expected to be biologically useful in the retardation of tooth decay, osteoporosis, and anemia through enhancement of mineral uptake and bioavailability [99]. Several phospho-peptides containing the cluster series of 3 phosphoserine followed by 2 Glu residue, i.e., SerP-SerP-SerP-Glu-Glu have been found in bovine

CN. These sequences offer the peptides with the distinctive capability to keep calcium, phosphorus, and other minerals in a soluble state at GI pH [38].

Phospho-peptide fractions may significantly differ in their binding capacity with Ca, owing to variation in AA composition around the phosphorylated region. The binding of minerals with phospho-peptides results in rapid absorption into the intestinal epithelium [135] and capture of definite minerals such as Fe in a peptide; therefore, reduced intra luminal availability of the metal and depressed microbial growth in the intestine [227].

Wang et al. [229] reported a direct connection between the degree of phosphorylation and Zn binding, with a decline in Zn-bound forms of peptide corresponding to the lesser phosphorylated residues of CPPs. Presence of CPPs in ileostomy fluid in human distal ileum, proves the capacity of CPPs to survive the GI tract [70].

It is suggested that the bound form of Fe is ideal vehicle for fortification of foods [93]. Bouhallab and Bougle [17] revealed that phospho-peptide derived from β-CN [f(1–25)] has a beneficial effect on Fe bioavailability *in vivo*. Worldwide nutritional problem of Fe deficiency can be overcome by the use of CPPs. CPPs form soluble complex with iron and prevent the formation of poorly absorbed high MW ferric hydroxides.

CPPs have an anti-cariogenic property by enhancing both re-calcification of dental enamel and by preventing the attachment of the organisms responsible for plaque formation [183]. CPP-$CaPO_4$ complexes appeared to be anti-tumerogenic and re-mineralize initial phases of tooth decay in animal as well as human trials. These peptides can provide amorphous calcium phosphate (ACP) and help attach it with the tooth enamel [197].

CPPs form nanoclusters with ACP thus supplying Ca and PO_4, which keep their supersaturation in saliva. As CPP-ACP stabilizes Ca and PO_4 in the aqueous medium, it can assist in the buffering of plaque pH and so that the levels of Ca and PO_4 content in the plaque are elevated. Hence, Ca and PO_4 levels within the subsurface lesions are kept high, which bring about remineralization.

Donida et al. [37] reported that CPPs acts as regulator of bone cell activity, most likely sustained by their capability as Ca transporters. Bouhallab et al. [18] revealed that the physiological activity of the various phospho-peptide residues is mainly reliant on their sequence and consequently on their comparative attraction towards Fe atoms. CPPs have the possibility of becoming a food-based dietary ingredients and augmenting the immune system [30]. Thus, CPPs find useful application in the areas of antihypertensive medicine, sports medicine, immune-enhancement, and immune-modulation [36, 219]. They are ideally suited for food fortification, since iron bound to CPPs showed excellent bioavailability with enhanced solubility and the retardation of interaction with other minerals [6].

CN and its BPs, in particular CPPs, are most researched and have powerful mineral-binding activities [50]. Such peptides have been reported from α_S-CN due to its higher phosphorylation. Miquel et al. [145] found the distribution of Ca, Fe, and Zn in CPP fractions isolated using the anion exchange column after GI digestion of infant formula. These specific sequences occur at f(66–70) in α_{S1}-CN, f(8–12) and f(56–60) in α_{S2}-CN. Mellander [141] reported that CPP enhances vitamin D independent bone mineralization in infants suffering from rickets. The phospho-peptides derived from α_S-CN exhibited higher calcium-binding capacity than those from β-CN [163]; the Ca-binding ability was 1.75 and 0.51 mmol/mg of α_S-CN and β-CN, respectively. The lower Ca-binding capacity of β-CN was probably due to lesser number of phosphoryl residues. Ferranti et al. [51] isolated 45 CPPs from Grana Padano cheese of which 24, 16 and 5 were from β-, α_{S1}- and α_{S2}-caseins, respectively.

Kitts [95] found that CPPs obtained from spray-dried whole tryptic digests of cow milk CN consist of unidentified BPs with MWs <6 kDa and an ability to sequester Fe^{2+}. Considering their importance, CPPs derived from milk protein have commercial potential as ingredients in foods, carbonated drinks and chewing gum, as well as toothpastes and mouthwashes to prevent tooth decay [180].

2.6.6 OPIOID PEPTIDES (OPS)

The opioid peptides (OPs) are opioid receptor ligands and perform various physiological activities including neuromodulator, hormone-like function, analgesia, sedative effect, induces sleep, reduces hypertension, ACE-inhibition, and more. OPs are found to be present in the nervous, endocrine, immune, and GI systems of humans. They are also derived from enzymatic hydrolysis of human and bovine β-CN *in vitro* [20]. Several hydrolysates of milk proteins like α-La, β-Lg, β-CN, κ-CN, and lactoferrin have been reported to exert both opioid agonist and antagonist behavior.

OPs derived from dietary proteins are known as exogenous opioids, while they have a Tyr residue at their N-termini, except α_{S1}-CN exorphin, casoxin 6 and lactoferroxin. They differ from the endogenous OPs by lacking an enkephalin sequence (Tyr-Gly-Gly-Phe-Met/Leu) [215], potentially with one more aromatic AA, Phe or Tyr, at the third or fourth place [59]. They react with the endogenous ligands, and exogenous and/or antagonist opioids and can regulate the central nervous system (CNS), which is engaged in blood pressure reduction, anorexia, variation in body temperature and modification of sexual behaviors [39, 147]. Endogenous agonist OPs may regulate the growth and activity of cells engaged in the CNS, while β-casomorphins (BCMs) are carried across mucus membranes of infants that control biological reactions bring calmness in neonates [22, 208].

BCMs interact with opiate receptors in intestinal epithelial cells and play an important role in some actions such as regulation of electrolyte transport, production of insulin, and absorption of food [220]. Opioid antagonists can inhibit the agonist action of enkephalin. Most important agonistic OPs called Casoxin-C have been derived from bovine κ-CN receptor [137]. Many scientific studies have demonstrated that the 2 casoxins (-A and -B) are opioid receptor ligands derived from bovine κ-CN that show low antagonistic activity [136]; and fractions of κ-CN represented by f(35–41) and f(58–61), respectively account for casoxin A and casoxin B. However, casoxin C is a powerful antagonist OP of κ-CN [f (25–34)], exhibiting maximum biological activity [239].

It has been reported that casomorphins, as opioid ligands, possess anti-secretory activity [31], stimulate analgesic behavior [129] and secretion of insulin and somatostatin [134]. α_S-CN derived exorphins isolated from bovine α_{S1}-CN [f(90–95) and f(90–96)] have opioid-like activities that are suppressed with naloxone [56].

2.6.6.1 β-CASOMORPHINS

The β-Casomorphins (BCMs) are four to eleven AA residues, present in a dormant form in β-casein, and are liberated during digestion both *in vivo* and *in vitro*. Among the BCMs, highly potent BPs are BCM7 and BCM5 corresponding to fractions f (60–66) and f(60–64), respectively [100]. These peptides can pass through the GI barrier and reach the cerebrospinal fluid in healthy human beings [211]. The biological effects of this mechanism have not yet been fully known; however, it is reported that an interrelation occurs between BCMs and autism. BCMs stimulate fos-type immune reactivity in regions of brain pertinent to autism, and increased concentrations of BCM7 appeared in individuals with such a circumstance

[211]. Likewise, BCM7 and opioid receptors might be correlated to the incidence of schizophrenia in individuals with a small number of receptors of opioid [193]. Regardless of elevated concentrations of BCM7 in these patients [211], BCM7 is not able to provide the soothing effect as it does in person with autism. It is reported that supplementation of milk may bring the capability to lessen the crying of infants appreciably. In neonates, OPs can pass the GI barrier to exert opioid effects over the entire body [31].

BCMs suppress the propagation of human lamina propria-derived lymphocytes *in vitro* and such an anti-proliferative action is inhibited by adding opiate receptor antagonist, naloxone to the culture [43]. BCM7 [f (60–66)] and BCM10 [f(193–202)] can exert bipolar regulatory action on proliferation of human peripheral blood lymphocytes. These peptides evaluated in *in-vitro* cultures with mitogen-stimulated T lymphocytes inhibited propagation at low levels, but stimulated propagation when ingested at higher levels [96]. However, several researchers have suggested that there is a connection between BCM7 (linked to intake of A1-milk) and various nervous disorders.

Bovine BCM7 has been reported to be a potential threat for human diseases such as ischemic heart disease, DM, coronary-artery disease, and crib death, in which symptoms appears to be connected with elevated circulating levels of exorphins [181]. However, the cause-effect relationship between intake of A1 milk and incidence of diseases has not been proved.

2.6.7 ANTI-CANCER PEPTIDES

Traditional anti-cancer treatments like chemotherapy and radiation are very expensive and have severe adverse side effects in the people. Therefore, natural antitumor substrates such as BPs are better alternate for prevention and control of cancer. Anti-tumor bio-peptides are also hidden in the AA sequence of milk proteins and are derived during proteolysis namely, GI digestion and fermentation. Most of the studies for antitumor activity of BPs are mainly on lunasin, the peptide derived from cereals or soybeans [231]. It is efficient in treating chemical and viral oncogene stimulated tumors and acts by suppressing histone acetyltransferase (HAT), which hinders the acetylation of H3 and H4, causing suppression of cell cycle multiplication, and inducing apoptosis of cells [75]. Jeong et al. [88] demonstrated that with the ingestion of rye, there is an increase in absorption of lunasin and the tissue-derived isolates preserve the antitumor HAT-inhibition of the parent molecule.

2.6.7.1 CASEIN-DERIVED ANTITUMOR PEPTIDES

Different studies showed that casomorphins have anticancer activities (Table 2.3). BCMs are opioid-like BPs obtained from β-CN hydrolysates and they could reach a considerable amount in the stomach as they are resistant to proteolytic action owing to the presence of proline AA in its sequence [212].

TABLE 2.3 Antitumor Peptides Identified from Bovine Casein

Casein	Peptide Sequence	Anticancer Properties	References
α_{S1}-CN	Tyr-Val-Pro-Phe-Pro[f(158–162)] (Casomorphin)	Suppressed proliferation of T47D human breast tumor cells	[90]
α_{S1}-CN	Arg-Pro-Lys [f(1–3)] Leu-Lys-Lys [f(101–103)] Tyr-Lys [f(104–105)] (Casecidins)	Induced necrosis of leukemic T- and B-cell lines	[159]
β-CN	Pro-Gly-Pro-Ile-Pro-Asn [f(63–68)]	Suppressed proliferation of SKOV3 human ovarian tumor cells partly through stimulating apoptosis by hampering BCL2 pathway	[232]
β-CN	Ile-Asn-Lys-Lys-Ile [f(41–45)]	Induced cytotoxicity to B16F10 melanoma cells	[9]
κ-CN	Phe-Phe-Ser-Asp-Lys [f(17–21)] (Casecidin)	Exhibited cytotoxicity to few mammalian cells, including human leukemic cell lines.	[126]

Casomorphins, purified from α-CN via peptic hydrolysate, mainly contained two peptides isolated from α_{S1}-CN, i.e., f (90 to 95) and f (90 to 96). The residues obtained by tryptic hydrolysate of β-CN demonstrated to have antitumor activity in an animal study and *in vitro* antiproliferative action. Also, κ-casecidin is isolated from κ-CN showed antitumor activity. A study suggested that BPs generated from CN can show their antitumor effect by a process of partially including opioid recipients [210]. Several caso morphin liberated from α- and β-CNs reduce the propagation of prostate carcinoma cell lines in such a manner.

CPPs are attributed to bind and solubilize Ca owing to the presence of numerous phosphorylated residues [13]. This characteristic is responsible for their anticancer properties against GI malignant cells (HT-29) via modulating propagation and apoptosis of cancer cells. Activation of voltage-activated L-type Ca channels in malignant colon cells (HT-29 and AZ-97) was found to be connected with apoptosis, and their obstruction may increase the proliferation of malignant cells [242]. Perego et al. [166] showed that CPPs protect differentiated intestinal cells against lethality because excess of Ca offsets apoptosis and aids proliferation, and in the meanwhile stimulating apoptosis in the undifferentiated malignant cells. This effect is due to the result of the binding of CPPs with extracellular Ca in a definite dose-response interrelation, which brings about a decline in the rate of cell proliferation and apoptosis.

2.7 COMMERCIAL BIOACTIVE PRODUCTS

It is now well reported that BPs can be produced during the fermentation of milk with the selective strains customarily engaged by the dairy industry [99, 223]. Consequently, peptides with several bioactive properties can be present in the final products such as varieties of cheese and fermented products. These conventional milk products under certain situations bring definite health benefits when administered as part of daily diet. A growing number of ingredients having precise BPs isolated from hydrolysates of milk proteins have been commercially introduced in the recent years, while some are currently under progress by global food enterprises. Such BPs have anticarcinogenic, antihypertensive, metal-binding, and stress-relieving properties, etc. Some of these commercially available bioactive peptides and their uses are mentioned in Table 2.4.

TABLE 2.4 Examples of Commercial Products Containing Bioactive Peptides [99]

Brand Name	Product Type	Bioactive Peptides	Functionality Claimed	Manufacturer
Calpis AMEEL S(Japan)/ Calpico (Europe)	Fermented product	VPP and IPP from β-CN and κ-CN	Hypotensive	Calpis Co., Japan
Evolus	Fermented milk, Ca-enriched	VPP and IPP from β-CN and κ-CN	Hypotensive	Valio, Finland
BioZate	Hydrolyzed WPI	β-Lg fragments	Antihypertensive	Davisco, USA
HCP102/ HCP105	Hydrolysate ingredient	CPP	Mineral absorption	Tatua, New Zealand
Capolac	Hydrolysate component	CPP	Mineral absorption	Arla foods ingredients, Sweden
Recaldent	Toothpaste and chewing gum	ACP-CPP	Remineralization of teeth	BreezeCare oral health company, Australia
BioPure-GMP	WPI	κ-CN f(106–109)	Prevention of tooth decay, controls the clotting blood	Davisco, USA
PRODIET F200/Lactium	Flavoured milk, confectionery, capsules	α_{S1}-CN f(91–100)	Decline in stress effects	Ingredia, France
Vivinal Alpha	Ingredient/ hydrolysate	Peptides from WP	Aids relaxation and sleep	Borculo domo ingredients
PeptoPro	Ingredient/ hydrolysate	Peptides from CN	Improves athletic performance and muscle recovery	DSM foods specialties, Netherlands
Cysteine peptide	Ingredient/ hydrolysate	Milk protein-derived peptide	Aids in enhancing the energy level and sleep	DMV International, Netherlands

2.8 SUMMARY

BPs have been used in producing several health benefiting functional foods. Besides, with the advent of novel technologies, the use of BPs has been greatly augmented. Membrane filtration processes are commonly applied for enrichment of BPs with precise MW. The NF and UF methods offer bioactive peptides, particularly from CN or WP hydrolysates [57]. In particular, isolation, and identification of novel substances from natural sources, including milk, exhibiting positive impact and perhaps with lesser adverse results can stand for a substitute to drugs. Further, *in vivo* and *in vitro* studies must be carried out to exhibit their use on a commercial scale for the prevention and treatment of several diseases. Such a work increases the market of value-added products or functional foods globally. However, several challenges faced while incorporating bioactive food protein hydrolysates and peptides into functional foods are their limited bioavailability, bitterness, hygroscopicity, hydrophobicity, reaction with the food matrix, incompatibility, and biostability. Different approaches can be used for delivering these peptides, namely: microemulsions, magnetic microcapsules, microencapsulation, and nano-encapsulation in the form of liposomes, nanofibers, niosomes, etc. Such techniques have become relevant for enhancing the consumption of food-derived BPs for human health endorsement.

KEYWORDS

- **bioactive peptides**
- **casein derived peptides**
- **characterization**
- **functional benefits**
- **isolation**
- **production**

REFERENCES

1. Abubakar, A., Saito, T., Kitazawa, H., Kawai, Y., & Itoh, T., (1998). Structural analysis of new antihypertensive peptides derived from cheese whey protein by proteinase K digestion. *Journal of Dairy Science, 81*(12), 3131–3138.
2. Abuja, P. M., & Albertini, R., (2001). Methods for Monitoring Oxidative Stress, Lipid Peroxidation and Oxidation Resistance of Lipoproteins. *Clinicachimica Acta, 306*(1, 2), 1–17.
3. Addeo, F., Garro, G., Intorcia, N., Pellegrino, L., Resmini, P., & Chianese, L., (1995). Gel electrophoresis and immunoblotting for the detection of casein proteolysis in cheese. *Journal of Dairy Research, 62*(2), 297–309.
4. Ahtesh, F., Stojanovska, L., Shah, N., & Mishra, V. K., (2016). Effect of flavoenzyme on angiotensin-converting enzyme inhibitory peptides formed in skim milk and whey protein concentrate during fermentation by *Lactobacillus helveticus*. *Journal of Food Science, 81*(1), M135–M143.

5. Akuzawa, R., Miura, T., & Kawakami, H., (2009). Bioactive components in caseins, caseinates, and cheeses. Chapter 8. In: Park, Y. W., (ed.), *Bioactive Components in Milk and Dairy Products* (pp. 215–233). USA: Wiley-Blackwell.
6. Amaro, A. L. M., & Camara, F. M., (2004). Iron availability: An updated review. *International Journal of Food Sciences and Nutrition, 55*(8), 597–606.
7. Aneja, R. P., Mathur, B. N., Chandan, R. C., & Banerjee, A. K., (2002). *Technology of Indian Milk Products* (p. 462). Dairy India Publication, Delhi-India.
8. Antonelli, A., Cocchi, M., Fava, P., Foca, G., Franchini, G. C., Manzini, D., & Ulrici, A., (2004). Automated evaluation of food color by means of multivariate image analysis coupled to a wavelet-based classification algorithm. *Analytica Chimica Acta, 515*(1), 3–13.
9. Azevedo, R. A., Ferreira, A. K., Auada, A. V. V., Pasqualoto, K. F. M., Marques-Porto, R., Maria, D. A., & Lebrun, I., (2012). Antitumor effect of cationic INKKI peptide from bovine β-casein on melanoma B16F10. *Journal of Cancer Therapy, 3*(4), 237–244.
10. Baumy, J. J., & Brule, G., (1988). Effect of pH and ionic strength on the binding of bivalent cations to β-casein. *Le Lait, 68*(4), 409–417.
11. Beevers, G., Lip, G. Y., & O'Brien, E., (2001). The patho physiology of hypertension. *British Medical Journal, 322*(7291), 912–916.
12. Benkerroum, N., (2010). Antimicrobial peptides generated from milk proteins: A survey and prospects for application in the food industry: A review. *International Journal of Dairy Technology, 63*(3), 320–338.
13. Berrocal, R., Chanton, S., Juillerat, M. A., Pavillard, B., Scherz, J. C., & Jost, R., (1989). Tryptic phosphopeptides from whole casein. II. Physicochemical properties related to the solubilization of calcium. *Journal of Dairy Research, 56*(3), 335–341.
14. Bhandari, D., Rafiq, S., Gat, Y., Gat, P., Waghmare, R., & Kumar, V., (2020). Review on bioactive peptides: Physiological functions, bioavailability, and safety. *International Journal of Peptide Research and Therapeutics, 26*(1), 139–150.
15. Bicket, D. P., (2002). Using ACE inhibitors appropriately. *American Family Physician, 66*(3), 461–468.
16. Bougatef, A., Hajji, M., Balti, R., Lassoued, I., Triki-Ellouz, Y., & Nasri, M., (2009). Antioxidant and free radical-scavenging activities of smooth hound muscle protein hydrolysates obtained by gastrointestinal proteases. *Food Chemistry, 114*(4), 1198–1205.
17. Bouhallab, S., & Bouglé, D., (2004). Biopeptides of milk: Caseino-phospho peptides and mineral bioavailability. *Reproduction, Nutrition, Development, 44*(5), 493–498.
18. Bouhallab, S., Cinga, V., Aít-Oukhatar, N., Bureau, F., Neuville, D., Arhan, P., Maubois, J. L., & Bouglé, D., (2002). Influence of various phosphopeptides of caseins on iron absorption. *Journal of Agricultural and Food Chemistry, 50*(24), 7127–7130.
19. Bouhallab, S., Leonil, J., & Maubois, J. L., (1991). Study of beta-casein phosphopeptide (1–25) iron complex: Action of alcalase and acid phosphatase. *Lait, 71*(4), 435–443.
20. Brantl, V., (1984). Novel opioid peptides derived from human beta-casein: Human beta-casomorphins. *European Journal of Pharmacology, 106*(1), 213–214.
21. Butylina, S., Luque, S., & Nyström, M., (2006). Fractionation of whey-derived peptides using a combination of ultrafiltration and nanofiltration. *Journal of Membrane Science, 280*(1, 2), 418–426.
22. Calvo, C. F., Cesselin, F., Gelman, M., & Glowinski, J., (2000). Identification of an opioid peptide secreted by rat embryonic mixed brain cells as a promoter of macrophage migration. *European Journal of Neuroscience, 12*(8), 2676–2684.

23. Celus, I., Brijs, K., & Delcour, J. A., (2007). Enzymatic hydrolysis of brewers spent grain proteins and techno-functional properties of the resulting hydrolysates. *Journal of Agricultural and Food Chemistry, 55*(21), 8703–8710.
24. Cervato, G., Cazzola, R., & Cestaro, B., (1999). Studies on the antioxidant activity of milk caseins. *International Journal of Food Sciences and Nutrition, 50*(4), 291–296.
25. Chabeaud, A., Vandanjon, L., Bourseau, P., Jaouen, P., Chaplain-Derouiniot, M., & Guérard, F., (2009). Performances of ultrafiltration membranes for fractionating a fish protein hydrolysate: Application to the refining of bioactive peptidic fractions. *Separation and Purification Technology, 66*(3), 463–471.
26. Chen, H. M., Muramoto, K., Yamauchi, F., Fujimoto, K., & Nokihara, K., (1998). Antioxidative properties of histidine-containing peptides designed from peptide fragments found in the digests of a soybean protein. *Journal of Agricultural and Food Chemistry, 46*(1), 49–53.
27. Chen, J., Lindmark-Månsson, H., Gorton, L., & Åkesson, B., (2003). Antioxidant capacity of bovine milk as assayed by spectrophotometric and amperometric methods. *International Dairy Journal, 13*(12), 927–935.
28. Clare, D. A., & Swaisgood, H. E., (2000). Bioactive milk peptides: A prospectus. *Journal of Dairy Science, 83*(6), 1187–1195.
29. Contreras, M., Carrón, R., Montero, M. J., Ramos, M., & Recio, I., (2009). Novel casein-derived peptides with antihypertensive activity. *International Dairy Journal, 19*(10), 566–573.
30. Cross, K. J., Huq, N. L., Palamara, J. E., Perich, J. W., & Reynolds, E. C., (2005). Physicochemical characterization of casein phosphopeptide-amorphous calcium phosphate nanocomplexes. *Journal of Biological Chemistry, 280*(15), 15362–15369.
31. Daniel, H., Vohwinkel, M., & Rehner, G., (1990). Effect of casein and β-casomorphins on gastrointestinal motility in rats. *The Journal of Nutrition, 120*(3), 252–257.
32. Dasgupta, C., & Zhang, L., (2011). Angiotensin II receptors and drug discovery in cardiovascular disease. *Drug Discovery Today, 16*(1, 2), 22–34.
33. Dashper, S. G., Liu, S. W., & Reynolds, E. C., (2007). Antibacterial peptides and their potential as oral therapeutic agents. *International Journal of Peptide Research and Therapeutics, 13*(4), 505–516.
34. Davalos, A., Miguel, M., Bartolome, B., & Lopez-Fandino, R., (2004). Antioxidant activity of peptides derived from egg white proteins by enzymatic hydrolysis. *Journal of Food Protection, 67*(9), 1939–1944.
35. Deak, N. A., Murphy, P. A., & Johnson, L. A., (2006). Effects of NaCl concentration on salting-in and dilution during salting-out on soy protein fractionation. *Journal of Food Science, 71*(4), C247–C254.
36. Devi, M. C., VasanthaBai, M., Vijayan-Lal, A., Uma-shankar, P. R., & Krishnan, L. K., (2002). An improved method for isolation of anti-viper venom antibodies from chicken egg yolk. *Journal of Biochemical and Biophysical Methods, 51*(2), 129–138.
37. Donida, B. M., Mrak, E., Gravaghi, C., Villa, I., Cosentino, S., Zacchi, E., Perego, S., et al., (2009). Casein phosphopeptides promote calcium uptake and modulate the differentiation pathway in human primary osteoblast-like cells. *Peptides, 30*(12), 2233–2241.
38. Dzau, V. J., (2001). Tissue angiotensin patho-biology of vascular disease: A unifying hypothesis. *Hypertension, 37*(4), 1047–1052.

39. Dziuba, J., Minkiewicz, P., Nałecz, D., & Iwaniak, A., (1999). Database of biologically active peptide sequences. *Die Nahrung, 43*(3), 190–195.
40. Eichler, H. G., Korn, A., Gasic, S., Pirson, W., & Businger, J., (1984). The effect of a new specific α-amylase inhibitor on post-prandial glucose and insulin excursions in normal subjects and type 2 (non-insulin-dependent) diabetic patients. *Diabetologia, 26*(4), 278–281.
41. Elbarbary, H. A., Abdou, A. M., Nakamura, Y., Park, E. Y., Mohamed, H. A., & Sato, K., (2012). Identification of novel antibacterial peptides isolated from a commercially available casein hydrolysate by autofocusing technique. *Biofactors, 38*(4), 309–315.
42. Elias, R. J., Kellerby, S. S., & Decker, E. A., (2008). Antioxidant activity of proteins and peptides. *Critical Reviews in Food Science and Nutrition, 48*(5), 430–441.
43. Elitsur, Y., & Luk, G. D., (1991). Beta-casomorphin (BCM) and human colonic lamina propria lymphocyte proliferation. *Clinical and Experimental Immunology, 85*(3), 493–497.
44. Ena, J. M., Van, B. E. C. H., Robben, A. J. P. M., & Schmidt, D. G., (1995). Whey protein antigenicity reduction by fungal proteinases and a pepsin/pancreatin combination. *Journal of Food Science, 60*(1), 104–110.
45. Erasmus, L. J., Rulquin, H., & Hermansen, J. E., (2001). Nutritional and management factors affecting milk protein content and composition. In: *Bulletin of the International Dairy Federation* (pp. 49–61).
46. Erdmann, K., Cheung, B. W., Immenschuh, S., & Schröder, H., (2008). Heme oxygenase-1 is a novel target and antioxidant mediator of s-adenosyl methionine. *Biochemical and Biophysical Research Communications, 368*(4), 937–941.
47. Evans, C. R., & Jorgenson, J. W., (2004). Multi-dimensional LC-LC and LC-CE for high-resolution separations of biological molecules. *Analytical and Bioanalytical Chemistry, 378*(8), 1952–1961.
48. Evert, A. B., Boucher, J. L., Cypress, M., Dunbar, S. A., & Franz, M. J., (2014). Nutrition therapy recommendations for the management of adults with diabetes. *Diabetes Care, 36*(11), 3821–3842.
49. Fang, X., Xie, N., Chen, X., Yu, H., & Chen, J., (2012). Optimization of antioxidant hydrolysate production from flying squid muscle protein using response surface methodology. *Food and Bioproducts Processing, 90*(4), 676–682.
50. Farooq, I., Moheet, I. A., Imran, Z., & Farooq, U., (2013). A review of novel dental caries preventive material: Casein phosphopeptide-amorphous calcium phosphate (CPP-ACP) complex. *Journal of Dental Sciences, 4*(2), 47–51.
51. Ferranti, P., Barone, F., Chianese, L., Addeo, F., Scaloni, A., Pellegrino, L., & Resmini, P., (1997). Phosphopeptides from grana Padano cheese: Nature, origin, and changes during ripening. *Journal of Dairy Research, 64*(04), 601–615.
52. FitzGerald, R. J., & Meisel, H., (2000). Milk protein-derived peptide inhibitors of angiotensin-I-converting enzyme. *The British Journal of Nutrition, 84*(S1), S33–S37.
53. FitzGerald, R. J., Murray, B. A., & Walsh, D. J., (2004). Hypotensive peptides from milk proteins. *The Journal of Nutrition, 134*(4), 980S–988S.
54. Fox, P. F., & McSweeney, P. L., (2003). *Advanced Dairy Chemistry—1 Proteins, Parts A and B* (3rd edn., p. 1318). Kluwer Academic/Plenum Publishers, New York.
55. Gaucheron, F., Le Graët, Y., Piot, M., & Boyaval, E., (1996). Determination of anions of milk by ion chromatography. *Le Lait, 76*(5), 433–443.

56. Gill, H. S., Doull, F., Rutherfurd, K. J., & Cross, M. L., (2000). Immuno-regulatory peptides in bovine milk. *The British Journal of Nutrition, 84*(S1), S111–S117.
57. Girgih, A. T., He, R., Malomo, S., Offengenden, M., Wu, J., & Aluko, R. E., (2014). Structural and functional characterization of hemp seed (*Cannabis sativa* L.) protein-derived antioxidant and antihypertensive peptides. *Journal of Functional Foods, 6*, 384–394.
58. Gobbetti, M., Minervini, F., & Rizzello, C. G., (2004). Angiotensin I-converting enzyme-inhibitory and antimicrobial bioactive peptides. *International Journal of Dairy Technology, 57*(2, 3), 173–188.
59. Gobbetti, M., Minervini, F., & Rizzello, C. G., (2007). Bioactive peptides in dairy products. Chapter 70. In: Hui, Y. H., (ed.), *Handbook of Food Products Manufacturing* (pp. 489–517). New Jersey-USA: John Wiley & Sons Inc.
60. Godinho, R., Mega, C., Teixeira-de-Lemos, E., Carvalho, E., Teixeira, F., Fernandes, R., & Reis, F., (2015). The place of dipeptidyl peptidase-4 inhibitors in type-2 diabetes therapeutics: A "me too" or "the special one" antidiabetic class? *Journal of Diabetes Research* (p. 28). Article ID: 806979.
61. Gómez-Ruiz, J. Á., López-Expósito, I., Pihlanto, A., Ramos, M., & Recio, I., (2008). Antioxidant activity of ovine casein hydrolysates: Identification of active peptides by HPLC-MS/MS. *European Food Research and Technology, 227*(4), 1061–1067.
62. Gooding, K. M., & Regnier, F. E., (2002). *HPLC of Biological Macro Molecules* (2nd edn., Vol. 87, p. 773). New York-USA: Marcel Dekker.
63. Groziak, S. M., & Miller, G. D., (2000). Natural bioactive substances in milk and colostrum: Effects on the arterial blood pressure system. *The British Journal of Nutrition, 84*(S1), S119–S125.
64. Guerra, A., Etienne-Mesmin, L., Livrelli, V., Denis, S., Blanquet-Diot, S., & Alric, M., (2012). Relevance and challenges in modeling human gastric and small intestinal digestion. *Trends in Biotechnology, 30*(11), 591–600.
65. Gulcicek, E. E., Shifman, M., Miller, P., & Colangelo, C. M., (2006). High probability MS identification of phosphopeptides by their mass defect. In: *54*th *ASMS Conference Proceedings* (p. 8).
66. Guttman, A., Varoglu, M., & Khandurina, J., (2004). Multidimensional separations in the pharmaceutical arena. *Drug Discovery Today, 9*(3), 136–144.
67. Halliwell, B., (2000). The antioxidant paradox. *The Lancet, 355*(9210), 1179–1180.
68. Haque, E., Chand, R., & Kapila, S., (2008). Biofunctional properties of bioactive peptides of milk origin. *Food Reviews International, 25*(1), 28–43.
69. Harnedy, P. A., & Fitz-Gerald, R. J., (2012). Bioactive peptides from marine processing waste and shellfish: A review. *Journal of Functional Foods, 4*(1), 6–24.
70. Hartmann, R., & Meisel, H., (2007). Food-derived peptides with biological activity: From research to food applications. *Current Opinion in Biotechnology, 18*(2), 163–169.
71. He, R., Girgih, A. T., Malomo, S. A., Ju, X., & Aluko, R. E., (2013). Antioxidant activities of enzymatic rapeseed protein hydrolysates and the membrane ultrafiltration fractions. *Journal of Functional Foods, 5*(1), 219–227.
72. Hernández-Ledesma, B., Amigo, L., Ramos, M., & Recio, I., (2004). Angiotensin converting enzyme inhibitory activity in commercial fermented products. Formation of peptides under simulated gastrointestinal digestion. *Journal of Agricultural and Food Chemistry, 52*(6), 1504–1510.

73. Hernández-Ledesma, B., Dávalos, A., Bartolomé, B., & Amigo, L., (2005). Preparation of antioxidant enzymatic hydrolysates from α-lactalbumin and β-lactoglobulin. Identification of active peptides by HPLC-MS/MS. *Journal of Agricultural and Food Chemistry, 53*(3), 588–593.
74. Hernández-Ledesma, B., Del, M. C. M., & Recio, I., (2011). Antihypertensive peptides: Production, bioavailability, and incorporation into foods. *Advances in Colloid and Interface Science, 165*(1), 23–35.
75. Hernández-Ledesma, B., Hsieh, C. C., & De Lumen, B. O., (2009). Lunasin, a novel seed peptide for cancer prevention. *Peptides, 30*(2), 426–430.
76. Hill, R. D., Lahav, E., & Givol, D., (1974). A rennin-sensitive bond in alpha-S1 B-casein. *Journal of Dairy Research, 41*(1), 147–153.
77. Hiller, B., & Lorenzen, P. C., (2009). Functional properties of milk proteins as affected by enzymatic oligomerization. *Food Research International, 42*(8), 899–908.
78. Holder, A., Birke, A., Eisele, T., Klaiber, I., Fischer, L., & Hinrichs, J., (2013). Selective isolation of angiotensin-I-converting enzyme-inhibitory peptides from micellar casein and β-casein hydrolysates via ultrafiltration. *International Dairy Journal, 31*(1), 34–40.
79. Holt, C., (1997). The milk salts and their interaction with casein. In: Fox, P. F., (ed.), *Advanced Dairy Chemistry, Volume 3: Lactose, Water, Salts, and Vitamins* (2nd edn., pp. 233–256). Boston-MA: Springer.
80. Holt, C., Timmins, P. A., Errington, N., & Leaver, J., (1998). A core-shell model of calcium phosphate nanoclusters stabilized by β-casein phospho-peptides derived from sedimentation equilibrium and small-angle x-ray and neutron-scattering measurements. *European Journal of Biochemistry, 252*(1), 73–78.
81. Hsu, K. C., Tung, Y. S., Huang, S. L., & Jao, C. L., (2013). Dipeptidyl peptidase-IV inhibitory activity of peptides in porcine skin gelatin hydrolysates. Chapter 8. In: Hernández-Ledesma, B., (ed.), *Bioactive Food Peptides in Health and Disease* (pp. 205–218). London-UK: IntechOpen.
82. Hurrell, R., (2002). How to ensure adequate iron absorption from iron-fortified food. *Nutrition Reviews, 60*(7.2), S7–S15.
83. Irshad, I., Kanekanian, A., Peters, A., & Masud, T., (2015). Antioxidant activity of bioactive peptides derived from bovine casein hydrolysate fractions. *Journal of Food Science and Technology, 52*, 1–9.
84. Iwaniak, A., Darewicz, M., & Minkiewicz, P., (2018). Peptides derived from foods as supportive diet components in the prevention of metabolic syndrome. *Comprehensive Reviews in Food Science and Food Safety, 17*(1), 63–81.
85. IzzoJr, J. L., & Weir, M. R., (2011). Angiotensin-converting enzyme inhibitors. *The Journal of Clinical Hypertension, 13*(9), 667–675.
86. Jan, F., Kumar, S., & Jha, R., (2016). Effect of boiling on the antidiabetic property of enzyme treated sheep milk casein. *Veterinary World, 9*(10), 1152–1156.
87. Jeon, Y. J., Byun, H. G., & Kim, S. K., (1999). Improvement of functional properties of cod frame protein hydrolysates using ultrafiltration membranes. *Process Biochemistry, 35*(5), 471–478.
88. Jeong, H. J., Lee, J. R., Jeong, J. B., Park, J. H., Cheong, Y. K., & De Lumen, B. O., (2009). The cancer preventive seed peptide lunasin from rye is bioavailable and bioactive. *Nutrition and Cancer, 61*(5), 680–686.
89. Jones, F. S., & Simms, H. S., (1930). The bacterial growth inhibitor (Lactenin) of milk. *Journal of Experimental Medicine, 51*(2), 327–339.

90. Kampa, M., Loukas, S., Hatzoglou, A., Martin, P., Martin, P. M., & Castanas, E., (1996). Identification of a novel opioid peptide (Tyr-Val-pro-Phe-pro) derived from human αs1 casein (αs1-casomorphin, and αs1-casomorphin amide). *Biochemical Journal, 319*(3), 903–908.
91. Kayser, H., & Meisel, H., (1996). Stimulation of human peripheral blood lymphocytes by bioactive peptides derived from bovine milk proteins. *FEBS Letters, 383*(1, 2), 18–20.
92. Keller, J. L., Lanou, A. J., & Barnard, N. D., (2002). The consumer cost of calcium from food and supplements. *Journal of the American Dietetic Association, 102*(11), 1669–1671.
93. Kibangou, I. B., Bouhallab, S., Henry, G., Bureau, F., Allouche, S., Blais, A., Guérin, P., et al., (2005). Milk proteins and iron absorption: Contrasting effects of different caseino-phosphopeptides. *Pediatric Research, 58*(4), 731–734.
94. Kitts, D., (2005). Antioxidant properties of casein-phosphopeptides. *Trends in Food Science and Technology, 16*(12), 549–554.
95. Kitts, D. D., & Weiler, K., (2003). Bioactive proteins and peptides from food sources. Applications of bioprocesses used in isolation and recovery. *Current Pharmaceutical Design, 9*(16), 1309–1323.
96. Kondrashina, A., Brodkorb, A., & Giblin, L., (2020). Dairy-derived peptides for satiety. *Journal of Functional Foods, 66*, 8. Article ID: 103801.
97. Konrad, B., Anna, D., Marek, S., Marta, P., Aleksandra, Z., & Józefa, C., (2014). The Evaluation of dipeptidyl peptidase (DPP)-IV, α-glucosidase and angiotensin converting enzyme (ACE) inhibitory activities of whey proteins hydrolyzed with serine protease isolated from asian pumpkin (*Cucurbita ficifolia*). *International Journal of Peptide Research and Therapeutics, 20*(4), 483–491.
98. Korhonen, H., (2009). Milk-derived bioactive peptides: From science to applications. *Journal of Functional Foods, 1*(2), 177–187.
99. Korhonen, H., & Pihlanto, A., (2006). Bioactive peptides: Production and functionality. *International Dairy Journal, 16*(9), 945–960.
100. Kostyra, E., Sienkiewicz-Szłapka, E., Jarmołowska, B., Krawczuk, S., & Kostyra, H., (2004). Opioid peptides derived from milk proteins. *Polish Journal of Food and Nutrition Sciences, 54*(1), 25–35.
101. Kuba, M., Tanaka, K., Tawata, S., Takeda, Y., & Yasuda, M., (2003). Angiotensin-converting enzyme inhibitory peptides isolated from tofuyo fermented soybean food. *Bioscience, Biotechnology, and Biochemistry, 67*(6), 1278–1283.
102. Lacroix, I. M., & Li-Chan, E. C., (2013). Inhibition of dipeptidyl peptidase (DPP)-IV and α-glucosidase activities by pepsin-treated whey proteins. *Journal of Agricultural and Food Chemistry, 61*(31), 7500–7506.
103. Lahl, W. J., & Braun, S. D., (1995). Enzymatic production of protein hydrolysates for food use. *Food Technology, 48*(10), 68–71.
104. Lahov, E., Edelsten, D., Sode-Morgensen, M. T., & Sofer, E., (1971). Properties of basic glycopeptides released from cow milk protein by heat. *Milchwissenschaft, 26*, 489–495.
105. Lahov, E., & Regelson, W., (1996). Antibacterial and immuno-stimulating casein-derived substances from milk: Casecidin, isracidin peptides. *Food and Chemical Toxicology, 34*(1), 131–145.
106. Lankisch, M., Layer, P., Rizza, R. A., & Di Magno, E. P., (1998). Acute postprandial gastrointestinal and metabolic effects of wheat amylase inhibitor (WAI) in normal, obese, and diabetic humans. *Pancreas, 17*(2), 176–181.

107. Latif, S., & Anwar, F., (2011). Aqueous enzymatic sesame oil and protein extraction. *Food Chemistry, 125*(2), 679–684.
108. Lee, S. H., & Song, K. B., (2009). Purification of an iron-binding Nona-peptide from hydrolysates of porcine blood plasma protein. *Process Biochemistry, 44*(3), 378–381.
109. Li, C. H., Matsui, T., Matsumoto, K., Yamasaki, R., & Kawasaki, T., (2002). Latent production of angiotensin I-converting enzyme inhibitors from buck wheat protein. *Journal of Peptide Science, 8*(6), 267–274.
110. Li, S., Hu, Q., Chen, C., Liu, J., He, G., Li, L., Wu, J., & Ren, D., (2020). Formation of bioactive peptides during simulated gastrointestinal digestion is affected by αs1-casein polymorphism in buffalo milk. *Food Chemistry, 313*, 7. Article ID: 126159.
111. Li, Y. W., Li, B., He, J., & Qian, P., (2011). Structure-activity relationship study of anti-oxidative peptides by QSAR modeling: Amino acid next to C-terminus affects the activity. *Journal of Peptide Science, 17*(6), 454–462.
112. Li, Y., Jiang, B., Zhang, T., Mu, W., & Liu, J., (2008). Antioxidant and free radical-scavenging activities of chickpea protein hydrolysate (CPH). *Food Chemistry, 106*(2), 444–450.
113. López-Expósito, I., Gómez-Ruiz, J. Á., Amigo, L., & Recio, I., (2006). Identification of antibacterial peptides from ovine αs2-casein. *International Dairy Journal, 16*(9), 1072–1080.
114. López-Expósito, I., Quirós, A., Amigo, L., & Recio, I., (2007). Casein hydrolysates as a source of antimicrobial, antioxidant, and antihypertensive peptides. *Le Lait, 87*(4, 5), 241–249.
115. López-Expósito, I., & Recio, I., (2008). Protective effect of milk peptides: Antibacterial and antitumor properties. In: Bosze, Z., (ed.), *Bioactive Components of Milk* (Vol. 606, pp. 271–293). New York: Springer Science + Business Media.
116. Lopez-Exposito, I., & Recio, I., (2006). Antibacterial activity of peptides and folding variants from milk proteins. *International Dairy Journal, 16*(11), 1294–1305.
117. Maeno, M., Yamamoto, N., & Takano, T., (1996). Identification of an antihypertensive peptide from casein hydrolysate produced by a proteinase from *Lactobacillus helveticus* CP790. *Journal of Dairy Science, 79*(8), 1316–1321.
118. Malkoski, M., Dashper, S. G., O'Brien-Simpson, N. M., Talbo, G. H., Macris, M., & Cross, K. J., (2001). Kappacin: Novel antibacterial peptide from bovine milk. *Antimicrobial Agents and Chemotherapy, 45*(8), 2309–2315.
119. Mandal, B., Roe, B. E., & Fein, S. B., (2014). Work and breastfeeding decisions are jointly determined for higher socioeconomic status US mothers. *Review of Economics of the Household, 12*(2), 237–257.
120. Mant, C. T., & Hodges, R. S., (1996). Analysis of peptides by high-performance liquid chromatography. *Methods in Enzymology, 271*, 3–50.
121. Mant, C. T., Kondejewski, L. H., Cachia, P. J., Monera, O. D., & Hodges, R. S., (1997). Analysis of synthetic peptides by high-performance liquid chromatography. *Methods in Enzymology, 289*, 426–469.
122. Mao, X. Y., Cheng, X., Wang, X., & Wu, S. J., (2011). Free-radical-scavenging and anti-inflammatory effect of yak milk casein before and after enzymatic hydrolysis. *Food Chemistry, 126*(2), 484–490.
123. Martínez-Maqueda, D., Hernández-Ledesma, B., Amigo, L., Miralles, B., & Gómez-Ruiz, J. Á., (2013). Extraction/ fractionation techniques for proteins and peptides and

protein digestion. Chapter 2. In: Tolldrá, F., & Nollet, L. M. L., (eds.), *Proteomics in Foods: Principles and Applications* (pp. 21–50). Boston-MA: Springer.
124. Maruyama, S., & Suzuki, H., (1982). A peptide inhibitor of angiotensin I converting enzyme in the tryptichydrolysate of casein. *Agricultural and Biological Chemistry, 46*(5), 1393–1394.
125. Matchar, D. B., McCrory, D. C., Orlando, L. A., & Patel, M. R., (2008). Systematic review: Comparative effectiveness of angiotensin-converting enzyme inhibitors and angiotensin II receptor blockers for treating essential hypertension. *Annals of Internal Medicine, 148*(1), 16–29.
126. Matin, A., & Otani, H., (2002). Cytotoxic and antibacterial activities of chemically synthesized κ-casecidin and its partial peptide fragments. *Journal of Dairy Research, 69*(2), 329–334.
127. Matin, M. A., Monnai, M., & Otani, H., (2000). Isolation and characterization of a cytotoxic pentapeptide κ-casecidin from bovine κ-casein digested with bovine trypsin. *Animal Science Journal, 71*(2), 197–207.
128. Matsufuji, H., Matsui, T., Seki, E., Osajima, K., Nakashima, M., & Osajima, Y., (1994). Angiotensin I-converting enzyme inhibitory peptides in an alkaline protease hydrolyzate derived from sardine muscle. *Bioscience, Biotechnology, and Biochemistry, 58*(12), 2244–2245.
129. Matthies, H., Stark, H., Hartrodt, B., Ruethrich, H. L., Spieler, H. T., Barth, A., & Neubert, K., (1984). Derivatives of beta-casomorphins with high analgesic potency. *Peptides, 5*(3), 463–470.
130. McCann, K. B., Shiell, B. J., Michalski, W. P., Lee, A., Wan, J., Roginski, H., & Coventry, M. J., (2006). Isolation and characterization of a novel antibacterial peptide from bovine αs1-casein. *International Dairy Journal, 16*(4), 316–323.
131. McDonagh, D., & FitzGerald, R. J., (1998). Production of caseino-phosphopeptides (CPPs) from sodium caseinate using a range of commercial protease preparations. *International Dairy Journal, 8*(1), 39–45.
132. McLachlin, D. T., & Chait, B. T., (2001). Analysis of phosphorylated proteins and peptides by mass spectrometry. *Current Opinion in Chemical Biology, 5*(5), 591–602.
133. Megías, C., Yust, M. D. M., Pedroche, J., Lquari, H., Girón-Calle, J., Alaiz, M., Millán, F., & Vioque, J., (2004). Purification of an ACE inhibitory peptide after hydrolysis of sunflower (*Helianthus annuus* L.) protein isolates. *Journal of Agricultural and Food Chemistry, 52*(7), 1928–1932.
134. Meisel, H., & Schlimme, E., (1990). Milk proteins: Precursors of bioactive peptides. *Trends in Foods Science &Technology, 1*, 41–43.
135. Meisel, H., (1997). Biochemical properties of regulatory peptides derived from milk proteins. *Peptide Science, 43*(2), 119–128.
136. Meisel, H., (1998). Overview on milk protein-derived peptides. *International Dairy Journal, 8*(5, 6), 363–373.
137. Meisel, H., & Fitzgerald, R. J., (2000). Opioid peptides encrypted in intact milk protein sequences. *The British Journal of Nutrition, 84*(S1), S27–S31.
138. Meisel, H., (2001). Bioactive peptides from milk proteins: A perspective for consumers and producers. *Australian Journal of Dairy Technology, 56*(2), 83–92.
139. Meisel, H., & FitzGerald, R., (2003). Biofunctional peptides from milk proteins: Mineral binding and cytomodulatory effects. *Current Pharmaceutical Design, 9*(16), 1289–1296.

140. Meisel, H., (2004). Multifunctional peptides encrypted in milk proteins. *Biofactors, 21*(1), 55–61.
141. Mellander, O., (1950). The physiological importance of the casein phosphopeptide calcium salts. Part II: Peroral calcium dosage of infants. Some aspects of the pathogenesis of rickets. *Acta Societatis Botanicorum Poloniae, 55*(5, 6), 247–255.
142. Mendis, E., Rajapakse, N., Byun, H. G., & Kim, S. K., (2005). Investigation of jumbo squid skin gelatin peptides for their *in vitro* antioxidant effects. *Life Sciences, 77*(17), 2166–2178.
143. Miguel, M., Aleixandre, M., Ramos, M., & López-Fandiño, R., (2006). Effect of simulated gastrointestinal digestion on the antihypertensive properties of ACE-inhibitory peptides derived from ovalbumin. *Journal of Agricultural and Food Chemistry, 54*(3), 726–731.
144. Miguel, M., Gómez-Ruiz, J. Á., Recio, I., & Aleixandre, A., (2010). Changes in arterial blood pressure after single oral administration of milk-casein-derived peptides in spontaneously hypertensive rats. *Molecular Nutrition and Food Research, 54*(10), 1422–1427.
145. Miquel, E., Alegría, A., Barberá, R., & Farré, R., (2006). Casein phosphopeptides released by simulated gastrointestinal digestion of infant formulas and their potential role in mineral binding. *International Dairy Journal, 16*(9), 992–1000.
146. Miquel, E., & Farré, R., (2007). Effects and future trends of casein phosphopeptides on zinc bioavailability. *Trends in Food Science and Technology, 18*(3), 139–143.
147. Molina, P. E., & Abumrad, N. N., (1994). Metabolic effects of opiates and opioid peptides. *Advances in Neuroimmunology, 4*(2), 105–116.
148. Motoi, H., & Kodama, T., (2003). Isolation and characterization of angiotensin I-converting enzyme inhibitory peptides from wheat gliadin hydrolysate. *Die Nahrung (The Food), 47*(5), 354–358.
149. Moure, A., Domínguez, H., & Parajó, J. C., (2006). Antioxidant properties of ultrafiltration recovered soy protein fractions from industrial effluents and their hydrolysates. *Process Biochemistry, 41*(2), 447–456.
150. Murray, B., & Fitz, G. R., (2007). Angiotensin converting enzyme inhibitory peptides derived from food proteins: Biochemistry, bioactivity, and production. *Current Pharmaceutical Design, 13*(8), 773–791.
151. Nakamura, Y., Yamamoto, N., Sakai, K., Okubo, A., Yamazaki, S., & Takano, T., (1995). Purification and characterization of angiotensin I-converting enzyme inhibitors from sour milk. *Journal of Dairy Science, 78*(4), 777–783.
152. Nakano, D., Ogura, K., Miyakoshi, M., Ishii, F., Kawanishi, H., Kurumazuka, D., Kwak, C. J., et al., (2006). Antihypertensive effect of angiotensin I-converting enzyme inhibitory peptides from a sesame protein hydrolysate in spontaneously hypertensive rats. *Bioscience, Biotechnology, and Biochemistry, 70*(5), 1118–1126.
153. Natesh, R., Schwager, S. L., Evans, H. R., Sturrock, E. D., & Acharya, K. R., (2004). Structural details on the binding of antihypertensive drugs captopril and enalaprilat to human testicular angiotensin I-converting enzyme. *Biochemistry, 43*(27), 8718–8724.
154. Nongonierma, A. B., & FitzGerald, R. J., (2013). Dipeptidyl peptidase IV inhibitory and antioxidative properties of milk protein-derived dipeptides and hydrolysates. *Peptides, 39*, 157–163.
155. Okada, Y., & Okada, M., (1998). Scavenging effect of water-soluble proteins in broad beans on free radicals and active oxygen species. *Journal of Agricultural and Food Chemistry, 46*(2), 401–406.

156. Oliveira, T. V. D., Polêto, M. D., & De Oliveira, M. R., (2020). Casein-derived peptides with antihypertensive potential: Production, identification, and assessment of complex formation with angiotensin I-converting enzyme (ACE) through molecular docking studies. *Food Biophysics, 15*, 162–172.
157. O' Mahony, F., (1988). *Rural Dairy Technology: Experiences in Ethiopia* (p. 65). ILCA Manual No. 4; Ethiopia: International Livestock Center.
158. Ondetti, M. A., Williams, N. J., Sabo, E., Pluscec, J., Weaver, E. R., & Kocy, O., (1988). Angiotensin-converting enzyme inhibitors from the venom of *Bothrops jararaca*. Isolation, elucidation of structure, and synthesis. *Biochemistry, 10*(22), 4033–4039.
159. Otani, H., & Suzuki, H., (2003). Isolation and characterization of cytotoxic small peptides, α-casecidins, from bovine αs1-casein digested with bovine trypsin. *Animal Science Journal, 74*(5), 427–435.
160. Panchaud, A., Affolter, M., & Kussmann, M., (2012). Mass spectrometry for nutritional peptidomics: How to analyze food bioactives and their health effects. *Journal of Proteomics, 75*(12), 3546–3559.
161. Panyam, D., & Kilara, A., (1996). Enhancing the functionality of food proteins by enzymatic modification. *Trends in Food Science and Technology, 7*(4), 120–125.
162. Pardo, M. F., & Natalucci, C. L., (2002). Electrophoretic analysis (tricine-SDS-PAGE) of bovine caseins. *Acta Farmacéutica Bonaerense, 21*(1), 57–60.
163. Park, O., Swaisgood, H., & Allen, J., (1998). Calcium binding of phosphopeptides derived from hydrolysis of a αs-casein or β-casein using immobilized trypsin. *Journal of Dairy Science, 81*(11), 2850–2857.
164. Park, Y. W., (2009). *Bioactive Components in Milk and Dairy Products* (p. 426). Iowa-USA: Wiley-Blackwell.
165. Patil, P., Mandal, S., Tomar, S. K., & Anand, S., (2015). Food protein-derived bioactive peptides in management of type-2 diabetes. *European Journal of Nutrition, 54*(6), 863–880.
166. Perego, S., Cosentino, S., Fiorilli, A., Tettamanti, G., & Ferraretto, A., (2012). Casein phosphopeptides modulate proliferation and apoptosis in HT-29 cell line through their interaction with voltage-operated L-type calcium channels. *The Journal of Nutritional Biochemistry, 23*(7), 808–816.
167. Peres, J. M., Bouhallab, S., Bureau, F., Maubois, J. L., Arhan, P., & Bouglé, D., (1997). Digestive absorption of iron bound to the 1–25 caseino-phosphopeptide of beta-casein. *Le Lait, 4*(77), 433–440.
168. Phelan, M., Aherne, A., FitzGerald, R. J., & O'Brien, N. M., (2009). Casein-derived bioactive peptides: Biological effects, industrial uses, safety aspects and regulatory status. *International Dairy Journal, 19*(11), 643–654.
169. Picariello, G., Ferranti, P., Fierro, O., Mamone, G., Caira, S., Di Luccia, A., Monica, S., & Addeo, F., (2010). Peptides surviving the simulated gastrointestinal digestion of milk proteins: Biological and toxicological implications. *Journal of Chromatography B, 878*(3), 295–308.
170. Pihlanto, A., (2006). Antioxidative peptides derived from milk proteins. *International Dairy Journal, 16*(11), 1306–1314.
171. Pihlanto-Leppälä, A., Rokka, T., & Korhonen, H., (1998). Angiotensin I converting enzyme inhibitory peptides derived from bovine milk proteins. *International Dairy Journal, 8*(4), 325–331.

172. Pintado, M. E., & Malcata, F. X., (2000). Hydrolysis of ovine, caprine, and bovine whey proteins by trypsin and pepsin. *Bioprocess Engineering, 23*(3), 275–282.
173. Poitou, B. C., Ciangura, C., Coupaye, M., Czernichow, S., Bouillot, J., & Basdevant, A., (2007). Nutritional deficiency after gastric bypass: Diagnosis, prevention, and treatment. *Diabetes and Metabolism, 33*(1), 13–24.
174. Pouliot, Y., & Gauthier, S. F., (2006). Milk growth factors as health products: Some technological aspects. *International Dairy Journal, 16*(11), 1415–1420.
175. Power, O., Jakeman, P., & FitzGerald, R., (2013). Antioxidative peptides: Enzymatic production, *in vitro* and *in vivo* antioxidant activity and potential applications of milk-derived antioxidative peptides. *Amino Acids, 44*(3), 797–820.
176. Qian, B., Xing, M., Cui, L., Deng, Y., Xu, Y., Huang, M., & Zhang, S., (2011). Antioxidant, antihypertensive, and immunomodulatory activities of peptide fractions from fermented skim milk with *Lactobacillus delbrueckii* ssp. *Bulgaricus* LB340. *Journal of Dairy Research, 78*(1), 72–79.
177. Quirós, A., Contreras, M. D. M., Ramos, M., Amigo, L., & Recio, I., (2009). Stability to gastrointestinal enzymes and structure-activity relationship of β-casein-peptides with antihypertensive properties. *Peptides, 30*(10), 1848–1853.
178. Ramchandran, L., & Shah, N. P., (2008). Proteolytic profiles and angiotensin-I converting enzyme and alpha-glucosidase inhibitory activities of selected lactic acid bacteria. *Journal of Food Science, 73*(2), 75–81.
179. Recio, I., & Visser, S., (1999). Identification of two distinct antibacterial domains within the sequence of bovine αs2-casein. *Biochimicaet Biophysica Acta (BBA)-General Subjects, 1428*(2, 3), 314–326.
180. Reema, S. D., Lahiri, P. K., & Roy, S. S., (2014). Review of casein phosphopeptides-amorphous calcium phosphate. *The Chinese Journal of Dental Research, 17*(1), 7–14.
181. Reichelt, K. L., & Stensrud, M., (1998). Increase in Urinary peptides prior to the diagnosis of schizophrenia. *Schizophrenia Research, 34*(3), 211–213.
182. Rejtar, T., Chen, H. S., Andreev, V., Moskovets, E., & Karger, B. L., (2004). Increased identification of peptides by enhanced data processing of high-resolution MALDITOF/TOF mass spectra prior to database searching. *Analytical Chemistry, 76*(20), 6017–6028.
183. Reynolds, E. C., Cai, F., Shen, P., & Walker, G. D., (2003). Retention in plaque and remineralization of enamel lesions by various forms of calcium in a mouth rinse or sugar-free chewing gum. *Journal of Dental Research, 82*(3), 206–211.
184. Rival, S. G., Boeriu, C. G., & Wichers, H. J., (2001). Caseins and casein hydrolysates, II: Antioxidative properties and relevance to lipoxygenase inhibition. *Journal of Agricultural and Food Chemistry, 49*(1), 295–302.
185. Robert, M. C., Razaname, A., Mutter, M., & Juillerat, M. A., (2004). Identification of angiotensin-I-converting enzyme inhibitory peptides derived from sodium caseinate hydrolysates produced by *Lactobacillus helveticus* NCC 2765. *Journal of Agricultural and Food Chemistry, 52*(23), 6923–6931.
186. Roblet, C., Amiot, J., Lavigne, C., Marette, A., Lessard, M., Jean, J., Ramassamy, C., et al., (2012). Screening of *in vitro* bioactivities of a soy protein hydrolysate separated by hollow fiber and spiral-wound ultrafiltration membranes. *Food Research International, 46*(1), 237–249.
187. Ruiz, J. Á. G., Ramos, M., & Recio, I., (2004). Angiotensin converting enzyme-inhibitory activity of peptides isolated from Manchego cheese. Stability under simulated gastrointestinal digestion. *International Dairy Journal, 14*(12), 1075–1080.

188. SabeenaFarvin, K. H., Baron, C. P., Nielsen, N. S., Otte, J., & Jacobsen, C., (2010). Antioxidant activity of yoghurt peptides: Part 2-characterization of peptide fractions. *Food Chemistry, 123*(4), 1090–1097.
189. Saito, T., Nakamura, T., Kitazawa, H., Kawai, Y., & Itoh, T., (2000). Isolation and structural analysis of antihypertensive peptides that exist naturally in gouda cheese. *Journal of Dairy Science, 83*(7), 1434–1440.
190. Salam, S. M. A., Kagawa, K. I., & Kawashiro, K., (2005). α-chymotrypsin-catalyzed peptide synthesis using N-protected D-amino acid carbamoyl methyl esters as acyl donors. *Biotechnology Letters, 27*(16), 1199–1203.
191. Sales, P. M., Souza, P. M., Simeoni, L. A., Magalhães, P. O., & Silveira, D., (2012). α-amylase inhibitors: A review of raw material and isolated compounds from plant source. *Journal of Pharmacy and Pharmaceutical Sciences, 15*(1), 141–183.
192. Sarmadi, B. H., & Ismail, A., (2010). Antioxidative peptides from food proteins: A review. *Peptides, 31*(10), 1949–1956.
193. Scarr, E., Money, T. T., Pavey, G., Neo, J., & Dean, B., (2012). Mu-opioid receptor availability in people with psychiatric disorders who died by suicide: A case control study. *BMC Psychiatry, 12*(1), 126–132.
194. Schägger, H., & Von, J. G., (1987). Tricine-sodium dodecyl sulfate-polyacrylamide gel electrophoresis for the separation of proteins from 1 to 100 kDa. *Analytical Biochemistry, 166*(2), 368–379.
195. Schanbacher, F. L., Talhouk, R. S., & Murray, F. A., (1997). Biology and origin of bioactive peptides in milk. *Livestock Production Science, 50*(1, 2), 105–123.
196. Schlimme, E., & Meisel, H., (1995). Bioactive peptides derived from milk proteins. Structural, physiological, and analytical aspects. *Nahrung, 39*(1), 1–20.
197. Schüpbach, P., Neeser, J., Golliard, M., Rouvet, M., & Guggenheim, B., (1996). Incorporation of caseinoglyco macro peptide and caseino phosphopeptide into the salivary pellicle inhibits adherence of mutans streptococci. *Journal of Dental Research, 75*(10), 1779–1788.
198. Seppo, L., Jauhiainen, T., Poussa, T., & Korpela, R., (2003). A fermented milk high in bioactive peptides has a blood pressure lowering effect in hypertensive subjects. *The American Journal of Clinical Nutrition, 77*(2), 326–330.
199. Shahidi, F., & Zhong, Y., (2008). Bioactive peptides. *Journal of AOAC International, 91*(4), 914–931.
200. Shai, Y., (1999). Mechanism of the binding, insertion, and destabilization of phospholipid bilayer membranes by α-helical antimicrobial and cell non-selective membrane-lytic peptides. *Biochimicaet Biophysica Acta (BBA)-Biomembranes, 1462*(1, 2), 55–70.
201. Sharma, S., Singh, R., & Rana, S., (2011). Bioactive peptides: A review. *International Journal of Bioautomation, 15*(4), 223–250.
202. Sharp, J. S., Becker, J. M., & Hettich, R. L., (2004). Analysis of protein solvent accessible surfaces by photochemical oxidation and mass spectrometry. *Analytical Chemistry, 76*(3), 672–683.
203. Shazly, A. B., Mu, H., Liu, Z., El-Aziz, M. A., Zeng, M., Qin, F., & Chen, J., (2019). Release of antioxidant peptides from buffalo and bovine caseins: Influence of proteases on antioxidant capacities. *Food Chemistry, 274*, 261–267.
204. Sienkiewicz-Szłapka, E., Jarmołowska, B., Krawczuk, S., Kostyra, E., Kostyra, H., & Bielikowicz, K., (2009). Transport of bovine milk-derived opioid peptides across caco-2 monolayer. *International Dairy Journal, 19*(4), 252–257.

205. Silva, S. V., & Malcata, F. X., (2005). Caseins as source of bioactive peptides. *International Dairy Journal, 15*(1), 1–15.
206. Silvestre, M., (1997). Methods for the analysis of protein hydrolysates. *Food Chemistry, 60*(2), 263–271.
207. Slattery, H., & Fitzgerald, R., (1998). Functional properties and bitterness of sodium caseinate hydrolysates prepared with a *Bacillus* proteinase. *Journal of Food Science, 63*(3), 418–422.
208. Sturner, R. A., & Chang, K. J., (1988). Opioid peptide content in infant formulas. *Pediatric Research, 23*, 4–10.
209. Suetsuna, K., Ukeda, H., & Ochi, H., (2000). Isolation and characterization of free radical scavenging activities peptides derived from casein. *The Journal of Nutritional Biochemistry, 11*(3), 128–131.
210. Sultan, S., Huma, N., Butt, M. S., Aleem, M., & Abbas, M., (2018). Therapeutic potential of dairy bioactive peptides: A contemporary perspective. *Critical Reviews in Food Science and Nutrition, 58*(1), 105–115.
211. Sun, Z., Cade, J. R., Fregly, M. J., & Privette, R., (1999). M. β-casomorphin induces FOS-like immunoreactivity in discrete brain regions relevant to schizophrenia and autism. *Autism, 3*(1), 67–83.
212. Sun, Z., Zhang, Z., Wang, X., Cade, R., Elmir, Z., & Fregly, M., (2003). Relation of β-casomorphin to apnea in sudden infant death syndrome. *Peptides, 24*(6), 937–943.
213. Tahrani, A. A., Bailey, C. J., Del, P. S., & Barnett, A. H., (2011). Management of type-2 diabetes: New and future developments in treatment. *The Lancet, 378*(9786), 182–197.
214. Takemoto, Y., (2012). Amino acids that centrally influence blood pressure and regional blood flow in conscious rats. *Journal of Amino Acids, 2012*, 9. Article ID: 831759.
215. Teschemacher, H., (2003). Opioid receptor ligands derived from food proteins. *Current Pharmaceutical Design, 9*(16), 1331–1344.
216. Thornberry, N. A., & Gallwitz, B., (2009). Mechanism of action of inhibitors of dipeptidyl-peptidase-4 (DPP-4). *Best Practice & Research Clinical Endocrinology & Metabolism, 23*(4), 479–486.
217. Timón, M. L., Andrés, A. I., Otte, J., & Petrón, M. J., (2019). Antioxidant peptides (< 3 kDa) identified on hard cow milk cheese with rennet from different origin. *Food Research International, 120*, 643–649.
218. Ting, B., Gauthier, S., & Pouliot, Y., (2007). Fractionation of β-lactoglobulin tryptic peptides using spiral wound nanofiltration membranes. *Separation Science and Technology, 42*(11), 2419–2433.
219. Tobita, K., Kawahara, T., & Otani, H., (2006). Bovine β-casein (1–28), a casein phosphopeptide, enhances proliferation and IL-6 expression of mouse CD19+ cells via toll-like receptor 4. *Journal of Agricultural and Food Chemistry, 54*(21), 8013–8017.
220. Tomé, D., & Debabbi, H., (1998). Physiological effects of milk protein components. *International Dairy Journal, 8*(5, 6), 383–392.
221. Tu, M., Wang, C., Chen, C., Zhang, R., Liu, H., Lu, W., & Du, M., (2018). Identification of a novel ACE-inhibitory peptide from casein and evaluation of the inhibitory mechanisms. *Food Chemistry, 256*, 98–104.
222. Turgeon, S. L., & Gauthier, S. F., (1990). Whey peptide fractions obtained with a two-step ultrafiltration process: Production and characterization. *Journal of Food Science, 55*(1), 106–110.

223. Udenigwe, C. C., (2014). Bioinformatics approaches, prospects, and challenges of food bioactive peptide research. *Trends in Food Science and Technology, 36*(2), 137–143.
224. Udenigwe, C. C., & Aluko, R. E., (2012). Food protein-derived bioactive peptides: Production, processing, and potential health benefits. *Journal of Food Science, 77*(1), R11–R24.
225. Uenishi, H., Kabuki, T., Seto, Y., Serizawa, A., & Nakajima, H., (2012). Isolation and identification of casein-derived dipeptidyl-peptidase 4(DPP-4)-inhibitory peptide LPQNIPPL from gouda-type cheese and its effect on plasma glucose in rats. *International Dairy Journal, 22*(1), 24–30.
226. Uluko, H., Li, H., Cui, W., Zhang, S., Liu, L., Chen, J., Sun, Y., et al., (2013). Response surface optimization of angiotensin converting enzyme inhibition of milk protein concentrate hydrolysates *in vitro* after ultrasound pretreatment. *Innovative Food Science & Emerging Technologies, 20*, 133–139.
227. Vegarud, G. E., Langsrud, T., & Svenning, C., (2000). Mineral-binding milk proteins and peptides; occurrence, biochemical, and technological characteristics. *The British Journal of Nutrition, 84*(S1), S91–S98.
228. Vermeirssen, V., Camp, J. V., & Verstraete, W., (2004). Bioavailability of angiotensin I converting enzyme inhibitory peptides. *The British Journal of Nutrition, 92*(3), 357–366.
229. Wang, B., Matsuoka, S., Ballif, B. A., Zhang, D., Smogorzewska, A., Gygi, S. P., & Elledge, S. J., (2007). Abraxas and RAP80 form a BRCA1 protein complex required for the DNA damage response. *Science, 316*(5828), 1194–1198.
230. Wang, H. J. Y., Shyu, Y. S., Wang, Y. T., & Hsu, C. K., (2010). Anti-oxidative properties of protein hydrolysate from defatted peanut kernels treated with esperase. *LWT-Food Science and Technology, 43*(2), 285–290.
231. Wang, J., Hu, J., Cui, J., Bai, X., Du, Y., Miyaguchi, Y., & Lin, B., (2008). Purification and Identification of a ACE inhibitory peptide from oyster proteins hydrolysate and the antihypertensive effect of hydrolysate in spontaneously hypertensive rats. *Food Chemistry, 111*(2), 302–308.
232. Wang, W., Gu, F., Wei, C., Tang, Y., Zheng, X., Ren, M., & Qin, Y., (2013). PGPIPN, a therapeutic hexapeptide, suppressed human ovarian cancer growth by targeting BCL2. *PloS One, 8*(4), 9. Article ID: 60701.
233. Whitney, R., (1977). McL. Milk proteins. In: Graham, H. D., (ed.), *Food Colloids* (pp. 66–151). Westport-Connecticut, USA: AVI Publishing Co.
234. Wong, P. Y. Y., & Kitts, D. D., (2003). A comparison of the buttermilk solids functional properties to nonfat dried milk, soy protein isolate, dried egg white, and egg yolk powders. *Journal of Dairy Science, 86*(3), 746–754.
235. Wu, H., Liu, Z., Zhao, Y., & Zeng, M., (2012). Enzymatic preparation and characterization of iron-chelating peptides from anchovy (*Engraulis japonicus*) muscle protein. *Food Research International, 48*(2), 435–441.
236. Wu, J., Aluko, R. E., & Nakai, S., (2006). Structural requirements of angiotensin I-converting enzyme inhibitory peptides: Quantitative structure-activity relationship study of di-and tripeptides. *Journal of Agricultural and Food Chemistry, 54*(3), 732–738.
237. Wu, Z., Pan, D., Zhen, X., & Cao, J., (2013). Angiotensin-I converting enzyme inhibitory peptides derived from bovine casein and identified by MALDI-TOF-MS/MS. *Journal of the Science of Food and Agriculture, 93*(6), 1331–1337.
238. Xia, Y., Bamdad, F., Gänzle, M., & Chen, L., (2012). Fractionation and characterization of antioxidant peptides derived from barley glutelin by enzymatic hydrolysis. *Food Chemistry, 134*(3), 1509–1518.

239. Xu, R. J., (1998). Bioactive peptides in milk and their biological and health implications. *Food Reviews International, 14*(1), 1–16.
240. Xue, L., Wang, X., Hu, Z., Wu, Z., Wang, L., Wang, H., & Yang, M., (2018). Identification and characterization of an angiotensin converting enzyme inhibitory peptide derived from bovine casein. *Peptides, 99*, 161–168.
241. Yamamoto, N., Maeno, M., & Takano, T., (1999). Purification and characterization of an antihypertensive peptide from a yogurt-like product fermented by *Lactobacillus helveticus* CPN4. *Journal of Dairy Science, 82*(7), 1388–1393.
242. Zawadzki, A., Liu, Q., Wang, Y., Melander, A., Jeppsson, B., & Thorlacius, H., (2008). Verapamil inhibits L-type calcium channel mediated apoptosis in human colon cancer cells. *Diseases of the Colon and Rectum, 51*(11), 1696–1702.
243. Zayas, J. F., (1997). Solubility of proteins. In: Zayas, J. F. (ed.), *Functionality of Proteins in Food* (pp. 6–75). Berlin-Heidelberg: Springer.
244. Zhang, Y., Chen, R., Ma, H., & Chen, S., (2015). Isolation and identification of dipeptidyl peptidase IV-inhibitory peptides from trypsin/chymotrypsin-treated goat milk casein hydrolysates by 2D-TLC and LC-MS/MS. *Journal of Agricultural and Food Chemistry, 63*(40), 8819–8828.
245. Zheng, L., Xu, Q., Lin, L., Zeng, X. A., Sun, B., & Zhao, M., (2019). *In vitro* metabolic stability of a casein-derived dipeptidyl peptidase-IV (DPP-IV) inhibitory peptide VPYPQ and its controlled release from casein by enzymatic hydrolysis. *Journal of Agricultural and Food Chemistry, 67*(38), 10604–10613.
246. Zhu, L., Chen, J., Tang, X., & Xiong, Y. L., (2008). Reducing, radical scavenging, and chelation properties of *in vitro* digests of alcalase-treated zein hydrolysate. *Journal of Agricultural and Food Chemistry, 56*(8), 2714–2721.
247. Zucht, H. D., Raida, M., Adermann, K., Mägert, H. J., & Forssmann, W. G., (1995). Casocidin-I: Casein-alpha-s2 derived peptide exhibits antibacterial activity. *FEBS Letters, 372*(2, 3), 185–188.

CHAPTER 3

WHEY PROTEIN DERIVED BIOACTIVE PEPTIDES: ISOLATION, CHARACTERIZATION, AND FUNCTIONAL ASPECTS

RONIT MANDAL and PAYAL KARMAKAR

ABSTRACT

Milk is an excellent source of several functional and biologically active compounds. Bioactive peptides are examples of such biologically active compounds that are obtained from the milk proteins during the gastrointestinal digestion, hydrolysis, or fermentation of milk. They are generally the fractions of proteins amino acid sequence. These bioactive peptides are very helpful in uplifting the human health by modulating several biological functions ranging from digestive, immune, cardiovascular, and nervous system based on the sequence of their amino acid residues. Recently, there have been many researches about the production and isolation of biologically active peptides from whey proteins. Therefore, the intent of this chapter is to explain to the readers about the whey protein-derived bioactive peptides. The manufacturing and isolation techniques of the peptides have been discussed. The health implications, technological aspects, and applications are also given in the chapter.

3.1 INTRODUCTION

Milk is undoubtedly the nature's most perfect food. Being a very good source of biologically active nutrients like fat, protein, carbohydrates,

Functional Dairy Ingredients and Nutraceuticals: Physicochemical, Technological, and Therapeutic Aspects. Megh R. Goyal, N. Veena, PhD & Santosh K. Mishra (Eds.)

minerals, vitamins, it modulates several biological functions in our body. Among these, milk proteins show a decisive in the promotion of several physiological functions. The proteins in milk which are assimilated as the building blocks of the body, present as the nutritional source apart from affecting the functional and sensory properties of milk products. Around 3.5% of milk proteins are present in bovine milk which comprise about 80% of casein and 20% whey proteins [64]. Among the caseins, α_{s1}-, α_{s2}-, β- and κ-caseins are typically available in the ratio of 38:11:38:13 [29]. On the other hand, whey proteins consist of β-lactoglobulin (β-Lg, ~65%), α-lactalbumin (α-La, ~25%), bovine serum albumins (BSA), immunoglobulins (Igs), and lactoferrin.

Milk proteins have long been attributed to deliver their biological functionalities apart from nutritional role, by virtue of the derived bioactive peptides (BPs). BPs are defined as the fragments of proteins (comprising of usually 3–20 amino acid residues) which are encrypted in the source proteins and become active to carry out their physiological functions when they are released [22]. Several potential BPs have been identified from not only plants but also animal sources. However, proteins present in milk have been the major and most studied source of BPs. Therefore, milk from various milch animals like cow, buffalo, goat, donkey, yak, etc., have been extensively investigated for the production and isolation of such beneficial BPs. The BPs are released from the proteins mainly by: (1) gastrointestinal enzyme action, (2) hydrolysis with proteolytic enzymes, and (3) milk fermentation by lactic starter cultures. Other methods like recombinant DNA technology and chemical synthesis methods have also been explored. BPs are claimed to have various health-promoting functions upon consumption, absorption, and release in the body such as antioxidative, antihypertensive, stimulatory, immune, and opioid effects. These effects of such health promoting BPs have been well studied and documented [23, 24, 54]. Biological activity derived from these peptides is rooted in the sequence of amino acid residues, composition, and hydrophobicity [29].

Whey is the main byproduct in the dairy, especially the cheese industry. Although the casein-derived BPs have been characterized and studied, the BPs produced from whey proteins are also shown to be associated with several of the mentioned biological functions. There exists enormous literature pertaining to these whey proteins BPs. Some of the examples of whey protein-derived BPs include α-lactorphin, β-lactotensin, β-lactorphin, lactoferricin, serorphins, and albutensin A [29]. Recent advances in the field of recovery, isolation, and purification technologies like membrane processes, chromatography, etc., and validation technologies includes matrix-assisted laser desorption/ionization mass spectrometry (MALDI-MS), etc., were introduced and employed to cater to the production and commercialization applications.

This chapter discusses the current developments of BPs derived from whey proteins in milk. The production process, recovery, isolation, and characterization are presented. Lastly, a section on the health implications, technological, and commercial aspects are also included.

3.2 WHEY PROTEIN DERIVED BIOACTIVE PEPTIDES

Whey or serum proteins were shown to be amongst the highest quality natural protein apart from egg proteins [4]. Whey proteins tend to play an active role upon consumption by virtue of the presence of nutrients with immense health benefits. This enormous nutritional value of whey proteins is linked to their protein quality and their diverse amino acid composition. They are globular and water-soluble in nature and contain well-balanced acidic/basic and hydrophilic/hydrophobic amino acids in their structures. It is due to these reasons, that whey, which is the major by-product of dairy/cheese industry needs to be utilized as a source of specialized ingredients, sports nutrition, etc.

Whey proteins as the reservoir of biologically active peptides were well studied and characterized by researchers based on the number of literatures that have accumulated recently. BPs from whey proteins have a huge potential from health and nutritional point of view. Whey proteins like β-Lg, α-La, lactoferrin, BSA have been treated or hydrolyzed under controlled conditions *in vitro* and *in vivo* for liberation of various BPs. They act as carrier ligands and involve in several biological functions. The β-Lg and α-La proteins were also studied in detail for their functional peptides. Some of the BPs includes α-lactorphin, β-lactorphin, β-lactotensin, etc.

Whey proteins attribute a little opioid activity due to the BPs like α-lactorphin and serorphin (from BSA) [44, 57]. A major biological activity of BPs relies in their angiotensin I-converting enzyme (ACE) inhibiting capacity which modulates blood pressure. Several BPs like lactokinins, β-lactorphin from β-lactoglobulin as well as the BPs from other whey proteins have been attributed to possess antihypertensive properties [11, 26, 36, 42, 43]. Lactoferrin and its derived BPs like lactoferricin, lactoferrampin have been shown to possess antimicrobial properties [32, 62, 63]. Table 3.1 presents BPs from several whey proteins.

TABLE 3.1 Different BPs Derived from Whey Proteins and their Biological Activity

Name	Fragment	Amino Acid Sequence	Biological Activity	Production Method	References
α-Lactalbumin					
α-lactorphin	50–53	YGLF	Opioid agonist ACE-inhibitory	Hydrolysis by gastric and pancreatic enzyme (E:S = 0.3:20 v/v), 37°C for 3 h	[44]
	–	GL	Antidiabetic activity	Enzymatic hydrolysis with elastase enzyme (E:S = 1–2% w/w)	[40]
	21–26	VSLPEW	ACE-inhibitory	Enzymatic hydrolysis with thermolysin enzyme (E:S = 1% w/w), 40°C for 3 h	[42]
Immunopeptide	18–20	YGG	Immunomodulatory	Synthetically prepared	[32]
	115–122	LDQWLCEK	Immunomodulatory	Fermentation of milk by *Lb. helveticus* LH2 added at 1:10 w/w, 39°C for 26 h	[60]
	123–129	WLAHKAL	ACE-inhibitory; antidiabetic activity	Enzymatic hydrolysis with thermoase PC10F enzyme (E:S = 3:10 w/w), 65°C for 3 h; followed by hydrolysis with Accelerzyme® CGP, Peptidase R or ProteAX at 1:25 w/w enzyme to hydrolysate ratio	[26]
	35–42	KGYGGVSL	ACE-inhibitory	Enzymatic hydrolysis with *Cynara cardunculus*proteases enzyme (E:S = 1.6%v/v)	[58]
	135–138	DQWL	Anti-inflammatory	Enzymatic hydrolysis with alcalase enzyme (E:S = 3:50 w/w), 55°C for 6 h	[30]

TABLE 3.1 *(Continued)*

Name	Fragment	Amino Acid Sequence	Biological Activity	Production Method	References
β-Lactoglobulin					
β-lactorphin	102–105	TLLF	ACE-inhibitory	Enzymatic hydrolysis with trypsin enzyme	[36]
β-lactotensin	146–149	HIRL	Hypercholesteremic activity	Synthetically prepared with FMOC solid-phase synthesis	[36]
βL3	83–91	IDALNENK	Antimicrobial against *E. coli*, *S. aureus*, *Lb. acidophilus* and *Lb. casei*	Enzymatic hydrolysis with plasmin enzyme (E:S = 1:150 v/v), 30°C for 44 h	[53]
β-lactosin A	78–80	IPA	Antidiabetic	Hydrolysis by proteinase A enzyme	[61]
	19–29	WYSLAMAASDI	Antioxidant activity	Hydrolysis by corolase PP enzyme (E:S = 1:20 w/w), 37°C for 24 h	[21]
Lactokinin	142–148	ALPMHIR	ACE-inhibitory	Synthetically prepared with FMOC solid-phase synthesis	[37]
	71–75	IIAEK	Hypercholesteremic activity	Enzymatic hydrolysis with trypsin enzyme (E:S = 0.4% w/w), 37°C for 3 h	[38]
	148–153	RLSFNP	ACE-inhibitory	Fermentation of sour milk by *L. helveticus* LB10 added at 4% v/w, 39°C	[43]
	17–19	GTW	ACE-inhibitory	Fermentation of low fat milk by lactic acid bacteria added at 0.1% w/v, 42°C for 30 h, After 5 h prozyme 6 added (E:S = 1:100 w/w)	[11]

TABLE 3.1 *(Continued)*

Name	Fragment	Amino Acid Sequence	Biological Activity	Production Method	References
	36–42	SAPLRVY	ACE-inhibitory	Enzymatic hydrolysis with protease N Amano enzyme (E:S = 0.002), 45/55°C for 8 h	[41]
	46–55	LKPTPEGDLE	Antidiabetic activity	Enzymatic hydrolysis with thermoase PC10F enzyme (E:S = 3:10 w/w), 65°C for 3 h; followed by hydrolysis with Accelerzyme® CGP, Peptidase R or ProteAX at 1:25 w/w enzyme to hydrolysate ratio	[26]
	114–118	DYKKY	Anti-inflammatory	Enzymatic hydrolysis with alcalase enzyme (E:S = 3:50 w/w), 55°C for 6 h	[30]
	92–100	VLVLDTDYK	Antimicrobial activity against *E. coli* k12	Gastrointestinal digestion of whey proteins using gastric juice (pH 2.5) for 30 min at 37°C; followed by duodenal juice (pH 8) at 37°C for 30 min	[2]
Bovine Serum Albumin					
Albutensin A	208–216	ALKAWSVAR	ACE-inhibitory	Enzymatic hydrolysis with proteinase enzyme	[17]
Serorphin	399–404	YGFQDA	Opioid activity	Enzymatic hydrolysis with pepsin enzyme	[57]
	179–190	PKIDAMREKVLA	ACE-inhibitory	Isolated from Idiazabal-type sheep (raw milk) cheese	[52]

TABLE 3.1 *(Continued)*

Name	Fragment	Amino Acid Sequence	Biological Activity	Production Method	References
Lactoferrin					
Lactoferricin	17–41/42	FKCRRWQWRMKKLGAPSICU RRAF/A	Antimicrobial against enterotoxigenic *E. coli* and *Listeria monocytogenes*	Enzymatic hydrolysis with pepsin enzyme	[32]
	232–238	LNNSRAP	ACE-inhibitory	Hydrolysis by pepsin enzyme (E:S = 3% w/w), 37°C for 4 h	[51]
Lactoferrampin	268–284	LLSKAQEKFGKNKSR	Antimicrobial activity against *Candida albicans*, *Streptococcus mutans*, *E. coli*, *Bacillus subtilis* and *Pseudomonas aeruginosa*	Synthetically prepared with FMOC solid-phase synthesis	[62, 63]
Lactoferroxin A/B/C	318–323/ 536–540/ 673–679	YLGSGY/ RYYGY/ KYLGPQY	Opioid activity	Enzymatic hydrolysis with pepsin enzyme(E:S = 1:50 w/w), 37°C for 6 h	[56]

Note: E:S: Enzyme-substrate ratio; FMOC: Fluorenylmethyloxycarbonyl peptide synthesis method.

Recently, a research group from Oregon State University have compiled numerous research and review related to milk protein-derived BPs into a database (http://mbpdb.nws.oregonstate.edu/) [39]. For more detailed information of various whey proteins BPs, readers can refer to this link.

3.3 PRODUCTION OF BIOACTIVE PEPTIDES

As the consumer awareness grows, the demand for BPs has increased. Such high demand and applications in pharmacological, physical, chemical, and clinical studies have led to their commercial and industrial production. However, their production faces some bottlenecks such as lack of an appropriate optimized production technology and insufficient yield. Different methods have been studied by researchers for production of BPs. As the BPs sequences lie encrypted in the parent protein's amino acid residue sequence, suitable cleaving techniques like hydrolysis using proteolytic enzymes based on top-down approach (*in vivo* by gastrointestinal enzymes, *in vitro* using fermentation, enzymes from plants and microorganisms or their combination) has been applied. Another method based on bottom-up approach process like chemical synthesis has also been widely studied. Some of the commonly used methods for production of BPs (Figure 3.1) are discussed in this section.

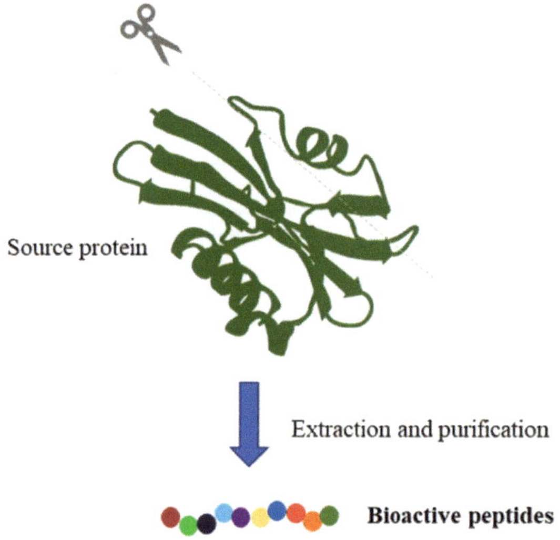

FIGURE 3.1 Production of BPs from proteins.

3.3.1 ENZYMATIC HYDROLYSIS WITH GASTROINTESTINAL ENZYMES

The BPs may be formed and released *in vivo* during the digestion process within the body. This method employs the common gastrointestinal enzymes like pepsin, trypsin or chymotrypsin in the digestive juices which hydrolyze the proteins. The BPs produced this way after consuming the dairy proteins were assumed to be associated with several biological activities like antimicrobial, antihypertensive, immunomodulatory, or opioid activity [16, 18, 65]. This method however has a limitation and rarely used for production since the isolation of the BPs is difficult and requires removal of intestinal contents. This method has been modified into a simulated *in vitro* digestion of the proteins of interest by treating them with digesting enzymes.

Simulated conditions should mimic the pH and temperatures of the gastrointestinal conditions to exact the BPs production. Other proteolytic enzymes like alcalase, thermolysin have been combined with pepsin and trypsin to release several BPs. The BPs produced this way show ACE-inhibitory effects. For instance, α-lactorphin from α-La (f (50–53)) was produced by Pihlanto-Leppälä [44] by enzymatic hydrolysis with gastric and pancreatic enzyme carried out at 37°C for 3 h was shown to have ACE-inhibitory effects.

3.3.2 FERMENTATION BY MICROBIAL STARTER CULTURES

Fermentation is also an efficient method of production of BPs (Figure 3.2). Fermented milks, for example, yogurt and cheese are very good reservoirs of beneficial BPs. The microbial cultures used for the production of yogurt, cheese, etc., carry out fermentation and produce metabolites due to their metabolic pathways and thus affect the properties of fermented dairy products. They produce BPs from the proteins by their proteolytic enzymes like bacterial proteases to utilize them as the nitrogen sources for their growth. Proteolytic systems of such starter cultures include a series of steps: firstly, the proteinases bound to their cell walls lead to initial proteolysis of milk proteins into oligopeptides. Specific transporters like the oligopeptide transporters (Opp), di-, and tripeptide transporters (Dtp) transfer those oligopeptides into the cytoplasm of the cells where the intracellular peptidases, endopeptidases, aminopeptidases, di-and tri-peptidases break them down into smaller peptides and free amino acid molecules [10, 49].

Type of microorganisms used and the fermentation condition affect the composition and activities of released BPs. Several lactic acid bacteria (LAB) are utilized to prepare BPs in the literature, such as *L. lactis* ssp. *cremoris*, *Lb. helveticus*, *L. lactis* ssp. *diacetylactis*, *Lb. delbrueckii* ssp. *bulgaricus*. Other microflora like yeasts have also been employed as starter cultures like in kefir, where they produce unique BPs from milk proteins [4]. A number of studies have shown that BPs produced due to fermentation have ACE-inhibitory activity, for instance, the BPs produced from β-Lg by fermentation using *Lb. helveticus* was attributed to have ACE-inhibitory effects [43]. Fermentation is a cheaper alternative to enzymatic methods, as starters are usually cheaper than enzymatic systems. Also, purification, and isolation of the BPs is easier.

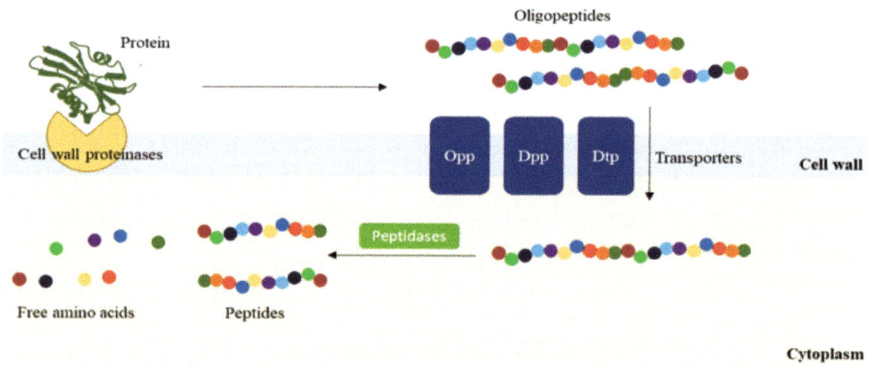

FIGURE 3.2 Production of BPs during microbial fermentation.

3.3.3 PROTEOLYSIS USING ENZYMES FROM DIFFERENT SOURCES

The *in vitro* synthesis of BPs employs proteolytic enzymes either as free enzyme or immobilized form for catalyzing the protein hydrolysis reactions. This method is the most studied, practiced, and reliable method for producing BPs due to lack of chemical residues as end by-products in the process. It is also faster and easy to control and predict the end product yields. Regio-selectivity and stereo-selectivity of enzymes is also advantageous to catalyze the proteins hydrolysis with very high specificity. Also, the use of different proteases, protein sources, and proteolytic conditions yield bioactive peptides from whey proteins having different biological functions.

As an example, based on protease specificity, peptides having antioxidant activity was identified owing to specific amino acid residues like tryptophan, cysteine, and methionine [28].

Proteases of plants, animals, and microbes are widely used for the manufacture of BPs from the protein sources. Some of the examples of proteases include pronase E (isolated from *Streptomyces griseus*), Flavourzyme (isolated form *Aspergillus oryzae*) [1], trypsin, chymotrypsin, and bromelain (from pineapples), papain (from papaya), which are gaining attention. It was reported that protease from the extracts of fennel, celery, parsley, and ficin (from figs) [12] have been great potential to release beneficial BPs from milk proteins.

Whey proteins were hydrolyzed by proteases from *Cynara cardunculus* to produce peptides having ACE-inhibitory and anti-ulcerative activities [55]. Thermolysin from microorganisms like *Bacillus thermoproteolyticus* has been utilized for the production of BPs having ACE-inhibitory and antioxidant activities [20, 42].

Although enzymatic methods are largely used in BP production, yet the use of recombinant DNA technology has opened up enormous possibilities. The use of cloning and gene expression for proteolytic enzymes in the production of peptides of biological importance has been well characterized. One of the bottlenecks of the employment of recombinant DNA technology is the need of skilled laborers and high cost of research and development. However, if properly established, such processes could help in boosting the production of BPs to a large scale. Nevertheless, *E. coli* based bacterial expression system has been the most studied techniques for expression of BPs. The *E. coli* cells are used for expression of genes that lead to huge production of cellular peptides. These peptides can be isolated and added to food matrices.

In a study by Prak and Utsumi [47], a hypercholesteremic peptide from β-Lg was produced (f (71–75), IIAEK) after gene expression in *E. coli*. In another study, shorter novel antimicrobial peptide derivates of the peptide lactoferricin (Lfcin, f (17–41)) were developed, namely, LfcinB-W10 and LfcinB-W4,10 [14]. These peptides were produced by altering the gene encoding them by replacing amino acids at position 10 and 4, 10 with tryptophan which were successfully expressed in *E. coli* BL21. These peptides showed antimicrobial activity against *S. aureus* ATCC25923.

Additionally, another peptide LFT33 has been produced by genes encoding thanatin (an antimicrobial peptide from insect) and bovine lactoferricin [15]. Another approach is to use genetically modified LABs used directly in food products which produce specific peptides in the food itself. However, studies related to the use of genetically modified LAB to produce BPs from the whey proteins are lacking.

3.3.4 CHEMICAL SYNTHESIS

Due to the advancement of chemical reaction techniques, improvement in process efficiencies and yields, chemical synthesis of peptides has been studied. Chemical synthesis is carried out mainly by two different approaches: in solution or in solid-state. In the solution method, the peptides of interest are obtained by successive recombination of amino acids by condensation reactions. The solid-state synthesis is preferred over solution method. In this process, a polymeric resin lined with linkers (functional groups) is used; the amino acids are linked to the linkers by their C-terminal by acylation and N-terminal derived protector group. This allows attaching of subsequent amino acid to the previous and the cycle is continued until the required sequence is obtained. Suitable cleaving agents are used to remove the amino acid chains after peptide formation from the resins.

The solid-state synthesis is categorized into several types by the way of reaction: sequential (addition of amino acid one by one) and convergent (synthesis of smaller fragments simultaneously which are joined later). Another way of classification is based on the use of protection group: tert-butyloxycarbonyl (BOC) and 9-fluorenylmethyloxycarbonyl (FMOC). FMOC method has been used by researchers to produce BPs having ACE-inhibitory, hypocholesteremic, and antimicrobial activities based on β-Lg and lactoferrin [36, 37, 62, 63].

3.4 ISOLATION, PURIFICATION, AND CHARACTERIZATION

3.4.1 ISOLATION OF BIOACTIVE PEPTIDES

The BPs produced from the mentioned processes are crude in nature. They are often found mixed with other fractions of proteins, microbial cells and enzymes. Purity is always desired when the BPs are to be tested for their applications in the industry. It is desired that the purity of BPs to be $\geq$ 95% for applications in clinical and therapeutic studies [9]. Isolation of the produced BPs is the first preparative step after production. This is carried out by washing of the product mixtures with deionized water and subjecting to centrifugation [9]. This removes the microbial cells, reagents, reaction products and by-products. The isolated peptides are still not totally purified and may be subjected to lyophilization before further purification.

3.4.2 PURIFICATION OF BIOACTIVE PEPTIDES

The purification process of the crude BPs is done using either chromatography methods or membrane processes. Mostly used method for purification of BPs is the reversed phase-high pressure liquid chromatography (RP-HPLC). This helps in a rapid purification and complete detection of the eluted BPs. The BPs are eluted through silica column using a specific solvent of some gradient and detected by array of detectors. Other chromatographic methods include size exclusion chromatography (SEC) or the gel filtration chromatography. This involves elution of BPs through a gel media which separates the BPs based on their molecular weights.

Ion exchange chromatography (IEC) is another method employed for BPs purification which involves the adsorption of cationic or anionic forms of BPs on the ion exchange resins. BPs of interest thus are retained on the resin and further characterized. Affinity chromatography and capillary electrophoresis techniques have also emerged due to ongoing research and development efforts. Gel electrophoresis techniques like the sodium dodecyl sulphate-Polyacrylamide gel electrophoresis (SDS-PAGE) have also been employed as an analytical method for BPs purification. One of the major drawbacks of using analytical purification techniques is their high cost and requirement of skilled personnel for operation [34].

Most widely used membrane separation method is the ultrafiltration. This method is used industrially for purification and enrichment of BPs due to lower processing cost and times, absence of phases. Ultrafiltration systems are based on the separation of protein fractions based on the molecular weights. These systems include a membrane of pore size of around 0.01 μm, through which the BPs solutions are passed and they get enriched in the BPs of interest either in the permeate or retentate streams. The ultrafiltration system works under operating pressures of 2–30 bar [9]. Ultrafiltration process has been introduced for enrichment of ACE-inhibitory and opioid peptides (α- and β-lactorphin) from purified hydrolysates of β-Lg and α-La [44, 45]. Enzymatic membrane reactors also have been introduced for production and enrichment of BPs [23].

3.4.3 CHARACTERIZATION OF BIOACTIVE PEPTIDES

The peptide sequences corresponding to the molecular masses of the produced BPs are mostly determined using mass spectrometry (MS) techniques. MS involves measuring the molecular masses of the compounds within the sample. This in a way helps in characterizing the BPs present within the sample. The commonly used techniques in MS for characterization of BPs are electrospray ionization (ESI)-MS, fast atom bombardment (FAB)-MS, and MALDI-MS.

3.5 FUNCTIONAL BENEFITS

3.5.1 ANTIHYPERTENSIVE ACTIVITY

Antihypertensive properties of the BPs are their most studied property. High blood pressure, also called hypertension, is related to many cardiovascular diseases like atherosclerosis, stroke, and myocardial infarction [6]. It is one of the major conditions that affect many adults worldwide every year. The enzyme ACE (a peptidyl dipeptide hydrolase enzyme) takes part in the modulation of the blood pressure through arteries. It basically catalyzes the formation of angiotensin-II, a vasoconstrictor from the angiotensin-I, its inactive form, into, and inactivation of bradykinin, a vasodilator thus releasing aldosterone and increasing sodium concentration in blood. The inactivation of ACE leads to decrease in blood pressure and control in hypertension. Many synthetic drugs and medicines have been developed to counter hypertension which has many side effects.

Many BPs have been associated with lowering hypertension and act as an alternative to the pharmaceutical drugs. Such BPs usually di or tri-peptides, have antihypertensive activity owing to their lysine, proline, or arginine residues in their C-terminal. Size, hydrophobicity, and sequence of amino acids are also important for such activity. It has been observed that the BPs compete with the ACE to bind with the substrate or change the conformation of the enzyme. This action is attributed to the amino acids having branched-chain structures like valine, leucine, and isoleucine that can interact with the nonpolar amino acids of ACE enzyme active site [6]. Some other mechanism of regulation of cardiovascular health is by blocking the release of endothelin-I by endothelial cells, bradykinin stimulation, and enhancement of nitric oxide synthase [23]. Nevertheless, administration of BPs or production of BPs in the gastrointestinal condition is not sufficient.

The BPs should be able to be absorbed in the blood stream. For instance, a peptide from β-Lg [f (142–148)] showed ACE-inhibition in laboratory conditions but not *in vivo*. Lactokinins are a proven group of ACE-inhibitory BPs resulting from the serum proteins in milk. It was also shown that the consumption of fermented products (fermentation by *Lb. helveticus* LBK-16H) led to high antihypertensive effect on the test group [33]. In another study, Chen et al. [11] identified ACE-inhibitory peptide [f(17–19)] from β-Lg.

Goudarzi et al. [19] optimized the hydrolysis conditions of whey protein isolate for producing a hydrolysate possessing maximum ACE-inhibitory activity. Tavares et al. [59] demonstrated antihypertensive effects of BPs from β-Lg and α-La in rats. In another study, consumption of white bread (with 4% whey protein hydrolysate) showed antihypertensive activity in rats [27]. These studies suggest that BPs demonstrate a tremendous potential in the development of functional food products with antihypertensive properties.

3.5.2 ANTIMICROBIAL ACTIVITY

Some biologically active peptides also possess the capacity to negatively affect the growth of several Gram-positive (G +ve) and negative (G –ve) bacteria like *E. coli*, *S. aureus*, *S. typhi*, *B. cereus*, *S. enteritidis*. The main cell inhibition mechanisms can be attributed to the interaction of the BPs with the microorganisms' outer cellular membrane [6]. This leads to pore formation or permeabilization and cell lysis. Some of the studied model for pore formation include barrel-stave, toroidal pore, carpet [6]. Another mechanism includes entry of the BPs into the cytoplasm of the cell, which interferes with the cellular activities like DNA, RNA synthesis or protein synthesis, or interacts with cellular components.

Factors like amino acid composition, cationic charge, ambipathicity, etc., lead to the interaction of BPs and the microbial cells. Lactoferrin in whey is a good source of peptides having antimicrobial characteristics. Some of the important antimicrobial BPs from lactoferrin includes lactoferricin and lactoferrampin. Lactoferricin [f(17–41)] shows antibacterial activity against numerous G +ve and G –ve bacteria and some yeasts. It binds iron that is required for the microorganisms and thus leads to their death. It has shown a potential inhibition of *E. coli* and *L. monocytogenes* [32]. Some derivatives also show antiviral properties against hepatitis B, C, and G virus, herpes, poliovirus. Lactoferrampin shows inhibition against bacteria like *Strepto-coccus* spp., *E. coli*, *B. subtilis*, *Pseudomonas aeruginosa* [34].

Some antimicrobial peptides were purified from hydrolysates of whey protein recently [33]. BPs from β-Lg [f(14–18), f(50–54), f(143–146), f(147–149)] and α-La [f(117–121)] were also shown to have antimicrobial action against *L. ivanovii* and *E. coli.*

3.5.3 OPIOID ACTIVITY

Opioid peptides have an action similar to the opiates in the brain. Some of these BPs have the ability to moderate brain activities like stimulation of endocrine secretions, regulation of analgesic effects, social, and emotional behavior moderation, response to stress. BPs with the opioid activity act as ligands to the opiate receptors within the nervous system, immune system, gastrointestinal tract to modulate antagonistic as well as agonistic effects. An important criteria to serve as opioid motif is to have a sequence of tyrosine-glycine-glycine-phenylalanine-(methionine or leucine) in the N-terminal of the BPs [9]. Casomorphins, casoxins are important BPs which have opioid activity. However, some BPs derived from whey proteins like α-lactorphin, and lactoferroxin from α-La and lactoferrin, respectively show potent opioid activity.

3.5.4 ANTIDIABETIC ACTIVITY

Diabetes and associated chronic health issues are prevalent metabolic disorders in the world. This is characterized by impaired insulin secretion by pancreatic cells. The main reasons for such conditions are the changing of healthy lifestyle, stress, lack of exercise. Absence of proper diet can also lead to diabetic conditions. In the modern working days, it is difficult for people to follow proper lifestyle. Thankfully, in the recent years, the peptides from milk proteins have demonstrated antidiabetic activities too. BPs in the form of drug supplements could be the solution to these prevalent problems in the world. The proposed modes of action for the antidiabetic activity of BPs are the insulinotropic activity, regulation of the blood glucose level by their inhibition of metabolic enzymes like α-amylase, α-glucosidase, and DPP-IV [33, 40].

Researchers have recently found that the peptides of whey protein possess better antidiabetic properties than the casein peptides, which was attributed to the higher solubility in gastric conditions, faster digestion, and availability of a greater number of amino acids having branched-chain structures in the former [33]. Overall, chain length, degree of hydrolysis of protein and rate of digestion affect the insulin response associated with the BPs.

Morifuji et al. [35], in their study observed that consuming whey proteins peptides composed mainly of branched-chain amino acid in the peptides like valine, leucine, isoleucine, threonine, lysine enhanced the plasma insulin level. In another study, consumption of whey protein hydrolysate rich in peptide (Isoleucine-proline-alanine) from β-lactoglobulin has shown a moderate inhibitory action on DPP-IV enzyme [61]. Hydrolysates from whey proteins are also important for the reduction of postprandial glycemia by secreting glucagon-like peptide 1, glucose-dependent insulinotropic polypeptide [33]. Thus, consumption of products like cheese could also lower the risk of diabetes due to the presence of antidiabetic BPs.

3.5.5 ANTIOXIDATIVE ACTIVITY

Formation of free radicals is a side effect of the respiration and oxidation metabolism in the body of aerobic organisms. Such free radicals are highly reactive and unstable. They can readily interact with other molecules in the organisms and lead to damage of cells and tissues and thus aging. They may also react with the DNA in the cells and cause mutations. Since these radicals are highly reactive species; the reactions they carry out may be responsible for the occurrence of diseases like atherosclerosis, cancer, rheumatoid arthritis, diabetes [6]. Antioxidants are the substance which is present in the body that can quench these free radicals and reduce their degenerative processes to a great extent. The BPs from milk proteins also possess such anti-oxidative property due to their action of binding the free radicals, making them unavailable to carry out any oxidative damages to the cells and tissues.

It has been established that milk protein peptides having 5–11 amino acid residues including proline, histidine, tryptophan, or tyrosine possess anti-oxidative properties. Consumption of whey proteins have also been attributed to produce glutathione in the body which plays a major role in scavenging free radicals [33]. Whey proteins like β-Lg, α-La have numerous cysteine residues which are very important for the glutathione synthesis in the body.

The peptides produced from β-Lg [f(19–29), f(42–46), f(145–149)] have also been shown to carry out anti-oxidative activities by scavenging the free radicals [21]. This was supposedly because of the existence of hydrophobic or aromatic amino acids in their peptide sequences like valine, tyrosine, tryptophan, methionine, isoleucine, leucine, phenylalanine which can easily donate proton to the radicals, thereby quenching them [31]. The inclusion of anti-oxidative BPs in food matrices could be a good strategy as the alternatives to many synthetic antioxidants used worldwide [31, 50].

3.5.6 IMMUNOMODULATORY AND ANTI-INFLAMMATORY ACTIVITIES

The diet always tends to govern the formation of body immune system and ability to fight diseases. Immunomodulatory peptides take part in a decisive role to improve body immune system by stimulating or suppressing the body immune functions. They have the tendency to boost immunity by regulating the formation of antibodies and cytokines. They inhibit the expression of cytokines like tumor necrosis factor-α (TNF-α), interleukins (IL-1 and 6). They also show a decisive role in carrying out anti-inflammatory activities by differential regulation of certain molecules like activator protein-1 and nuclear transcription factor-κB or blocking inflammatory reaction pathways like c-Jun N-terminal kinase (JNK) or mitogen-activated protein kinase (MAPK) [6].

BPs from whey proteins have been shown in many studies *in vivo* and *in vitro* to regulate the immune system. BPs from the β-Lg [f (15–20), f(55–60), f(139–148)] and α-La [f(10–16)] were demonstrated to inhibit or stimulate cytokine production and proliferation of human lymphocytes [4]. Commercially available whey protein isolates possess immunomodulatory BPs which are produced by hydrolysis by enzymes. Such products could be very beneficial to the development of pediatric and infant foods with balanced immunomodulatory BPs.

3.5.7 ANTICANCER ACTIVITY

Cancer has been a major cause of global mortality which occur due to uncontrolled growth and propagation of cancerous or abnormal cells in the body. It may be triggered due to exposure to radiation, chemicals, mutations, obesity, etc. However, due to the non-specificity of the abnormal cells, the treatment of cancer bears a lot of side effects. It has been recently established that BPs from the food sources could be used in the treatment of cancers [33]. The mode of action of anticancer BPs is by the killing of the tumor cells by the cell membrane lyses and mitochondrial lyses. Positively charged peptide reacts with the negatively charged surface phospholipids in tumor cells. On the contrary, the normal cells possess neutral change owing to the zwitterionic phospholipids [33].

Whey proteins when taken as dietary supplements showed anticancer effects on urogenital cancer [4]. Peptides from β-Lg have been demonstrated to suppress tumor growth by increasing glutathione concentration and free radical scavenging, which was due to the high level of cysteine residues. Apart from that, the apoptosis-inducing property of α-La has been shown on tumor cells. Lactoferricins from lactoferrin have also been shown to have cytotoxicity effects in melanoma, fibrosarcoma, and colon carcinoma, neuroblastoma, and breast cancer cells. Though these results are in their infancy, more clinical trials are warranted to use BPs as potential anticancer drugs.

3.6 TECHNOLOGICAL APPLICATIONS

As already mentioned, traditional or commercial fermented milk products like sour milk, yogurt, dahi, and various cheeses have been proven to be associated with health benefits. For instance, yogurt, and cheese consumption is related to lowering of the risk of high blood pressure [5, 7, 8]. In their study [8], comprising pooled health study cohorts ($n > 180{,}000$), participants consuming yogurt (≥ 5 serving/week vs. <1 serving/month) and cheese (1–4 servings/day vs. <1 serving/week) had lower risks of high blood pressure of 16% and 6% (95% confidence interval), respectively. Such biological activities could well be attributed to the BPs present in products formed due to fermentation [16, 18]. The specific activity, occurrence, and amount of such BPs depend on the product type, starter cultures and fermentation conditions used. Control and optimization of production of BPs in fermented products can be a venture of future research and development. Table 3.2 summarizes some of the BPs identified in dairy products like yogurt and cheese.

In recent years, hydrolyzed or fermented products containing BPs from whey proteins have been developed and commercialized. Based on the available evidence for safety and cytotoxicity, these BPs enriched products can be claimed to have therapeutic uses. These products are sold as functional foods or dietary supplements after approval from regulatory agencies. Valid documentation and recording of associated health benefits of the developed products is warranted as a part of future research and development [4, 23]. Table 3.3 lists some of the commercialized products in the market containing whey protein derived BPs.

TABLE 3.2 Identified BPs Sequences and Their Applications in Dairy Products

Product	Identified Peptide Sequence	Biological Activity	References
Freeze-dried yogurt-based diets	Whey protein peptides	Hypercholesteremic effect	[48]
Idiazabal-type sheep (raw milk) cheese	BSA (179–190)	ACE-inhibitory activity	[52]
Yogurt (commercial and traditional)	β-Lg-derived peptides containing W, Y, and M amino acids	Antioxidant activity	[3]
Yogurt (commercial)	β-Lg [(104–105), (122–124), (103–105), (142–146)] α-La [(52–53), (59–60), (104–105), (118–119), (25–26), (50–52), (104–108)]	ACE-Inhibitory activity	[25]
Yogurt (probiotic)	Whey protein peptides rich in C-terminal residues	Antioxidant and antidiabetic activity	[13]
Yogurt, ropy milk, and sour milk	β-Lg [(9–14), (15–20)] α-La (105–110)	ACE-inhibitory activity	[46]

3.7 SUMMARY

Dairy products consumption, especially fermented milk and cheese is related to many health benefits. Many milk proteins are excellent reservoirs of several peptides which play biologically active roles in the body. In the recent years, the bioactive peptides produced from the whey proteins were demonstrated to perform and regulate many important biological functions within our body. They are produced by hydrolysis by enzymes in the gastrointestinal system *in vivo* or enzymes from plants, animal, and microbial sources *in vitro*. Several *in vivo* and *in vitro* tests have shown their potential to act as antidiabetic, anti-oxidative, anti-carcinogenic, antihypertensive, or anti-inflammatory peptides. However, the administration of the peptides for drug and therapeutic use should be carried out after thorough clinical studies. Nevertheless, many commercially available products having whey protein-derived BPs have emerged in recent years, which promise immense benefits to the consumers.

TABLE 3.3 Selected Commercial Products Containing Whey Protein Derived BPs and Their Health Benefits

Brand	Food Product Type	Protein Source	Peptide	Health Claim	Manufacturer
Biozate 3 and 7	Ingredient	Whey protein hydrolysate (WPH)	β-Lg fragments	Antihypertensive activity	Davisco Foods (Minnesota)
BiPRO WPI	Infant formula; ingredient	Whey protein isolate (WPI)	Glycomacropeptide (106–109)	Anticarcinogenic; antithrombic activity	Davisco Foods (Minnesota)
Cysteine Peptide	Functional ingredient	WPH	Peptides from milk protein	Raising energy level and sleep	DMV International (Netherlands)
Hiprotal® whey protein 35 and Hiprotal® whey protein 45	Ingredient in infant formula	Whey peptide	WPC	Muscle development	FrieslandCampina (Netherlands)
Lacprodan® ALPHA-10; Lacprodan® ALPHA-20	Health food	α-La	α-La fragments	Improvement of sleep and memory	Arla Foods (Denmark)
Lacprodan® DI-3071/3021 Lacprodan® hydro-milk	Sports supplements and nutrition	Whey protein	WPH	Improvement of oral hygiene; vitamins and minerals uptake	Arla Foods (Denmark)
Lacprodan® CGMP-10	Food for infants	Whey peptide	CMP	Cognitive and nervous system maturation	Arla Foods (Denmark)
Peptigen® IF-3087/3090/ 3012/3080	Food for infants	Whey protein	WPH	Binding of mineral	Arla Foods (Denmark)

TABLE 3.3 *(Continued)*

Brand	Food Product Type	Protein Source	Peptide	Health Claim	Manufacturer
PeptoPro®	Ingredient	Milk proteins	Dipeptides and tripeptides	Boosting of muscle protein synthesis	DSM (Netherlands)
Prevention	Capsule form	WPH	Lactoferrin fragments	Acne reduction	DMV International, (Netherlands)
Vivinal Alpha	Functional ingredient	WPH	α-La fragments	Improvement of sleep and relaxation	Borculo Domo Ingredients (BDI) (Netherlands)

KEYWORDS

- **bioactive peptides**
- **fermented milks**
- **health benefits**
- **whey proteins**

REFERENCES

1. Ahtesh, F., Stojanovska, L., Shah, N., & Mishra, V. K., (2016). Effect of flavourzyme (®) on angiotensin-converting enzyme inhibitory peptides formed in skim milk and whey protein concentrate during fermentation by *Lactobacillus helveticus*. *Journal of Food Science, 81*(1), M135–M143.
2. Almaas, H., Eriksen, E., Sekse, C., Comi, I., Flengsrud, R., Holm, H., Jensen, E., et al., (2011). Antibacterial peptides derived from caprine whey proteins, by digestion with human gastrointestinal juice. *British Journal of Nutrition, 106*(6), 896–905.
3. Aloğlu, H. S., & Oner, Z., (2011). Determination of antioxidant activity of bioactive peptide fractions obtained from yogurt. *Journal of Dairy Science, 94*(11), 5305–5314.
4. Anusha, R., & Bindhu, O. P., (2016). Bioactive peptides from milk. Chapter 6. In: Gigli, I., (ed.), *Milk Proteins: From Structure to Biological Properties and Health Aspects* (pp. 101–140). London-United Kingdom: IntechOpen Ltd.
5. Ashar, M. N., & Chand, R., (2004). Fermented milk containing ACE-inhibitory peptides reduces blood pressure in middle aged hypertensive subjects. *Milchwissenschaft, 59*(7), 363–366.
6. Bersi, G., Barberis, S. E., Origone, A. L., & Adaro, M. O., (2018). Bioactive peptides as functional food ingredients. In: Grumezescu, A. M., & Holban, A. M., (eds.), *Role of Materials Science in Food Bioengineering* (pp. 147–186). London-United Kingdom: Academic Press.
7. Buendia, J. R., Yanping, L. I., Hu, F. B., Cabral, H. J., Bradlee, M. L., Quatromoni, P. A., Singer, M. R., et al., (2018). Long-term yogurt consumption and risk of incident hypertension in adults. *Journal of Hypertension, 36*(8), 1671–1679.
8. Buendia, J. R., Yanping, L. I., Hu, F. B., Cabral, H. J., Bradlee, M. L., Quatromoni, P. A., Singer, M. R., et al., (2018). Regular yogurt intake and risk of cardiovascular disease among hypertensive adults. *American Journal of Hypertension, 31*(5), 557–565.
9. Chauhan, V., & Kanwar, S. S., (2020). Bioactive peptides: Synthesis, functions, and biotechnological applications. In: Verma, M. L., & Chandel, A. K., (eds.), *Biotechnological Production of Bioactive Compounds* (pp. 107–137). Amsterdam-Netherlands: Elsevier.
10. Chaves-López, C., Serio, A., Paparella, A., Martuscelli, M., Corsetti, A., Tofalo, R., & Suzzi, G., (2014). Impact of microbial cultures on proteolysis and release of bioactive peptides in fermented milk. *Food Microbiology, 42*, 117–121.

11. Chen, G. W., Tsai, J. S., & Pan, B. S., (2007). Purification of angiotensin I-converting enzyme inhibitory peptides and antihypertensive effect of milk produced by protease-facilitated lactic fermentation. *International Dairy Journal, 17*(6), 641–647.
12. Di Pierro, G., O'keeffe, M. B., Poyarkov, A., Lomolino, G., & Fitzgerald, R. J., (2014). Antioxidant activity of bovine casein hydrolysates produced by *Ficus carica* L.-derived proteinase. *Food Chemistry, 156*, 305–311.
13. Ejtahed, H. S., Mohtadi-Nia, J., Homayouni-Rad, A., Niafar, M., Asghari-Jafarabadi, M., & Mofid, V., (2012). Probiotic yogurt improves antioxidant status in type 2 diabetic patients. *Nutrition, 28*(5), 539–543.
14. Feng, X., Liu, C., Guo, J., Bi, C., Cheng, B., Li, Z., Shan, A., & Li, Z., (2010). Expression and purification of an antimicrobial peptide, bovine lactoferricin derivative LfcinB-W10 in *Escherichia coli. Current Microbiology, 60*, 179–184.
15. Feng, X., Liu, C., Guo, J., Song, X., Li, J., Xu, W., & Li, Z., (2012). Recombinant expression, purification, and antimicrobial activity of a novel hybrid antimicrobial peptide LFT33. *Applied Microbiology and Biotechnology, 95*, 1191–1198.
16. FitzGerald, R. J., & Murray, B. A., (2006). Bioactive peptides and lactic fermentations. *International Journal of Dairy Technology, 59*(2), 118–125.
17. FitzGerald, R. J., & Meisel, H., (1999). Lactokinins: Whey protein-derived ACE inhibitory peptides. *Nahrung, 43*(3), 165–167.
18. Gobbetti, M., Minervini, F., & Rizzello, C. G., (2004). Angiotensin I-converting-enzyme-inhibitory and antimicrobial bioactive peptides. *International Journal of Dairy Technology, 57*(2, 3), 173–188.
19. Goudarzi, M., Madadlou, A., Mousavi, M. E., & Emam-Djomeh, Z., (2012). Optimized preparation of ACE-inhibitory and antioxidative whey protein hydrolysate using response surface method. *Dairy Science & Technology, 92*(6), 641–653.
20. Guinane, C. M., Kent, R. M., Norberg, S., O'connor, P. M., Cotter, P. D., Hill, C., Fitzgerald, G. F., et al., (2015). Generation of the antimicrobial peptide caseicin A from casein by hydrolysis with thermolysin enzymes. *International Dairy Journal, 49*, 1–7.
21. Hernández-Ledesma, B., Dávalos, A., Bartolomé, B., & Amigo, L., (2005). Preparation of antioxidant enzymatic hydrolysates from α-lactalbumin and β-lactoglobulin. Identification of active peptides by HPLC-MS/MS. *Journal of Agricultural and Food Chemistry, 53*(3), 588–593.
22. Kitts, D. D., & Weiler, K., (2003). Bioactive proteins and peptides from food sources. Applications of bioprocesses used in isolation and recovery. *Current Pharmaceutical Design, 9*(16), 1309–1323.
23. Korhonen, H., & Pihlanto, A., (2006). Bioactive peptides: Production and functionality. *International Dairy Journal, 16*(9), 945–960.
24. Korhonen, H., & Pihlanto, A., (2003). Food-derived bioactive peptides opportunities for designing future foods. *Current Pharmaceutical Design, 9*(16), 1297–1308.
25. Kunda, P. B., Benavente, F., Catalá-Clariana, S., Giménez, E., Barbosa, J., & Sanz-Nebot, V., (2012). Identification of bioactive peptides in a functional yogurt by micro liquid chromatography time-of-flight mass spectrometry assisted by retention time prediction. *Journal of Chromatography A, 1229*, 121–128.
26. Lacroix, I. M., Meng, G., Cheung, I. W., & Li-Chan, E. C., (2016). Do whey protein-derived peptides have dual dipeptidyl-peptidase IV and angiotensin I-converting enzyme inhibitory activities? *Journal of Functional Foods, 21*, 87–96.

27. Lafarga, T., Gallagher, E., Aluko, R. E., Auty, M. A., & Hayes, M., (2016). Addition of an enzymatic hydrolysate of bovine globulins to bread and determination of hypotensive effects in spontaneously hypertensive rats. *Journal of Agricultural and Food Chemistry, 64*(8), 1741–1750.
28. Lin, S., Tian, W., Li, H., Cao, J., & Jiang, W., (2012). Improving antioxidant activities of whey protein hydrolysates obtained by thermal preheat treatment of pepsin, trypsin, alcalase, and flavourzyme. *International Journal of Food Science and Technology, 47*(10), 2045–2051.
29. Lucarini, M., (2017). Bioactive peptides in milk: From encrypted sequences to nutraceutical aspects. *Beverages, 3*(3), 41.
30. Ma, Y., Liu, J., Shi, H., & Yu, L. L., (2016). Isolation and characterization of anti-inflammatory peptides derived from whey protein. *Journal of Dairy Science, 99*(9), 6902–6912.
31. Mann, B., Kumari, A., Kumar, R., Sharma, R., Prajapati, K., Mahboob, S., & Athira, S., (2015). Antioxidant activity of whey protein hydrolysates in milk beverage system. *Journal of Food Science and Technology, 52*(6), 3235–3241.
32. Meisel, H., & Bockelmann, W., (1999). Bioactive peptides encrypted in milk proteins: Proteolytic activation and thropho-functional properties. *Antonie Van Leeuwenhoek, 76*, 207–215.
33. Mohammadian, M., Salami, M., Emam-Djomeh, Z., & Alavi, F., (2017). Nutraceutical properties of dairy bioactive peptides. In: Watson, R. R., Collier, R. J., & Preedy, V. R., (eds.), *Dairy in Human Health and Disease Across the Lifespan* (pp. 325–342). London-United Kingdom: Academic Press.
34. Mohanty, D., Jena, R., Choudhury, P. K., Pattnaik, R., Mohapatra, S., & Saini, M. R., (2016). Milk derived antimicrobial bioactive peptides: A review. *International Journal of Food Properties, 19*(4), 837–846.
35. Morifuji, M., Ishizaka, M., Baba, S., Fukuda, K., Matsumoto, H., Koga, J., Kanegae, M., & Higuchi, M., (2010). Comparison of different sources and degrees of hydrolysis of dietary protein: Effect on plasma amino acids, dipeptides, and insulin responses in human subjects. *Journal of Agricultural and Food Chemistry, 58*(15), 8788–8797.
36. Mullally, M. M., Meisel, H., & FitzGerald, R. J., (1996). Synthetic peptides corresponding to α-lactalbumin and β-lactoglobulin sequences with angiotensin-I-converting enzyme inhibitory activity. *Biological Chemistry Hoppe-Seyler, 377*(4), 259–260.
37. Mullally, M. M., Meisel, H., & FitzGerald, R. J., (1997). Identification of a novel angiotensin-i-converting enzyme inhibitory peptide corresponding to a tryptic fragment of bovine beta-lactoglobulin. *FEBS Letter, 402*(2, 3), 99–101.
38. Nagaoka, S., Futamura, Y., Miwa, K., Awano, T., Yamauchi, K., Kanamaru, Y., Tadashi, K., & Kuwata, T., (2001). Identification of novel hypocholesterolemic peptides derived from bovine milk β-lactoglobulin. *Biochemical and Biophysical Research Communications, 281*(1), 11–17.
39. Nielsen, S. D., Beverly, R. L., Qu, Y., & Dallas, D. C., (2017). Milk bioactive peptide database: A comprehensive database of milk protein-derived bioactive peptides and novel visualization. *Food Chemistry, 232*, 673–682.
40. Nongonierma, A. B., Le Maux, S., Hamayon, J., & FitzGerald, R. J., (2016). Strategies for the release of dipeptidyl peptidase IV (DPP-IV) inhibitory peptides in an enzymatic hydrolyzate of α-lactalbumin. *Food & Function, 7*(8), 3437–3443.

41. Ortiz-Chao, P., Gómez-Ruiz, J. A., Rastall, R. A., Mills, D., Cramer, R., Pihlanto, A., Korhonen, H., & Jauregi, P., (2009). Production of novel ACE inhibitory peptides from β-lactoglobulin using protease N Amano. *International Dairy Journal, 19*(2), 69–76.
42. Otte, J., Shalaby, S. M., Zakora, M., & Nielsen, M. S., (2007). Fractionation and identification of ACE-inhibitory peptides from α-lactalbumin and β-casein produced by thermolysin-catalyzed hydrolysis. *International Dairy Journal, 17*(12), 1460–1472.
43. Pan, D., & Guo, Y., (2010). Optimization of sour milk fermentation for the production of ACE-inhibitory peptides and purification of a novel peptide from whey protein hydrolysate. *International Dairy Journal, 20*(7), 472–479.
44. Pihlanto-Leppälä, A., Koskinen, P., Paakkari, I., Tupasela, T., & Korhonen, H., (1996). Opioid whey protein peptides obtained by membrane filtration. *IDF Bulletin, 311*, 36–38.
45. Pihlanto-Leppälä, A., Koskinen, P., Piilola, K., Tupasela, T., & Korhonen, H., (2000). Angiotensin I-converting enzyme inhibitory properties of whey protein digests: Concentration and characterization of active peptides. *Journal of Dairy Research, 67*(1), 53–64.
46. Pihlanto-Leppälä, A., Rokka, T., & Korhonen, H., (1998). Angiotensin I-converting enzyme inhibitory peptides derived from bovine milk proteins. *International Dairy Journal, 8*(4), 325–331.
47. Prak, K., & Utsumi, S., (2009). Production of a bioactive peptide (IIAEK) in *Escherichia coli* using soybean proglycinin A1aB1b as a carrier. *Journal of Agricultural and Food Chemistry, 57*(9), 3792–3799.
48. Ramchandran, L., & Shah, N. P., (2011). Yogurt can beneficially affect blood contributors of cardiovascular health status in hypertensive rats. *Journal of Food Science, 76*(4), H131–H136.
49. Raveschot, C., Cudennec, B., Coutte, F., Flahaut, C., Fremont, M., Drider, D., & Dhulster, P., (2018). Production of bioactive peptides by *Lactobacillus* species: From gene to application. *Frontiers in Microbiology, 9*, 2354.
50. Ries, D., Ye, A., Haisman, D., & Singh, H., (2010). Antioxidant properties of caseins and whey proteins in model oil-in-water emulsions. *International Dairy Journal, 20*(2), 72–78.
51. Ruiz-Giménez, P., Salom, J. B., Marcos, J. F., Vallés, S., Martínez-Maqueda, D., Recio, I., Torregrosa, G., et al., (2012). Antihypertensive effect of a bovine lactoferrin pepsin hydrolysate: Identification of novel active peptides. *Food Chemistry, 131*(1), 266–273.
52. Sagardia, I., Iloro, I., Elortza, F., & Bald, C., (2013). Quantitative structure-activity relationship based screening of bioactive peptides identified in ripened cheese. *International Dairy Journal, 33*(2), 184–190.
53. Sedaghati, M., Ezzatpanah, H., Mashhadiakbar, B. M., Tajabadi, E. M., & Aminafshar, M., (2015). Plasmin-digest of β-lactoglobulin with antibacterial properties. *Food and Agricultural Immunology, 26*(2), 218–230.
54. Shimizu, M., (2004). Food-derived peptides and intestinal functions. *Biofactors, 21*(1–4), 43–47.
55. Silva, S. V., Pihlanto, A., & Malcata, F. X., (2006). Bioactive peptides in ovine and caprine cheese like systems prepared with proteases from *Cynara cardunculus*. *Journal of Dairy Science, 89*(9), 3336–3344.

56. Tani, F., Iio, K., Chiba, H., & Yoshikawa, M., (1990). Isolation and characterization of opioid antagonist peptides derived from human lactoferrin. *Agricultural and Biological Chemistry, 54*(7), 1803–1810.
57. Tani, F., Shiota, A., Chiba, H., & Yoshikawa, M., (1994). Serophin, an opioid peptide derived from serum albumin. In: Brantl, V., & Teschemacher, H., (eds.), *β-Casomorphins and Related Peptides: Recent Developments* (pp. 49–53). Weinheim-Germany: Wiley.
58. Tavares, T., Sevilla, M. Á., Montero, M. J., Carrón, R., & Malcata, F. X., (2012). Acute effect of whey peptides upon blood pressure of hypertensive rats, and relationship with their angiotensin I-converting enzyme inhibitory activity. *Molecular Nutrition & Food Research, 56*(2), 316–324.
59. Tavares, T., Del, M. C. M., Amorim, M., Pintado, M., Recio, I., & Malcata, F. X., (2011). Novel whey-derived peptides with inhibitory effect against angiotensin-converting enzyme: *In vitro* effect and stability to gastrointestinal enzymes. *Peptides, 32*(5), 1013–1019.
60. Tellez, A., Corredig, M., Brovko, L. Y., & Griffiths, M. W., (2010). Characterization of immune-active peptides obtained from milk fermented by *Lactobacillus helveticus*. *Journal of Dairy Research, 77*(2), 129–136.
61. Tulipano, G., Sibilia, V., Caroli, A. M., & Cocchi, D., (2011). Whey proteins as source of dipeptidyl dipeptidase IV (dipeptidyl peptidase-4) inhibitors. *Peptides, 32*(4), 835–838.
62. Van, D. K. M. I., Groenink, J., Nazmi, K., Veerman, E. C., Bolscher, J. G., & Amerongen, A. V. N., (2004). Lactoferrampin: A novel antimicrobial peptide in the n1-domain of bovine lactoferrin. *Peptides, 25*(2), 177–183.
63. Van, D. K. M. I., Nazmi, K., Teeken, A., Groenink, J., Van't, H. W., Veerman, E. C., Bolscher, J. G. M., & Amerongen, A. V. N., (2005). Lactoferrampin, an antimicrobial peptide of bovine lactoferrin, exerts its candidacidal activity by a cluster of positively charged residues at the C-terminus in combination with a helix-facilitating N-terminal part. *Biological Chemistry, 386*(2), 137–142.
64. Walstra, P., Wouters, J. T., & Geurts, T. J., (2005). *Dairy Science and Technology* (2nd edn., p. 4). Boca Raton-Florida: CRC Press, Taylor, and Francis Group.
65. Yamamoto, N., Ejiri, M., & Mizuno, S., (2003). Biogenic peptides and their potential use. *Current Pharmaceutical Design, 9*(16), 1345–1355.

PART II
Therapeutic Aspects of Dairy Ingredients

CHAPTER 4

COLOSTRUM AS A SOURCE OF NUTRACEUTICALS

ANKITKUMAR J. THESIYA, KUNAL K. AHUJA, ROHIT G. SINDHAV, and ASHVIN S. HARIYANI

ABSTRACT

Colostrum is the first mammary gland secretion obtained after parturition, which significantly boosts up the immunity of the newborn. Bovine colostrum contains less lactose and more protein as compared to normal milk. Thus, it may become suitable for the consumption of lactose-intolerant people. It is also rich in immunoglobulins, lactoferrin, lysozyme, cytokines, lactoperoxidase, hormones, thyroxine, nucleotides, polyamines, growth factors, and antimicrobial factors. Immunoglobulins, lactoferrin, and lactoperoxidase present in colostrum are well known to exhibit antimicrobial defense against harmful microbes. Lactoferrin also has antiviral, antioxidant, antifungal, anti-inflammatory, antiparasitic, and anticancer activities. Anti-stress, antiviral as well as antitumor properties have been reported for α-lactalbumin, whereas lysozyme has been proven to alleviate the symptoms of inflammation of tissue near teeth. The activity of the insulin-like growth factor is higher in colostrum in comparison with normal milk. The presence of such functional components in colostrum extends their therapeutic role by helping in the prevention or cure of various illnesses. This enables colostrum a unique source of components for pharmaceutical, nutraceuticals, and functional food products. This chapter explores the composition, application, and nutraceutical potential of colostrum.

Functional Dairy Ingredients and Nutraceuticals: Physicochemical, Technological, and Therapeutic Aspects. Megh R. Goyal, N. Veena, PhD & Santosh K. Mishra (Eds.)

4.1 INTRODUCTION

Colostrum is the foremilk obtained after parturition [3, 17, 56]. It is a yellowish fluid having thick consistency as compared to normal milk [6, 52], which is produced in higher amounts by a healthy cow, mostly extra to the calf's need [69]. It is one of the primary requisites for nutrition, health, and survival of the newborn [6, 17]. This complex milk is rich in immunoglobulins (Igs), lysozyme, growth hormone (GH), insulin-like growth factors (IGF) I and II, lactoferrin, lactoperoxidase, transforming growth factor (TGF), etc. [3, 70]. These biologically active components provide beneficial physiological effects and therefore suggested for newborns for the development of passive immunity [56, 64, 86].

Normally, colostrum is not mixed with normal milk and kept separate from bulk milk [55]. The addition of colostrum to normal milk increased protein and mineral content which might reduce the heat stability and harm certain dairy operations like UHT processing [55, 59, 85]. On the other hand, colostrum contains many valuable constituents which increase its use as a functional food ingredient [47].

Colostrum and colostrum-derived ingredients can give therapeutic benefits above basic satiety. Consumer's interest in healthy foods is increasing day by day which will exploit the opportunity for the development of food formulations using such compounds. However, the rights of newborns for colostrum from mother should be reserved before using colostrum for harvesting the bioactive components for the development of health foods.

This chapter focuses on the detailed composition of colostrum, its health benefits and various applications of colostrum.

4.2 COMPOSITION

The composition of colostrum is mainly affected by many factors like breed, individuality, pre-partum nutrition, dry period duration of milch animal and postpartum time [62, 73–25]. Colostrum contains high amounts of proteins, lipids, vitamins, and minerals [6]. It also contains a high amount of long-chain fatty acids and low amounts of phospholipids, trans fatty acids and cis-9 trans-11-conjugated linoleic acids as compared to normal milk [11, 12, 76]. The cholesterol present in colostrum with an average of 327 mg/100 g fat, which is higher than the value for normal milk (i.e., 285 mg/100 g fat) [79].

Among proteins, the casein content of colostrum is higher than milk [18, 54]. Colostrum contains a significant amount of α-lactalbumin and β-lactoglobulin as whey protein fractions as compared to mature milk [27,

55]. Colostrum is one of the rich sources of lactoferrin, immunoglobulin, and growth factors [3]. Immunoglobulin content of colostrum ranges from 40–200 mg/ml, which contributes up to 70–80% of the total protein [48, 51].

Colostrum contains three major classes of immunoglobulins namely, IgG, IgM, and IgA [48, 89]. The concentration of IgA, IgM, and IgG in colostrum from Murrah buffalo was 3.22, 5.22 and 54.0 mg/ml respectively, which were higher as compared to a cow colostrum [22]. Its amount decreases rapidly following parturition [48]. The significant amounts of immunoglobulins in colostrum grants passive immunity to the neonates [55, 90, 92]. It also impacts on endocrine systems, metabolism, and the nutritional state of neonates [13, 14, 33].

Colostrum also contains a significant amount of somatic cell count as compared to normal milk [34]. Colostrum also contains high amount of bovine serum albumin and nucleotides as compared to milk [83, 97]. The concentrations of these constituents are affected by many factors including environmental and genetic factors [77].

Fat and protein content remains higher whereas lactose remains lower in the colostrum on the first day (Table 4.1). The enzymes concentration was also higher in colostrum than milk [87]. Initial milk (7 days) of Pelibuey ewes was reported to have more total solids, fat, and energy than the milk of subsequent days [53]. The detailed composition colostrum is summarized in Table 4.1.

TABLE 4.1 Composition of Bovine Colostrum

Parameter	**References**					
	[98]	**[42]**	**[26]**	**[50]**	**[1]**	
Milch animal	**Red Holstein × Simmental cows**	**Holstein cows**	**Holstein cows**	**Holstein cows**	**Egyptian buffaloes**	**Holstein cows**
Country	**Switzerland**	**Pennsylvania**	**United States**	**Japan**	**Egypt**	**Egypt**
After parturition	2nd day	3rd day	1st day	0 h	At calving	At calving
Total solids	–	27.64 %	23.90 %	31.1 %	26.67%	24.19 %
Protein	52.0 g/L	14.92 %	14 %	15.1 %	13.46 %	13.45 %
Lactose	43.9 g/L	2.49 %	2.70 %	–	–	–
Ca	–	4716.1 mg/kg	0.26 %	265 mg/dl	279 mg/100 g	281.4 mg/100 g
Fat	70.5 g/L	6.7 %	6.70 %	–	9.59 %	8.04 %

TABLE 4.1 *(Continued)*

Parameter	References					
	[98]	[42]	[26]	[50]	[1]	
Milch animal	**Red Holstein × Simmental cows**	**Holstein cows**	**Holstein cows**	**Holstein cows**	**Egyptian buffaloes**	**Holstein cows**
Country	**Switzerland**	**Pennsylvania**	**United States**	**Japan**	**Egypt**	**Egypt**
Fe	–	5.33 mg/kg	0.2 mg/100 ml	1.8 ppm	2.36 mg/100 g	2.21 mg/100 g
K	44 mmol/L	1159.89 mg/kg	0.14 %	–	107 mg/100 g	179.5 mg/100 g
Mg	–	733.24 mg/kg	0.04 %	38.6 mg/dl	35.52 mg/100 g	42.86 mg/100 g
Na	27 mmol/L	1058.93 mg/kg	0.07 %	–	150.50 mg/100 g	212.80 mg/100 g
P	–	4452.10 mg/kg	–	218 mg/dl	58.00 mg/100 g	53.00 mg/100 g
Zn	–	15.90 mg/kg	1.22 mg/100 ml	25.8 ppm	0.23 mg/100 g	0.27 mg/100 g
Vitamin A	–	4.9 μg/g	2.8 μg/g	–	166.67 IU/100 ml	250 IU/100 ml
Vitamin E	–	77.17 μg/g of fat	84 μg/g of fat	–	342.00 IU/100 ml	234.00 IU/100 ml
IgG	28.3 g/L	–	32 mg/ml	–	33.20 mg/ml	32.33 mg/ml
IgA	–	1.66 mg/ml	–	–	–	–
IgG1	–	34.96 mg/ml	–	–	–	–
IgG2	–	6.00 mg/ml	–	–	–	–
IgM	–	4.32 mg/ml	–	–	3.00 mg/ml	3.20 mg/ml
Lactoferrin	–	0.82 mg/ml	–	–	1.085 mg/ml	1.196 mg/ml

Colostrum contains less amount of lactose in the initial days of milking, then after it changes in its concentration which follows a reverse pattern to fat, protein, and ash [15, 42, 72]. The oligosaccharides content of colostrum ranges from ~22–24 g/L, which is significantly higher than human milk [93]. Among all mineral, magnesium, sodium, copper, iron, zinc, and manganese were found higher in colostrum as compared with normal milk [1, 40, 42, 92].

Colostrum contains a high level of fat-soluble vitamins, which include vitamin A, E, D, and K (especially vitamin K_1, phylloquinone) as compared to normal milk [16, 23, 38, 39, 96]. Likewise, the level of various water-soluble vitamins like vitamin B_6, B_{12}, C, thiamine, folate, and riboflavin is more whereas, the concentration of biotin and pantothenic acid are found less in colostrum as compared to normal milk [46, 55].

4.2.1 GROWTH FACTOR

In addition to other constituents, colostrum too contains many growth factors like IGF-I and II [20], epidermal growth factor [95], TGF-β1 and β2 [21, 29], betacellulin [10], platelet-derived growth factor, fibroblast growth factor 1 and 2 [44], etc. The presence of such growth factors in colostrum became beneficial to neonates [88]. Initially, its concentration is higher which reduces rapidly following parturition [10, 20, 29]. They help to restore normal growth of skin, bone, cartilage, muscle, and nervous tissues. They also aid in the synthesis of nucleic acids, control of blood sugar, build, and maintain lean muscle.

The hepatocyte growth factor is also found at a high level in colostrum which is one of the important mitogen, repair factors, and stimulators of hepatocyte proliferation [66]. The concentration of IGF-I, insulin, prolactin, and tumor necrosis factor-α (TNF-α) was reported as 103, 4.55, 120, 5.0 μg/L [68]. Buffalo milk colostrum has a high level of IGF-I with respect to cow milk colostrum [1]. The concentration of growth factors is affected by stages of lactation, diseases in animal, aging, etc., and fall its level rapidly after parturition [65].

4.3 PHYSICOCHEMICAL PROPERTIES OF COLOSTRUM

As the lactation progresses, the milk properties like rennet clotting time, density, and titratable acidity were decreased however there was increased ethanol stability and pH [45, 57, 58, 81]. Catalase activity of colostrum was found to be higher than milk [24]. The specific gravity of colostrum tends to decrease as the lactation progresses [72]. Buffering capacity and osmotic pressure of colostrum was greater than that of normal milk and declines rapidly during the first four secretions [58]. The colostrum has poor ethanol stability during the initial few days following birth [92].

4.4 EFFECTS OF HEAT TREATMENT ON COLOSTRUM

Heating affects the various heat labile constituents of colostrum. For example, if colostrum powder prepared by high-heat pasteurization and drying which significantly denature the sensitive immunoglobulins and proline-rich polypeptides [85]. It was found that about 34% native IgG concentration decreased in the colostrum after heat treatment (63°C for 120 min) [59]. Likewise, there is a 28.4% native IgG concentration in the colostrum decreased when heated at 72°C temperature and holding at such temperature for 15 seconds by using high-temperature short-time pasteurization [32].

4.5 FUNCTIONAL BENEFITS OF COLOSTRUM

Colostrum is an exclusive source for biomolecules as it contains many naturally occurring, physiologically active components, including immunoglobulins, lactoferrin, leukocytes, cytokines, carotenoids, xanthophyll esters, apocarotenoids, oligosaccharides, several growth factors and hormones which provide health benefits other than the basic nutrition [15, 30, 31, 56, 99]. Thus, colostrum can become an important source of ingredient for the development of nutraceuticals [15, 31, 36].

Colostrum is one of the potential sources of peptides and proteins, bioactive components that provide health benefits to newborn [5, 15, 41, 91]. The health benefits of colostrum [19, 28, 37, 43, 49, 60, 80] are indicated in Figure 4.1.

Human colostrum contains carotenoids, xanthophyll esters, and apocarotenoids in three different forms namely β-citraurin, apo-8′-lycopenal and apo-8′-β-carotenal [99]. Colostrum also contains phospholipids which are associated as brain functions enhancer to improve memory [85]. Bovine colostrum plays an indispensable role in the overall improvement of health and longevity by providing anti-aging benefits [94]. It is widely used in the gut-related various clinical applications like short-bowel syndrome, gut injury, inflammatory bowel disease, chemotherapy-induced mucositis, necrotizing enterocolitis, etc. [85].

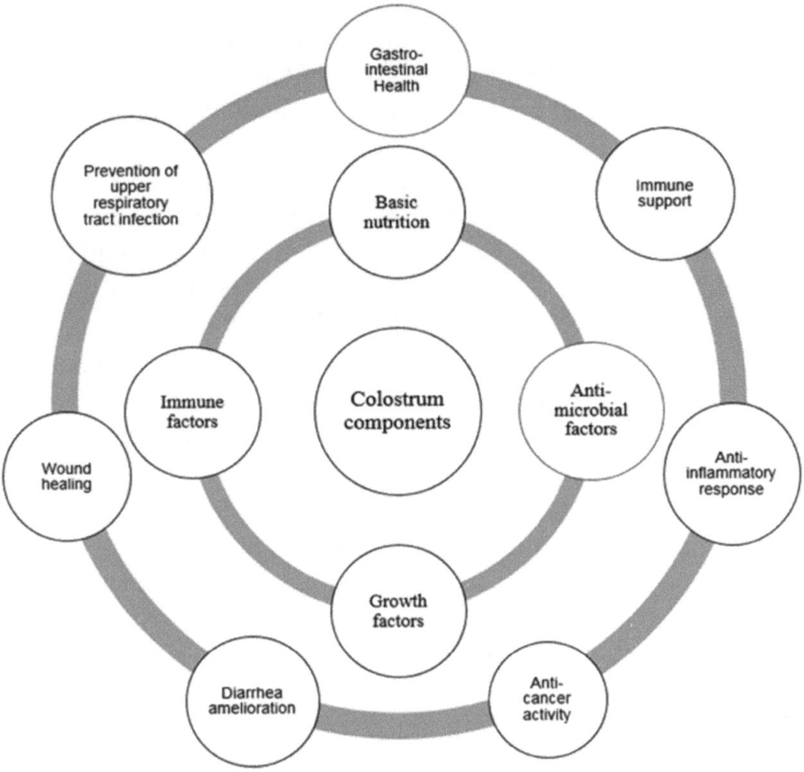

FIGURE 4.1 Health benefits associated with colostrum components.

4.5.1 NEONATAL PASSIVE IMMUNITY

Mortality is often observed due to immunity-related disorders [9]. When the significant levels of IgG or total protein are not attained within 24–28 hrs after birth, which may cause the failure of passive immunity transfer [88]. Initially, colostrum contains a significant number of immunoglobulins due to specific transfer from the serum into the mammary gland [48]. It provides immune protection until the activation of the immune system of newborn and builds natural immunity which reduces the chances of mortality [9, 35, 98]. Thus, colostrum is an important source of passive immunity and should be administered immediately after birth [30, 48, 91].

4.5.2 ANTIMICROBIAL PROTECTION

The disorders related to immunity are prominent reasons for the mortality [9, 15]. Colostrum is an important source of antimicrobial factors that gives protection and provides passive immunity during initial days [3, 15, 47, 71]. Among antimicrobial factors, immunoglobulins and lactoferrin plays an significant role in the improvement of natural immunity by protecting newborns from microbial infections [9, 15, 48, 84]. Lactoferrin is also known for its antiviral, antibacterial, antifungal, and antiparasitic properties [98]. These properties helped in the inhibition of the growth of rotavirus, adenovirus, and enterovirus [41]. Lactoperoxidase, lysozyme, and glycomacropeptide also possesses various antibacterial activities whereas α-lactalbumin has shown antiviral properties [41, 98]. These antimicrobial properties of colostrum enable its use for the improvement of immunity in the patients as well as healthy individuals [15].

4.5.3 ANTICARCINOGENIC PROPERTIES

Lactoferrin and α-Lactalbumin like compounds present in colostrum which have an advantageous effect in cancer treatment [3, 41, 98].

4.5.4 INTESTINAL HEALTH

Numerous compounds including bioactive molecules present in colostrum, improves the digestive tract in newborns [15, 25, 30]. Colostrum contains a significant amount of lactoferrin which unaffected towards tryptic digestion and thus, its greater amount emits in fecal which is important for its antiviral properties in the gastrointestinal tract [41]. Lactoferrin also aids in the regulation of iron absorption in the bowel [15]. Gastrointestinal infections can be controlled by IgA and other antimicrobial components present in colostrum [91].

4.5.5 SUITABILITY FOR LACTOSE INTOLERANCE PEOPLE

The lactose concentration is low in colostrum and hence, it may be suitable for lactose intolerant people [15]. Colostrum proteins were also found to have less side-effect profile as compared to milk [9].

4.5.6 DISEASE TREATMENT AND PREVENTION

Many studies have suggested that colostrum can be used as a supplement for the treatment, mitigation or prevention of many infectious diseases [4, 9, 61] like wound healing [49], prevention of upper respiratory tract infection [80], etc. It can be used as an immune support [28], diarrhea amelioration [37], anti-inflammatory response [19], anticancer activity [43]. It may also improve neuroendocrine systems and used as a health-promoting diet [4, 30]. It also became helpful in treating type-2 diabetes and gastrointestinal illnesses [60, 67].

α-Lactalbumin, hepatocyte growth factor and glycomacropeptide were reported to exhibit anti-stress, anti-allergic, antithrombotic activities, respectively [41, 64, 98]. α-Lactalbumin has been found to prevent diarrhea and lower the blood pressure in the rat [98]. Lysozyme can be used in the inhibition of tooth decay and periodontitis treatment [41, 98]. Colostrum can also have a positive effect on the prevention of inflammation by generating specific cytokines [6, 61].

Hyper-immune bovine colostrum can be developed by immunization of cows which could be more effective against diseases related to the gastro-intestinal tract [84]. Proline-rich polypeptide is proven for its immunotropic properties. Colostrinin chewable tablets which are rich in proline-rich poly-peptide indicated a positive influence on Alzheimer's disease patients [41]. A complex of α-lactalbumin and oleic acid was reported to have significant effect on cutaneous papillomas patients [98]. Lysozyme enriched milk was used in the feeding of concomitant diseases suffering from premature infants [98].

4.6 APPLICATIONS

The demand for colostrum is increasing in the market mainly due to the presence of valuable components in it and its significant health effect on the human [82]. For that researchers tried to produce dairy products that contained bovine colostrum. The incorporation of bovine colostrum and date syrup in yogurt enhanced the functional and nutritional quality of yogurt for children without affecting the quality [2]. It was also reported that the addition of bovine colostrum don't affect the specific microflora of yogurt and kefir like fermented product and on the other way it also increases the functional properties of products [8].

Ice cream containing bovine colostrum without compromising with the quality was also developed [63]. Bovine colostrum also used in Khess-a traditional Indian dessert [78]. Colostrum, colostrum powder, or its fractions can be used as an element in various food product formulas like drinks, yogurt mix preparation, chewing gums, etc. [7, 56]. Athletes widely use colostrum as a supplement to improve sports performance [17, 30, 61]. Colostrum-based immune milk products are prepared mainly for humans suffering from various infectious diseases [48]. Travelan, a commercial product for the prevention of enterotoxigenic *Escherichia coli* induced diarrhea [84]. Colostrum products are available in market which depicted in Table 4.2.

TABLE 4.2 Commercially Available Colostrum Products

Product	Manufacturer
Bovine colostrum capsules	Dr. Patkar's Healthcare India, Mumbai, India
Bovine colostrum powder, capsules, and tablets	Biostrum Nutritech Pvt. Ltd., Mahesana, Gujarat, India
Cow colostrum powder	Aadvik Food and Products Pvt. Ltd., Bikaner, Rajasthan, India
Goat milk colostrum capsules	Mt. CAPRA, WA, USA
Immulox® Colostrum powder and capsules	APS BioGroup, Phoenix Arizona, USA
Strawfit® (Milk flavoring straw with colostrums)	Bourgeon Food LLP, Indore, Madhya Pradesh, India
Sumul Immune Power (cow colostrums powder)	Surat District Cooperative Milk Producers' Union Limited, Surat, Gujarat, India

4.7 SUMMARY

Colostrum is an outstanding food for the newborn as it contains several bioactive components including growth factors. The admiration of colostrum is increasing day by day owing to its health-beneficial properties like antimicrobial protection, anti-carcinogenic properties, neonatal passive immunity, improved intestinal health, mitigation, or prevention of many infectious diseases, etc. Healthy components from bovine colostrum have a strong ability to be included as an ingredient in the formulation of nutraceutical food. However, more scientific data are required to support the specific health benefits associated with bovine colostrum components in human clinical studies.

KEYWORDS

- **colostrum**
- **functional foods**
- **growth factors**
- **health benefits**
- **nutraceuticals**

REFERENCES

1. El-Fattah, A. M. A., Abd, R. F. H., El-Dieb, S. M., & El-Kashef, H. A., (2012). Changes in composition of colostrum of Egyptian buffaloes and Holstein cows. *BMC Veterinary Research, 8*(1), 19.
2. Abdel-Ghany, A. S., & Zaki, D. A., (2018). Production of novel functional yogurt fortified with bovine colostrum and date syrup for children. *Alexandria Science Exchange Journal, 39*(4), 651–662.
3. Agarwal, P., & Gupta, R., (2016). A review on anticancer property of colostrum. *Research and Reviews Journal of Medical and Health Sciences, 5*(4), 1–9.
4. Alexieva, B., Markova, T., & Nikolova, E., (2004). Bovine colostrum-the promising nutraceutical. *Czech Journal of Food Sciences, 22*(1), 73–79.
5. Appukutty, M., Radhakrishnan, A. K., Ramasamy, K., Majeed, A. B. A., Noor, M. I., Safii, N. S., Chinna, K., & Koon, P. B., (2011). Modulation of interferon-gamma response through orally administered bovine colostrum in active adolescent boys. *Biomedical Research, 22*(1), 18–22.
6. Arya, R. K., Tahilramani, H., & Garg, A., (2018). Impact of human colostrum associated microbial population on neonatal health. *International Journal of Medical Science and Innovative Research, 3*(5), 86–90.
7. Aryana, K. J., Albers, E., & Cueva, O., (2008). Colostrum fortified probiotic fat free yogurt. *Milk Science International, 63*(1), 64–68.
8. Ayar, A., Siçrama, H., & Çetin, I., (2016). The effect of bovine colostrum on the lactic flora of yogurt and kefir. *JSM Biotechnology & Biomedical Engineering, 3*(4), 1062.
9. Bagwe, S., Tharappel, L. J., Kaur, G., & Buttar, H. S., (2015). Bovine colostrum: An emerging nutraceutical. *Journal of Complementary and Integrative Medicine, 12*(3), 175–185.
10. Bastian, S. E., Dunbar, A. J., Priebe, I. K., Owens, P. C., & Goddard, C., (2001). Measurement of betacellulin levels in bovine serum, colostrum, and milk. *The Journal of Endocrinology, 168*(1), 203–212.
11. Belyea, R. L., & Adams, M. W., (1990). Energy and nitrogen utilization of high versus low producing dairy cows. *Journal of Dairy Science, 73*(4), 1023–1030.
12. Bitman, J., & Wood, D. L., (1990). Changes in milk fat phospholipids during lactation. *Journal of Dairy Science, 73*(5), 1208–1216.

13. Blum, J. W., & Hammon, H. M., (2000). Colostrum effects on the gastrointestinal tract and on nutritional, endocrine, and metabolic parameters on neonatal calves. *Livestock Production Science, 66*(2), 151–159.
14. Blum, J. W., & Hammon, H. M., (2000). Colostrum—more than just an immunoglobulin supplier. *Schweizer Archiv fur Tierheilkunde, 142*(5), 221–228.
15. Buttar, H. S., Bagwe, S. M., Bhullar, S. K., & Kaur, G., (2017). Health benefits of bovine colostrum in children and adults. Chapter 1. In: Watson, R. R., Collier, R. J., & Preedy, V. R., (eds.), *Dairy in Human Health and Disease Across the Lifespan* (pp. 3–20). London, UK: Academic Press.
16. Calderón, F., Chauveau-Duriot, B., Martin, B., Graulet, B., Doreau, M., & Nozière, P., (2007). Variations in carotenoids, vitamins a and e and color in cow's plasma and milk during late pregnancy and the first three months of lactation. *Journal of Dairy Science, 90*(5), 2335–2346.
17. Carrillo, A. E., Koutedakis, Y., & Flouris, A. D., (2013). Exercise and exposure to heat following bovine colostrum supplementation: A review of gastrointestinal and immune function. *Cellular and Molecular Biology, 59*(1), 84–88.
18. Cerbulis, J., & Farrell, H. M., (1975). Composition of milks of dairy cattle. I. Protein, lactose, and fat contents and distribution of protein fraction. *Journal of Dairy Science, 58*(6), 817–827.
19. Chae, A., Aitchison, A., Day, A. S., & Keenan, J. I., (2017). Bovine colostrum demonstrates anti-inflammatory and antibacterial activity in *in vitro* models of intestinal inflammation and infection. *Journal of Functional Foods, 28*, 293–298.
20. Collier, R. J., Miller, M. A., Hildebrandt, J. R., Torkelson, A. R., White, T. C., Madsen, K. S., Vicini, J. L., et al., (1991). Factors affecting insulin-like growth factor-i concentration in bovine milk. *Journal of Dairy Science, 74*(9), 2905–2911.
21. Cox, D. A., & Burk, R., (1991). Isolation and characterization of milk growth factor, a transforming-growth-factor-beta 2-related polypeptide, from bovine milk. *European Journal of Biochemistry, 197*(2), 353–358.
22. Dang, A. K., Kapila, S., Purohit, M., & Singh, C., (2009). Changes in colostrum of murrah buffaloes after calving. *Tropical Animal Health and Production, 41*(7), 1213–1217.
23. Debier, C., Pottier, J., Goffe, C. H., & Larondelle, Y., (2005). Present knowledge and unexpected behaviors of vitamins A and E in colostrum and milk. *Livestock Production Science, 98*(1, 2), 135–147.
24. Farkye, N. Y., (2002). Enzymes indigenous to milk: Other enzymes. In: Roginski, H., Fuquay, J. W., & Fox, P. F., (eds.), *Encyclopedia of Dairy Sciences* (pp. 934–938). London, UK: Academic Press.
25. Fischer, A. J., Villot, C., Van, N. J. K., Yohe, T. T., Renaud, D. L., & Steele, M. A., (2019). Invited review: Nutritional regulation of gut function in dairy calves: From colostrum to weaning. *Applied Animal Science, 35*(5), 498–510.
26. Foley, J. A., & Otterby, D. E., (1978). Availability, storage, treatment, composition, and feeding value of surplus colostrum: A review. *Journal of Dairy Science, 61*(8), 1033–1060.
27. Georgiev, P., (2008). Differences in chemical composition between cow colostrum and milk. *Bulgarian Journal of Veterinary Medicine, 11*(1), 3–12.
28. Gephart, S. M., Weller, M., & Gephart, S., (2014). Colostrum as oral immune therapy to promote neonatal health. *Advances in Neonatal Care, 14*(1), 44–51.

29. Ginjala, V., & Pakkanen, R., (1998). Determination of transforming growth factor-β1 (TGF-β1) and insulin-like growth factor (IGF-1) in colostrum samples. *Journal of Immunoassay, 19*(2, 3), 195–207.
30. Glówka, N., & Woźniewicz, M., (2019). Potential use of colostrum bovinum supplementation in athletes: A review. *Acta Scientiarum Polonorum Technologia Alimentaria, 18*(2), 115–123.
31. Gosch, T., Apprich, S., Kneifel, W., & Novalin, S., (2014). A combination of microfiltration and high-pressure treatment for the elimination of bacteria in bovine colostrum. *International Dairy Journal, 34*(1), 41–46.
32. Green, L., Godden, S., & Feirtag, J., (2003). Effect of batch and high temperature-short time pasteurization on immunoglobulin g concentrations in colostrum. *Journal of Dairy Science, 86*(Suppl 1), 246.
33. Guilloteau, P., Le Huerou-Luron, I., Chayvialle, J. A., Toullec, R., Zabielski, R., & Blum, J. W., (1997). Gut regulatory peptides in young cattle and sheep. *Journal of Veterinary Medicine, 44*(1), 1–23.
34. Hallberg, J. W., Dame, K. J., Chester, S. T., Miller, C. C., Fox, L. K., Pankey, J. W., Nickerson, S. C., & Weaver, L. J., (1995). The visual appearance and somatic cell count of mammary secretions collected from primigravid heifers during gestation and early postpartum. *Journal Dairy Science, 78*(7), 1629–1636.
35. Hernandez-Castellano, L. E., Ozcelik, R., Hernandez, L. L., & Bruckmaier, R. M., (2018). Supplementation of colostrum and milk with 5-hydroxyl-tryptophan affects immune factors but not growth performance in newborn calves. *Journal of Dairy Science, 101*(1), 794–800.
36. Holland, P. T., Cargill, A., Selwood, A. I., Arnold, K., Krammer, J. L., & Pearce, K. N., (2011). Determination of soluble immunoglobulin G in bovine colostrum products by protein g affinity chromatography-turbidity correction and method validation. *Journal of Agricultural and Food Chemistry, 59*(10), 5248–5256.
37. Huppertz, H. I., Rutkowski, S., Busch, D. H., Eisebit, R., Lissner, R., & Karch, H., (1999). Bovine colostrum ameliorates diarrhea in infection with diarrheagenic *Escherichia coli*, Shiga toxin-producing *E. coli* and *E. coli* expressing intimin and hemolysin. *Journal of Pediatric Gastroenterology and Nutrition, 29*(4), 452–456.
38. Indyk, H. E., & Woolard, D. C., (1995). The endogenous vitamin K_1 content of bovine milk: Temporal influence of season and lactation. *Food Chemistry, 54*(4), 403–407.
39. Jensen, S. K., Johannsen, A. K., & Hermansen, J. E., (1999). Quantitative secretion and maximal secretion capacity of retinol, beta-carotene, and alpha-tocopherol into cows' milk. *Journal of Dairy Research, 66*(4), 511–522.
40. Jeong, S. G., Ham, J. S., Kim, D. H., Ahn, C. N., Chae, H. S., You, Y. M., Jang, A. R., et al., (2009). Physiochemical properties of colostrum by milking time of Gyeonggi province. *Korean Journal for Food Science of Animal Resources, 29*(4), 445–456.
41. Kanwar, J. R., Kanwar, R. K., Sun, X., Punj, V., Matta, H., Morley, S. M., Parratt, A., et al., (2009). Molecular and biotechnological advances in milk proteins in relation to human health. *Current Protein and Peptide Science, 10*(4), 308–338.
42. Kehoe, S. I., Jayarao, B. M., & Heinrichs, A. J., (2007). A survey of bovine colostrum composition and colostrum management practices on Pennsylvania dairy farms. *Journal of Dairy Science, 90*(9), 4108–4116.

43. Kim, Y., Kim, M. J., Han, K. S., Imm, J. Y., Oh, S., & Kim, S. H., (2009). Anticancer activity of lactoferrin isolated from caprine colostrum on human cancer cell lines. *International Journal of Dairy Technology, 62*(2), 277–281.
44. Kirihara, O., & Ohishi, H., (1995). Functional proteins in bovine milk. *Japanese Journal of Dairy and Food Science, 44*, 9–17.
45. Klimeš, J., Jagoš, P., Bouda, J., & Gajdůšek, S., (1986). Basic qualitative parameters of cow colostrum and their dependence on season and postpartum time. *Acta Veterinaria Brno, 55*(1, 2), 23–39.
46. Kon, S. K., & Watson, M. B., (1937). The vitamin C content of cow's milk. *Journal of Biochemistry, 31*(2), 223–226.
47. Korhonen, H., (1998). Colostrum immunoglobulins and the complement system-potential ingredients of functional foods. *Bulletin of the International Dairy Federation, 336*, 36–40.
48. Korhonen, H., Marnila, P., & Gill, H. S., (2000). Milk immunoglobulins and complement factors. *British Journal of Nutrition, 84*(S1), 75–80.
49. Kshirsagar, A. Y., Vekariya, M. A., Gupta, V., Pednekar, A. S., Mahna, A., Patankar, R., Shaikh, A., & Nagur, B., (2015). A comparative study of colostrum dressing versus conventional dressing in deep wounds. *Journal of Clinical and Diagnostic Research, 9*(4), PC01–PC04.
50. Kume, S. I., & Tanabe, S., (1993). Effect of parity on colostral mineral concentrations of Holstein cows and value of colostrum as a mineral source for newborn calves. *Journal of Dairy Science, 76*(6), 1654–1660.
51. Larson, B. L., (1992). Immunoglobulins of the mammary secretions. In: Fox, P. F., (ed.), *Advanced Dairy Chemistry: Proteins* (Vol. 1, pp. 231–254). London: Elsevier Applied Science.
52. Li, H., & Aluko, R. E., (2006). Bovine colostrum as a bioactive product against human microbial infections and gastrointestinal disorders. *Current Topics in Nutraceutical Research, 4*(3, 4), 227–237.
53. Macedo, R. J., Arredondo, V., García, F., Aguilar, M., Prado, O., & Rodríguez, R., (2012). Effect of supplemental yeast culture and physiological factors on colostrum and milk composition of pelibuey ewes. *Tropical Animal Health and Production, 44*(2), 349–354.
54. Madsen, B. D., Rasmussen, M. D., Nielsen, M. O., Wiking, L., & Larsen, L. B., (2004). Physical properties of mammary secretions in relation to chemical changes during transition from colostrum to milk. *Journal of Dairy Research, 71*(3), 263–272.
55. Marnila, P., & Korhonen, H., (2002). Colostrum. In: Roginski, H., Fuquay, J. W., & Fox, P. F., (eds.), *Encyclopedia of Dairy Sciences* (pp. 473–478.). London-UK: Academic Press.
56. Marnila, P., & Korhonen, H., (2011). Milk: Colostrum. In: Fuquay, J. W., Fox, P. F., & McSweeney, P. L. H., (eds.), *Encyclopedia of Dairy Sciences* (2nd edn., pp. 591–597). London-UK: Academic Press.
57. McCarthy, O. J., & Singh, K., (2009). Physico-chemical properties of milk. Chapter 15. In: McSweeney, P. L. H., & Fox, P. F., (eds.), *Advanced Dairy Chemistry, Volume 3: Lactose, Water, Salts, and Minor Constituents* (3rd edn., pp. 691–749). New York-USA: Springer.

58. McIntyre, R. T., Parrish, D. B., & Fountaine, F. C., (1952). Properties of the colostrum of the dairy cow. VII. pH, buffering capacity and osmotic pressure. *Journal of Dairy Science, 35*(4), 356–362.
59. McMartin, S., Godden, S. M., Metzger, L., Feirtag, J., Bey, R., Stabel, J., Goyal, S., et al., (2006). Heat treatment of bovine colostrum I. Effects of temperature on viscosity and immunoglobulin G level. *Journal of Dairy Science, 89*(6), 2110–2118.
60. Menchetti, L., Traina, G., Tomasello, G., Casagrande-Proietti, P., Leonardi, L., Barbato, O., & Brecchia, G., (2016). Potential benefits of colostrum in gastrointestinal diseases. *Frontiers in Bioscience, 8*(1), 331–351.
61. Mizelman, E., Duff, W., Kontulainen, S., & Chilibeck, P. D., (2017). The health benefits of bovine colostrum. Chapter 4. In: Watson, R. R., Collier, R. J., & Preedy, V. R., (eds.), *Nutrients in Dairy and their Implications on Health and Disease* (pp. 51–60). London-UK: Academic Press.
62. Moody, E. G., Wise, G. H., Parrish, D. B., & Atkeson, W., (1951). Properties of the colostrum of the dairy cow. VI. Creaming and rate of flow. *Journal of Dairy Science, 34*(2), 106–115.
63. Mouton, E., & Aryana, K. J., (2015). Influence of colostrum on the characteristics of ice cream. *Food and Nutrition Sciences, 6*(5), 480–484.
64. Munblit, D., Treneva, M., Dissanayeke, S., Chow, L. Y., Pampura, A., Boyle, R., & Warner, J., (2013). Environmental influences on hepatocyte growth factor in human colostrum: 1223. *Allergy: European Journal of Allergy and Clinical Immunology, 68*, 444.
65. Munblit, D., Treneva, M., Peroni, D., Colicino, S., Chow, L. Y., Dissanayeke, S., Pampura, A., et al., (2014). Difference in growth factor levels in colostrum and breast milk of mothers from Moscow, London, and Verona: 1712. *Allergy: European Journal of Allergy and Clinical Immunology, 69*, 592.
66. Munblit, D., Umar, B., Treneva, M., Chow, L. Y., Dissanayeke, S., Pampura, A., Warner, J., & Boyle, R., (2013). Hepatocyte growth factor in human colostrum and mature breast milk. *Clinical and Experimental Allergy, 43*(12), 1440.
67. Nikolic, I., Stojanovic, I., Vujicic, M., Fagone, P., Mangano, K., Stosic-Grujicic, S., Nicoletti, F., & Saksida, T., (2017). Standardized bovine colostrum derivative impedes development of type 1 diabetes in rodents. *Immunobiology, 222*(2), 272–279.
68. Ontsouka, C. E., Bruckmaier, R. M., & Blum, J. W., (2003). Fractionized milk composition during removal of colostrum and mature milk. *Journal of Dairy Science, 86*(6), 2005–2011.
69. Oyeniyi, O. O., & Hunter, A. G., (1978). Colostral constituents including immunoglobulins in the first three milkings postpartum. *Journal of Dairy Science, 61*(1), 44–48.
70. Pakkanen, R., (1998). Determination of transforming growth factor-β2 (TGF-β2) in bovine colostrum samples. *Journal of Immunoassay, 19*(1), 23–37.
71. Pandjaitan, M., & Sucianto, N. A., (2018). The study using of colostrum milk and edamame to increase human growth hormone. In: *International Conference on Innovation, Entrepreneurship, and Technology Proceeding* (Vol. 2, No. 2, pp. 121–127).
72. Parrish, D. B., Wise, G. H., Hughes, J. S., & Atkeson, F. W., (1950). Properties of the colostrum of the dairy cow. V. Yield, specific gravity and concentrations of total solids and its various components of colostrum and early milk. *Journal of Dairy Science, 33*(6), 457–465.

73. Parrish, D. B., Wise, G. H., & Hughes, J. S., (1947). Properties of the colostrum of the dairy cow. I. Tocopherol levels in colostrum and in early milk. *Journal of Dairy Science, 30*(11), 849–860.
74. Parrish, D. B., Wise, G. H., Hughes, J. S., & Atkeson, F. W., (1948). Properties of the colostrum of the dairy cow. II. Effect of prepartal rations upon nitrogenous constituents. *Journal of Dairy Science, 31*(10), 889–895.
75. Parrish, D. B., Wise, G. H., & Hughes, J. S., (1949). Properties of the colostrum of the dairy cow. IV. Effect of form of vitamin a and of tocopherol supplements on concentrations of vitamin a and carotenoids. *Journal of Dairy Science, 32*(5), 458–464.
76. Paszczyk, B., Żegarska, Z., & Borejszo, Z., (2005). The contents of trans fatty acids and CLA in cow colostrum and milk fat in the early lactation period. *Czech Journal of Food Sciences, 23*, 159–165.
77. Pecka-Kiełb, E., Zachwieja, A., Wojtas, E., & Zawadzki, W., (2018). Influence of nutrition on the quality of colostrum and milk of ruminants. *Mljekarstvo: Journal for Dairy Production and Processing Improvement, 68*(3), 169–181.
78. Poonia, A., & Dabur, R. S., (2012). Physico-chemical and sensory properties of khees obtained from buffalo and cow colostrum. *Journal of Dairying, Foods & Home Sciences, 31*(4), 256–258.
79. Precht, D., (2001). Cholesterol content in European bovine milk fats. *Nahrung, 45*(1), 2–8.
80. Saad, K., Abo-Elela, M. G. M., El-Baseer, K. A. A., Ahmed, A. E., Ahmad, F. A., Tawfeek, M. S. K., El-Houfey, A. A., et al., (2016). Effects of bovine colostrum on recurrent respiratory tract infections and diarrhea in children. *Medicine, 95*(37), e4560.
81. Sanchez-Macias, D., Moreno-Indias, I., Castro, N., Morales-delaNuez, A., & Arguello, A., (2014). From goat colostrum to milk: Physical, chemical, and immune evolution from partum to 90 days postpartum. *Journal of Dairy Science, 97*(1), 10–16.
82. Scammell, A. W., (2001). Production and uses of colostrum. *Australian Journal of Dairy Technology, 56*(2), 74–82.
83. Schlimme, E., Martin, D., & Tait, D., (2002). Nucleosides and Nucleotides in milk. In: Roginski, H., Fuquay, J. W., & Fox, P. F., (eds.), *Encyclopedia of Dairy Sciences* (pp. 2100–2107). London-UK: Academic Press.
84. Sears, K. T., Tennant, S. M., Reymann, M. K., Simon, R., Konstantopoulos, N., Blackwelder, W. C., Barry, E. M., & Pasetti, M. F., (2017). Bioactive immune components of anti-diarrheagenic enterotoxigenic *Escherichia coli* hyperimmune bovine colostrum products. *Clinical and Vaccine Immunology, 24*(8), e00186–16.
85. Seth, R., & Das, A., (2011). Colostrum powder and its health benefits. In: *Winter School Compendium on Chemical Analysis of Value-Added Dairy Products and their Quality Assurance* (pp. 59–67). National Dairy Research Institute; Karnal.
86. Shah, A. M., Shah, A. R., Hassan, M. F., Yousif, M., & Wang, Z., (2019). Colostrum composition and its importance to the health of animals: A review. *Punjab University Journal of Zoology, 34*(2), 197–206.
87. Shahani, K. M., Harper, W. J., Jensen, R. G., Parry, Jr. R. M., & Zittle, C. A., (1973). Enzymes in bovine milk: A review. *Journal of Dairy Science, 56*(5), 531–543.
88. Singh, A. K., Pandita, S., Vaidya, M. M., Singh, S. V., Chandra, G., Pampoori, Z. A., Huozha, R., et al., (2011). Bovine colostrum and neonate immunity: A review. *Agricultural Reviews, 32*(2), 79–90.

89. Smolenski, G., Haines, S., Kwan, F. Y. S., Bond, J., Farr, V., Davis, S. R., Stelwagen, K., & Wheeler, T. T., (2007). Characterization of host defense proteins in milk using a proteomic approach. *Journal of Proteome Research, 6*(1), 207–215.
90. Stelwagen, K., Carpenter, E., Haigh, B., Hodgkinson, A., & Wheeler, T. T., (2009). Immune components of bovine colostrum and milk. *Journal of Animal Science, 87*(13), 3–9.
91. Thapa, B. R., (2005). Health factors in colostrum. *The Indian Journal of Pediatrics, 72*(7), 579–581.
92. Tsioulpas, A., Grandison, A. S., & Lewis, M. J., (2007). Changes in physical properties of bovine milk from the colostrum period to early lactation. *Journal of Dairy Science, 90*(11), 5012–5017.
93. Urashima, T., Asakuma, S., Leo, F., Fukuda, K., Messer, M., & Oftedal, O. T., (2012). The predominance of type I oligosaccharides is a feature specific to human breast milk. *Advances in Nutrition, 3*(3), 473S–482S.
94. Wyatt, D. A., (2013). *Anti-aging Benefits of Bovine Colostrum.* Semantic Scholar. https://www.semanticscholar.org/paper/Anti-Aging-Benefits-of-Bovine-Colostrum-Wyatt/f59ed5baa9c33fb03f57722e074c4de5ddb958f3 (accessed on 15 November 2021).
95. Yagi, H., Suzuki, S., Noji, T., Nagashima, K., & Kuroume, T., (1986). Epidermal growth factor in cow's milk and milk formulas. *Acta Paediatrica Scandinavica, 75*(2), 233–235.
96. Yan, Y., Fang, C., Ye, W., & Zhou, Z., (1993). The lipid composition of bovine colostrum. *Acta Nutrimenta Sinica, 15*, 299–303.
97. Zhang, L. Y., Wang, J. Q., Yang, Y. X., Bu, D. P., Li, S. S., & Zhou, L. Y., (2011). Comparative proteomic analysis of changes in the bovine whey proteome during the transition from colostrum to milk. *Asian-Australasian Journal of Animal Sciences, 24*(2), 272–278.
98. Zimecki, M., & Artym, J., (2005). Therapeutic properties of proteins and peptides from colostrum and milk. *Postepy Higieny I Medycyny Doswiadczalnej (Online), 59*, 309–323.
99. Zoccali, M., Giuffrida, D., Granese, R., Salafia, F., Dugo, P., & Mondello, L., (2020). Determination of free apocarotenoids and apocarotenoid esters in human colostrum. *Analytical and Bioanalytical Chemistry, 412*(6), 1335–1342.

CHAPTER 5

STRUCTURAL AND FUNCTIONAL ASPECTS OF MILK OLIGOSACCHARIDES

ANINDITA DEBNATH, SRUSHTY PATIL, and VENUS BANSAL

ABSTRACT

Milk has been designated as a complete food owing to its nutritional and functional properties. From ages, dairy products have been the basic diet of society and are the main actor in the progress of mankind. The shift in paradigm from just-food to functional foods has led to explore the different components of milk that have certain health benefits. Oligosaccharides (OSs) are not apprehended from this as nutritionally and functionally enriched milk also contains OSs that have been explored in the past few years. Milk of all mammalian species comprises OSs; however, their proportion is very less as compared to that of human milk (HM). An assortment of beneficial effects related to health viz. immune modulation, anti-inflammation, mineral absorption, and the brain development activity have been reported to be conferred by oligosaccharides. The proven health benefits of oligosaccharides from human milk (HMOs) have driven the researchers' focus to a new arena to find out the OSs in other mammalian species in order to harness the functional properties of OSs for masses.

5.1 INTRODUCTION

Carbohydrates are an inherent part of our food system and a primary source of energy. Among different foods, milk that is an integral part of our diet

Functional Dairy Ingredients and Nutraceuticals: Physicochemical, Technological, and Therapeutic Aspects. Megh R. Goyal, N. Veena, PhD & Santosh K. Mishra (Eds.)

also contains carbohydrates in the varying proportion of 1.5 to 10% in different species of mammals. In milk, lactose is the major component of milk carbohydrates with a trace quantity of monosaccharides and oligosaccharides (OSs). Milk OSs can be defined as the carbohydrates consisting of three to ten monosaccharides. These monosaccharides are bound together by glycosidic bonds that are covalent in nature [4, 104].

The curiosity of researchers to explore the structure and health aspects of milk OSs has augmented in the past few years because of their promising biofunctional attributes. Several investigators have studied the effect of various mammalian milk OSs on the health promoting aspects of these constituents [65–67, 121]. However, OS content of mammalian species differ with respect to their concentration as well as structure that can affect their functional attributes. Among different mammals, HM contains proportionately more concentration of OSs with 2–2.3% in colostrum and 0.7–1.2% in mature milk [11, 18, 19, 75, 116]. In HM, OSs is the third prime constituent next to lipids and lactose (Figure 5.1) [11, 29, 62, 78, 108].

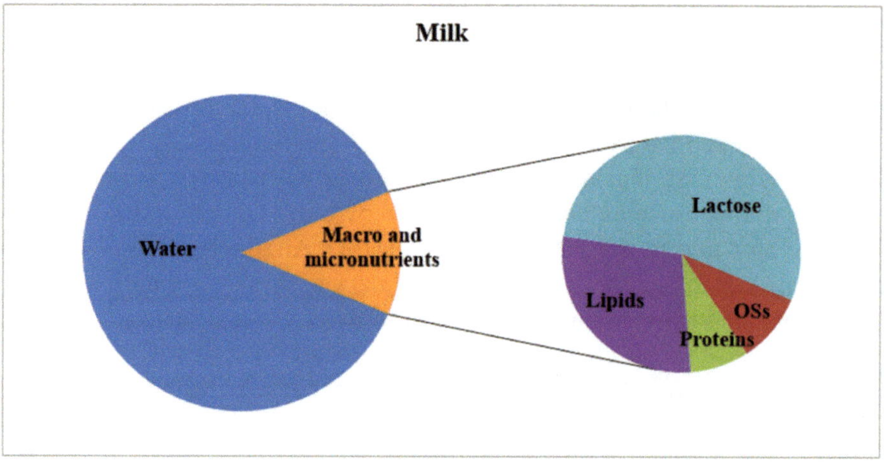

FIGURE 5.1 Fraction of human milk constituents.

The percentage of OSs in HM is comparatively more compared to milk from other mammalian species, i.e., cow milk, sheep milk, goat milk, and horse milk [110]. About 100–1000 folds lower concentration of milk OSs in domestic animals than HM has been reported [6]. Cow milk and colostrum contain about 0.05 g/L and 1 g/L OSs, respectively [1]. In goat milk, about 250–300 mg/L OSs has been reported [42, 71]. The OSs content of bovine milk is 4–5 times lower than goat milk, though caprine milk contains a lower percentage of OSs than human milk. Moreover, the level of OSs usually decreases at late lactation stage. At late lactation stage, about 5–10 times lower proportion of OSs in HM (Figure 5.1) as compared to early stage of lactation has been testified [63]. However, the fraction of sialylated OSs in bovine milk increases during late lactation [69].

Owing to a considerably higher proportion of OSs in human milk (HM), researchers have extensively explored the OSs of HM. With advancement in analytical chemistry, 200 types of HMOs have been recognized until now. These OSs have been reported to aid in the proliferation of good bacteria in the gut, reduce risk from infection and allergy, and even support brain development and improve cognitive function [6]. Considering these, researchers have focused their study to explore OSs of other mammalian species in order to harness the therapeutic properties of OSs in inhabitants other than breast-fed infants. However, the structure of bovine milk oligosaccharides (BMOs) may differ from that of HMOs and thus may not have similar therapeutic properties as that of HMOs. The monosaccharides which constitute the HMOs are D-galactose (Gal), D-glucose (Glu), L-Fucose (Fuc), N-acetylglucosamine (GlcNAc) and sialic acid (N-acetyl neuraminic acid (NANA)). In addition to these five monomers, the presence of other types of sialic acid (i.e., N-glycolylneuraminic acid (NeuGc)) as monomer in OSs of bovine milk colostrum has also been observed [11, 16, 72, 111, 123]. The structure of BMOs as compared to other animal milk OSs are more close to HMOs [71].

This chapter highlights the structure, classification of OSs in different mammalian milk with special focus on HM. The proven health aspects of various milk OSs have also been presented in order to comprehend therapeutic properties for the development of functional foods.

5.2 CLASSIFICATION AND STRUCTURE OF MILK OLIGOSACCHARIDES

On account of type of monosaccharides, OSs are divided into different categories viz. (i) core OSs that are made of Gal, Glc, and GlcNAc (ii) fucosyl-OSs (addition of Fuc to the core OS); (iii) sialyl-OSs (addition of NANA to the core OS) and (iv) sialylfucosyl-OSs which are formed after addition of both Fuc and NANA to the core OS [78]. Kunz et al. [62] and Sabharwal et al. [88] observed that Glc, Gal, Fuc, N-acetylgalactosamine

(GalNAc), GlcNAc, and sialic acids-NANA and NeuGc covalently linked together in varying manner to form a large varieties of milk OSs.

In most of the milk OSs, at their reducing end, a lactose unit remains attached to GlcNAc, Gal, Fuc, NANA or NeuGc residues. Generally, in human colostrum and milk, OSs NANA is attached to Gal or GlcNAc at α2–3/ 2–6 position and Fuc is linked to Gal, GlcNAc or reducing Glc of the main unit at α1–2/ 1–3/ 1–4 position. In some OSs N-acetyllactosamine unit (Gal(β1–4)GlcNAc) is present at reducing terminal, attached to Gal [2, 31].

HMOs mostly contain either type I unit, i.e., Lacto-N-biose unit (Gal (β1–3)GlcNAc) or type II unit, i.e., Gal(β1–4)GlcNAc unit attached to lactose core unit. Some BMOs also are of type I or type II [6, 2]. In HM, OSs with type I unit predominates over the type II unit associated OSs, while, in milk of other species type II unit is predominantly present.

The existence of about 130 OSs in colostrum and milk from humans has been identified by mass spectra, while analysis through microfluidic HPLC-chip mass spectroscopy (MS) revealed the presence of 200 OSs. Till now, the structure of 115 OSs has already been found out [79, 113]. The structure of 30–50 BMOs in mature bovine milk and colostrum has been identified [2, 4, 104, 105]. Tao et al. [104] and Urashima and Taufik [113] reported the existence of 39 OSs in bovine milk among which the structure of 25 OSs has been recognized.

The milk OSs, based on the presence of charged monosaccharide residue, can further be classified into two types such as neutral and acidic OSs [31, 61]. Neutral OSs contain no charged residue of monosaccharide while, acidic OSs contain negatively charged residue of monomer like sialic acid. The HM contains larger amounts of neutral OSs as compared to acidic OSs. It has been reported that about 50–80% of total HMOs structures are fucosylated and 10–30% are sialylated [6, 108].

In neutral fraction of animal milk OS, the amount of fucosylated-OSs are lesser as compared to HMOs. The acidic OSs of animal milk mainly contains sialic acid [10]. Bovine milk contains about 70% sialylated-OSs and 1–5% fucosylated-OSs [6, 105, 108]. These authors have also reported that caprine milk exhibits about 41% fucosylation and 18% sialylation in OS structures. Acidic HM or its colostrum shows the presence of only NANA [111]. Same authors also observed in bovine milk colostrum, 10 sialyl-OSs that contain both NANA and NeuGc. The common sialyl-OSs of bovine and human milk are 3'- and 6'-sialyllactose, and 3'- and 6'-galactosyllactose, respectively [11].

Table 5.1 represents the identified HMOs of different series and class on the basis of their group. Table 5.2 shows the identified bovine and caprine milk OSs which are also categorized based on their group.

TABLE 5.1 Classification of HMOs-Based on Their Groups

Class of Oligosaccharides	Name of Oligosaccharides	References
Lactose Series		
Core oligosaccharides	3'-Galactosyllactose (3'-GL)	[24]
	4'-Galactosyllactose (4'-GL)	[101]
	6'-Galactosyllactose (6'-GL)	[125]
	Lacto-N-triose II (LN triose II)	[23]
Fucosylated oligosaccharides	2'-Fucosyllactose (2'-FL)	[55]
	3'-Fucosyllactose (3'-FL)	[74]
	Difucosyllactose (DFL)	[59]
Sialylated oligosaccharides	3'-Sialyllactose (3'-SL)	[58]
Sialylated fucosylated oligosaccharides	3'-fucosyl-3′-sialyllactose (FSL)	[33]
Lacto-N-Tetraose Series		[52]
Core oligosaccharides	Lacto-N-fucopentaose I (LNFP I)	[54]
	Lacto-N-fucopentaose II (LNFP II)	[53]
	Lacto-N-fucopentaose V (LNFP V)	[30]
	Lacto-N-difucohexaose I (LNDFH I)	[60]
	Lacto-N-difucohexaose II (LNDFH II)	[60]
Sialylated oligosaccharides	3′-sialyllacto-N-tetraose (LSTa)	[58]
	6'-sialyllacto-N-tetraose (LSTb)	[58]
	Disialyllacto-N-tetraose (DS-LNT)	[32]
Sialylated fucosylated oligosaccharides	Fucosyl-sialyllacto-N-tetraose (F-LSTa)	[122]
	Fucosyl-sialyllacto-N-tetraose (F-LSTb)	[122]
	Fucosyl-disialyllacto-N-tetraose I (FDS-LNT I)	[46]
	Fucosyl-disialyllacto-N-tetraose II (FDS-LNT II)	[34]
Lacto-N-Neotetraose Series		[58]
Fucosylated oligosaccharides	Lacto-N-fucopentaose III (LNFP III)	[78]
Sialylated oligosaccharides	6′-sialyllacto-N-neotetraose (LSTc)	[58]
Sialylated fucosylated oligosaccharides	Fucosyl-sialyllacto-N-neotetraose (F-LSTc)	[96]

TABLE 5.1 *(Continued)*

Class of Oligosaccharides	Name of Oligosaccharides	References
Lacto-N-Hexaose Series		[49]
Fucosylated oligosaccharides	2-Fucosyl-lacto-N-hexaose I (F-LNH I)	[126]
	3-Fucosyl-lacto-N-hexaose II (F-LNH II)	[25]
	Difucosyl-lacto-N-hexaose (DF-LNH)	[25]
	Difucosyl-lacto-N-hexaose (DF-LNHa)	[126]
	Trifucosyl-lacto-N-hexaose (TF-LNH)	[86]
Sialylated oligosaccharides	6'-sialyl-lacto-N-hexaose (S-LNH)	[49]
	Disialyl-lacto-N-hexaose I (DS-LNH I)	[46]
	Disialyl-lacto-N-hexaose II (DS-LNH II)	[46]
	Trisialyl-lacto-N-hexaose (TS-LNH)	[28]
Sialylated fucosylated oligosaccharides	Fucosyl-sialyl-lacto-N-hexaose (FS-LNH)	[126]
	Fucosyl-sialyl-lacto-N-hexaose I (FS-LNH I)	[35]
	Fucosyl-sialyl-lacto-N-hexaose II (FS-LNH II)	[35]
	Fucosyl-sialyl-lacto-N-hexaose III (FS-LNH III)	[35]
	Fucosyl-sialyl-lacto-N-hexaose IV (FS-LNH IV)	[45]
	Difucosyl-sialyl-lacto-N-hexaose I (DFS-LNH I)	[35]
	Difucosyl-sialyl-lacto-N-hexaose II (DFS-LNH II)	[45]
	Fucosyl-disialyl-lacto-N-hexaose I (FDS-LNH I)	[127]
	Fucosyl-disialyl-lacto-N-hexaose II (FDS-LNH II)	[127]
	Fucosyl-disialyl-lacto-N-hexaose III (FDS-LNH III)	[45]
Lacto-N-Neohexaose (LNnH) Series		[50]
Fucosylated oligosaccharides	Fucosyl-lacto-N-neohexaose (F-LNnH)	[50]
	Difucosyl-lacto-N-neohexaose (DF-LNnH)	[37]
Sialylated oligosaccharides	6-sialyl-lacto-N-neohexaose I (S-LNnH I)	[50]
	6-sialyl-lacto-N-neohexaose II (S-LNnH II)	[106]
	Disialyl-lacto-N-neohexaose (DS-LNnH)	[35]

TABLE 5.1 *(Continued)*

Class of Oligosaccharides	Name of Oligosaccharides	References
Sialylated fucosylated oligosaccharides	Fucosyl-sialyl-lacto-N-neohexaose I (FS-LNnH I)	[33]
	Fucosyl-sialyl-lacto-N-neohexaose II (FS-LNnH II)	[50]
	Difucosyl-sialyl-lacto-N-neohexaose (DFS-LNnH)	[35]
	Fucosyl-disialyl-lacto-N-neohexaose FDS-LNnH)	[127]
Para-Lacto-N-Hexaose Series		[124]
Fucosylated oligosaccharides	3-fucosyl-para-lacto-N-hexaose I (F-paraLNH I)	[87]
	4-fucosyl-para-lacto-N-hexaose II (F-paraLNH II)	[14]
	Difucosyl-para-lacto-N-hexaose (DF-paraLNH)	[124]
	Trifucosyl-para-lacto-N-hexaose I (TF-paraLNH I)	[99]
	Trifucosyl-para-lacto-N-hexaose II (TF-paraLNH II)	[14]
	Difucosyl-para-lacto-N-hexaose sulfate I (DF-paraLNH sulfate I)	[36]
	Difucosyl-para-lacto-N-hexaose sulfate II (DF-paraLNH sulfate II)	[36]
	Trifucosyl-para-lacto-N-hexaose sulfate (TF-paraLNH sulfate)	[36]
Para-Lacto-N-Neohexaose Series		[124]
Fucosylated oligosaccharides	Difucosyl-para-lacto-N-neohexaose (DF-paraLNnH)	[124]
	Trifucosyl-para-lacto-N-neohexaose (TF-paraLNnH)	[14]
Lacto-N-Octaose Series		
Fucosylated oligosaccharides	3-fucosyl-lacto-N-octaose (F-LNO)	[103]
	Difucosyl-lacto-N-octaose I (DF-LNO I)	[103]
	Difucosyl-lacto-N-octaose II (DF-LNO II)	[103]
	Trifucosyl-lacto-N-octaose (TF-LNO)	[103]
Sialylated fucosylated oligosaccharides	Fucosyl-sialyl-lacto-N-octaose (FS-LNO)	[43]
	Difucosyl-sialyl-lacto-N-octaose (DFS-LNO)	[44]

TABLE 5.1 *(Continued)*

Class of Oligosaccharides	Name of Oligosaccharides	References
Lacto-N-Neooctaose Series		[103]
Fucosylated oligosaccharides	3-fucosyl-lacto-N-neooctaose (F-LNnO)	[128]
	Difucosyl-lacto-N-neooctaose I (DF-LNnO I)	[103]
	Difucosyl-lacto-N-neooctaose II (DF-LNnO II)	[103]
	Trifucosyl-lacto-N-neooctaose (TF-LNnO)	[103]
Iso-Lacto-N-Octaose Series		[98]
Fucosylated oligosaccharides	3-fucosyl-iso-lacto-N-octaose (F-isoLNO)	[51]
	Difucosyl-iso-lacto-N-octaose I (DF-isoLNO I)	[100]
	Difucosyl-iso-lacto-N-octaose II (DF-isoLNO II)	[100]
	Trifucosyl-iso-lacto-N-octaose I (TF-isoLNO I)	[98]
	Trifucosyl-iso-lacto-N-octaose II (TF-isoLNO II)	[51]
	Tetrafucosyl-iso-lacto-N-octaose (TetraF-isoLNO)	[37]
	Pentafucosyl-iso-lacto-N-octaose (PF-isoLNO)	[37]
Sialylated fucosylated oligosaccharides	Fucosyl-sialyl-iso-lacto-N-octaose (FS-isoLNO)	[44]
	Difucosyl-sialyl-iso-lacto-N-octaose I (DFS-isoLNO I)	[44]
	Difucosyl-sialyl-iso-lacto-N-octaose II (DFS-isoLNO II)	[44]
	Trifucosyl-sialyl-iso-lacto-N-octaose (TFS-isoLNO)	[43]
Lacto-N-Decaose Series		[3, 14]
Fucosylated oligosaccharides	3-fucosyl-lacto-N-decaose (F-LND)	[16]
Para-Lacto-N-Octaose Series		[37]
Fucosylated oligosaccharides	Tetrafucosyl-para-lacto-N-octaose (Tetra F-para LNO)	[37]

TABLE 5.2 Classification of Some Bovine and Caprine Milk Oligosaccharides

Class of Oligosaccharides	Name of Oligosaccharides	Source*	References
Neutral Oligosaccharides			
Core oligosaccharides	N-acetyllactosamine	b, c	[90]
	N-acetylgalactosaminylglucose	b, c	[90]
	α3'-N-acetylgalactosaminyl-lactose	b, c	[112]
	α3'-galactosyllactose	b, c	[109, 112]
	β3'-galactosyllactose	b, c	[89, 109]
	β4'-galactosyllactose	c	[41]
	β6'-galactosyllactose	b, c	[89, 109]
	lacto-N-novopentaose I	c	[112]
	lactose-3'-O-phosphate	b, c	[22]
Fucosylated oligosaccharides	2-fucosyllactose	c	[109]
	3-fucosyllactosamine	b, c	[89]
Acidic Oligosaccharides			
Sialylated oligosaccharides	3-N-acetylneuraminylgalactose	b, c	[57]
	3'-N-acetylneuraminyllactose	b, c	[93, 110]
	6'-N-acetylneuraminyllactose	b, c	[57, 110]
	3'-N-glycolylneuraminyllactose	b, c	[57]
	6'-N-glycolylneuraminyllactose	b, c	[110, 114]
	3'-N-acetylneuraminyl-3'-galactosyllactose	b, c	[80]
	3'-N-acetylneuraminyl-β6'-galactosyllactose	c	[117]
	6'-N-acetylneuraminyl-β3'-galactosyllactose	c	[117]
	6'-N-acetylneuraminyllactosamine	b, c	[57, 110]
	6'-N-glycolylneuraminyllactosamine	b, c	[114]
	di-N-acetylneuraminyllactose	b, c	[57]
	6'-N-acetylneuraminyllactosamine-1-O-phosphate	b, c	[80]
	6'-N-acetylneuraminyllactosamine-6-O-phosphate	b, c	[80]

Note: *b: bovine milk or colostrum; *c: caprine milk or colostrum.

5.3 FUNCTIONAL BENEFITS

The significance of milk OSs in the health promoting aspects has been explored in recent years by several researchers throughout the world. The prebiotic effect of milk OSs is known for 100 years. However, recent studies revealed that milk OSs have wide potential as functional ingredient. It promotes the health of an individual by enhancing the absorption of minerals, controlling the immune response system, restricting the pathogens from adhering on the epithelial cells, averting swelling and allergenic diseases and by cognitive development [5, 11, 31, 82].

5.3.1 PREBIOTIC EFFECT

The functional role of milk OSs as prebiotic is an area of great importance. Several researchers have already established the potency of OSs as prebiotic. The prebiotics are non-digestible ingredient of food that remain undigested in the gut of the host, provide food to beneficial host bacteria, aid in their proliferation and activity against pathogens in the colon. The growth and stimulation of host bacteria and metabolic end products of prebiotic food are recognized to inhibit the proliferation of pathogenic bacteria. Microbiota of healthy gut consists majorly of *Bifidobacterium* and *Lactobacillus* spp. These bacteria favored the host by boosting the immunity, alleviating the risk of several diseases like traveler's diarrhea in children, constipation, cancer, allergy, etc., through balancing the microbial ecosystem, acting against undesirable pathogenic bacteria, maintaining healthy mucosal system [20, 73, 81].

Milk OSs have already been proved as one of the influential factors to augment the number of beneficial bacteria, mainly of *Bifidobacterium* and *Lactobacillus species* in the infant's colon [20, 77, 82, 113]. Bifidobacteria metabolize OSs into short-chain fatty acids, which causes reduction of the colonic pH, and unsuitable condition for the proliferation of pathogens. Bifidobacteria, by this way inhibits the colonization of *Shigella flexneri* and *Escherichia coli* [31]. Lactobacilli prevent occurrence of urogenital infections, protect the vaginal mucosa through formation of a biofilm on the mucosal surface [83]. BMOs can also be used as prebiotic for human being as human lacks glycoside hydrolases enzymes [5, 11]. Bovine milk and colostrum OSs are therefore, recommended for medicinal purposes [113].

5.3.2 ANTI-PATHOGENIC

Milk OSs show anti-pathogenic properties against both bacteria and virus [38, 130]. Viruses and bacteria use to adhere to the mucosal epithelium cell of the respiratory tract and digestive tract of the host by developing a link with non-reducing terminal of sugar moieties of glycol-conjugates present in the cells. The configuration of non-reducing site of milk OS being similar to non-reducing end of sugar chains present in glycol-conjugate prevent pathogens from adhering to host cell and thus alleviate the risk of occurring pathogenic diseases, provide protection to infant [47, 113]. The HMOs are found to interfere with the adherence of *Campylobacter jejuni, Helicobacter pylori, Escherichia coli, Salmonella fyris and Vibrio cholera* on the outside of epithelial cells [21, 76]. Fucosylated-OSs exhibit promising role in reducing the risk of *Campylobacter jejuni* associated diarrhea and neurological paralysis in infants [85]. Sialylated-OSs protect the host from ulcerative colitis *by Helicobacter pylori* [7, 40].

Several researchers have studied the anti-infection properties of BMOs. Hakkarainen et al. [38] reported that the presence of neutral as well as acidic OSs at an amount of 1–2 g/L in milk helps to prevent meningitis and septicemia causative *Neisseria meningitides* from attaching to the host cell. The OS fraction obtained from milk of Swiss Brown cow inhibits the enterotoxigenic *E. coli* strains associated hemagglutination [70].

5.3.3 IMMUNOMODULATORY ACTIVITY

Milk OSs show the ability to boost the immune response by regulating the immune system [95]. Kurakevich et al. [64] reported that HMOs viz., 3'-sialyllactose (3'-SL) stimulates mesenteric lymph node CD11c+ dendritic cells. These cells promote T helper cells of type 1 (Th1) and Type 17 (Th17) immune responses by producing tumor necrosis factor-α (TNF-α), interleukin-12 (IL-12), interferon-γ (IFN-γ) and transforming growth factor-β1(TGF-β1) and thus induce immune response. Thomas et al. [107] noted the immunomodulatory functions of HMOs-LNFPIII which stimulate dendritic cell 2 (DC2) and induce Type 2 T helper (Th2) responses. The ability of HMOs-LNFP III and LNneoT to produce interleukin-10 (IL-10) was also observed in murine [115]. Eiwegger et al. [26] studied that HMOs improve postnatal immune maturation and prevent allergy by stimulating cord blood T cells that produce Th2 cytokine and IFN-γ.

5.3.4 MINERAL ABSORPTION

Milk OSs exhibits a pivotal role in the absorption of minerals in to intestine and colon. The improvement in mineral transport to small intestine and colon with OS consumption was observed by Scholz-Ahrens et al. [94]. Heuvel et al. [39] reported that bacteria ferments OSs in the colon and consequently yields fatty acids of short chains which facilitate the mineral assimilation into the gut through lowering of pH. The assimilation of calcium was found to rise in the rats that consumed milk OSs [17]. Boehm and Stahl [12] too observed the positive influence of milk OSs on absorption as well as retention of iron, calcium, magnesium, and zinc. Heuvel et al. [39] also reported that intake of OSs through milk or other diet, significantly increase calcium absorption in postmenopausal women.

5.3.5 ALLERGIC DISEASES

Nowadays, people are more prone to allergic diseases, and it has reached to a stage of an epidemic across the world. All type of foods, viz., milk, pulses, cereals, meat, and meat products and so forth contains allergenic substances. The most effective means of averting allergies is to eliminate the allergen foods from the diet. Recently, it has been documented that HMOs possess potent therapeutic value against allergenic diseases.

Castillo-Courtade et al. [15] carried out an *in vitro* study on mouse to gauge the impact of HMOs (2'-FL, 6′-SL) on ovalbumin associated anaphylaxis and found a positive result. The authors hypothesized that the potentiality of these HMOs to trigger Interleukin-10(+) T regulatory cells and stabilize mast cells might be the possible reason. In another study, these HMOs were shown to regulate the responses of human epithelial cell against allergic diseases [129]. The function of 6′-SL involves inhibition of chemokines release. As a consequence, resists the entry of inflammatory cells to the intestine and thus reducing the symptoms of food allergy.

5.3.6 BRAIN DEVELOPMENT

Sialylated-OSs of both HM and bovine milk play a vital role in brain and central nervous system development [119]. Sialylated-OSs aids in the biosynthesis of gangliosides and glycoproteins that take part in cognitive development [13, 62]. These constituents regulate structural and functional roles of frontal cortex of brain, play role in receptor binding and modulation, cell-to-cell contact, cell recognition, biosignal transduction and immunomodulation [119]. Sialic acids act as receptors for neurotransmitters in the central nervous system [37, 118, 119].

Lucas et al. [68] tested the relation of breast milk with intelligent quotient (IQ) of preterm infants. The authors concluded that infants fed with breast milk had higher IQ at the age of 7, while superior brain developmental scores were observed at 18th month. It was reported that within a time period amid a few months before parturition and few years after birth, the brain consists of more than double the amount of sialic acid [102]. The higher intelligence skill in babies consuming mother's milk than the babies intaking infant formula also support the influential role of milk OSs in brain development [13, 27, 97]. Also, the outcomes of experiments conducted on animal studies highlighted that consumption of sialic acid in their protein-bound form via food improves the cognitive skill at an early stage of development [37, 120]. Schaur and Kamerling [91] reported that this might be owing to better functioning of the central nervous system because of the ability of sialic acid to bind calcium to ganglioside.

5.3.7 ANTI-INFLAMMATION

Selectin-binding ligands are present in milk OSs that have anti-inflammation properties [37, 84]. Selectins are the cell adhesion molecules discharged as chemical signal from the damaged tissues in the course of the inflammation process. P-selectins, after making bond with P-selectin glycoprotein ligand 1 (PSGL-1), help in the formation of platelet neutrophil complexes (PNC) which cause phagocytosis and reactive oxygen species (ROS) production. The structures of acidic HMOs are analogous to PSGL-1 [9]. Bode et al. [9] reported that acidic HMOs reduce the binding of P-selectin to PSGL-1 by binding with P-selectin and thus restrict the occurrence of PNC formation during inflammation to 20%, while neutral HMOs exhibit no effect. The sialyl fraction of HMOs like 3'-SL and 3'-sialyl-3-fucosyl-lactose (FSL) has been noticed to act against inflammation in breast milk-fed infants [8]. Hence, only acidic milk oligosaccharides are more effective than neutral milk oligosaccharides [95].

5.4 SUMMARY

Milk contains several bioactive constituents that stimulate the growth of neonatal. OSs are one of such bioactive components that are constituent of milk of all mammalian species. The concentration of OSs is very less except human milk. This review indicates that OSs have unlimited perspective to be used as bioactive components in the food matrix. Indeed, OSs can be used in the formulation of pharmacological drugs as a natural ingredient for improving the well-being of individuals. However, there is a need to understand the structure-function relationship of different OSs in order to harness its therapeutic value for masses.

KEYWORDS

- **bioactive function**
- **bovine milk**
- **functional aspects**
- **human milk**
- **oligosaccharides**
- **structural aspects**

REFERENCES

1. Albrecht, S., Lane, J. A., Marino, K., Al Busadah, K. A., Carrington, S. D., Hickey, R. M., & Rudd, P. M., (2014). A Comparative study of free oligosaccharides in the milk of domestic animals. *British Journal of Nutrition, 111*(7), 1313–1328.
2. Aldredge, D. L., Geronimo, M. R., Hua, S., Nwosu, C. C., Lebrilla, C. B., & Barile, D., (2013). Annotation and structural elucidation of bovine milk oligosaccharides and determination of novel fucosylated structures. *Glycobiology, 23*(6), 664–676.
3. Amano, J., Osanai, M., Orita, T., Sugahara, D., & Osumi, K., (2009). Structural determination by negative-ion MALDI-QIT-TOFMSN after pyrene derivatization of variously fucosylated oligosaccharides with branched decaose cores from human milk. *Glycobiology, 19*(6), 601–614.
4. Barile, D., Marotta, M., Chu, C., Mehra, R., Grimm, R., Lebrilla, C. B., & German, J. B., (2010). Neutral and acidic oligosaccharides in Holstein-Friesian colostrum during the first 3 days of lactation measured by high performance liquid chromatography on a microfluidic chip and time-of-flight mass spectrometry. *Journal of Dairy Science, 93*(9), 3940–3949.

5. Barile, D., Tao, N., Lebrilla, C. B., Coisson, J. D., Arlorio, M., & German, J. B., (2009). Permeate from cheese whey ultrafiltration is a source of milk oligosaccharides. *International Dairy Journal, 19*(9), 524–530.
6. Bode, L., (2012). Human milk oligosaccharides: Every baby needs a sugar mama. *Glycobiology, 22*(9), 1147–1162.
7. Bode, L., (2006). Recent advances on structure, metabolism, and function of human milk oligosaccharides. *The Journal of Nutrition, 136*(8), 2127–2130.
8. Bode, L., Kunz, C., Muhly-Reinholz, M., Mayer, K., Seeger, W., & Rudloff, S., (2004). Inhibition of monocyte, lymphocyte, and neutrophil adhesion to endothelial cells by human milk oligosaccharides. *Thrombosis and Haemostasis, 92*(12), 1402–1410.
9. Bode, L., Rudloff, S., Kunz, C., Strobel, S., & Klein, N., (2004). Human milk oligosaccharides reduce platelet-neutrophil complex formation leading to a decrease in neutrophil beta 2 integrin expression. *Journal of Leukocyte Biology, 76*(4), 820–826.
10. Boehm, G., & Moro, G., (2008). Structural and functional aspects of prebiotics used in infant nutrition. *The Journal of Nutrition, 138*(9), 1818S–1828S.
11. Boehm, G., & Stahl, B., (2007). Oligosaccharides from milk. *The Journal of Nutrition, 137*(3), 847S–849S.
12. Boehm, G., & Stahl, B., (2003). Oligosaccharides. In: Mattila-Sandholm, T., & Saarela, M., (eds.), *Functional Dairy Products* (Vol. 1, pp. 203–243). Cambridge-UK; Woodhead publishers.
13. Bogoch, S., (1977). Recognins and their Chemoreciprocals. In: DeFeudis, F. V., & Delgado, J. M., (eds.), *Behavioral Neurochemistry* (p. 270). New York; Spectrum Publishers.
14. Bruntz, R., Dabrowski, U., Dabrowski, J., Ebersold, A., Peter-Katalinić, J., & Egge, H., (1988). Fucose-containing oligosaccharides from human milk from a donor of blood group 0 Le(a) nonsecretor. *Biological Chemistry Hoppe-Seyler, 369*(4), 257–273.
15. Castillo-Courtade, L., Han, S., Lee, S., Mian, F. M., Buck, R., & Forsythe, P., (2015). Attenuation of food allergy symptoms following treatment with human milk oligosaccharides in a mouse model. *Allergy, 70*(9), 1091–1102.
16. Chai, W., Piskarev, V. E., Zhang, Y., Lawson, A. M., & Kogelberg, H., (2005). Structural determination of novel lacto-N-decaose and its monofucosylated analogue from human milk by electrospray tandem mass spectrometry and 1H NMR spectroscopy. *Archives of Biochemistry and Biophysics, 434*(1), 116–127.
17. Chonan, O., & Watanuki, M., (1996). The effect of 6'-galactooligosaccharides on bone mineralization of rats adapted to different levels of dietary calcium. *Journal International of Vitamin and Nutrition Research, 66*(3), 244–249.
18. Coppa, G. V., Gabrielli, O., Pierani, P., Catassi, C., Carlucci, A., & Giorgi, P. L., (1993). Changes in carbohydrate composition in human milk over 4 months of lactation. *Pediatrics, 91*(3), 637–641.
19. Coppa, G. V., Gabrielli, O., Pierani, P., Zampini, L., Rottoli, A., Carlucci, A., & Giorgi, P. L., (1991). Qualitative and quantitative studies of carbohydrates of human colostrum and mature milk. *Rivista Italiana di Pediatria, 17*(3), 303–307.
20. Coppa, G. V., Pierani, P., Zampini, L., Bruni, S., Carloni, I., & Gabrielli, O., (2001). Characterization of oligosaccharides in milk and feces of breast-fed infants by high performance anion-exchange chromatography. Chapter 33. In: Newburg, D. S., (ed.), *Bioactive Components* of *Human Milk* (pp. 307–314). New York-USA; Springer Science + Business Media.

21. Coppa, G. V., Zampini, L., Galeazzi, T., Facinelli, B., Ferrante, L., Capretti, R., & Orazio, G., (2006). Human milk oligosaccharides inhibit the adhesion to caco-2 cells of diarrheal pathogens: *Escherichia coli, Vibriocholerae*, and *Salmonella fyris*. *Pediatric Research, 59*(3), 377–382.
22. Cumar, F. A., Ferchmin, P. A., & Caputto, R., (1965). Isolation and identification of a lactose phosphate ester from cow colostrum. *Biochemical and Biophysical Research Communications, 20*(1), 60–62.
23. Dabrowski, U., Egge, H., & Dabrowski, J., (1983). Proton-nuclear magnetic resonance study of peracetylated derivatives of ten oligosaccharides isolated from human milk. *Archives of Biochemistry and Biophysics, 224*(1), 254–260.
24. Donald, A. S., & Feeney, J., (1988). Separation of human milk oligosaccharides by recycling chromatography. First isolation of lacto-N-neo-difucohexaose II and 3′-galactosyllactose from this source. *Carbohydrate Research, 178*(1), 79–91.
25. Dua, V. K., Goso, K., Dube, V. E., & Bush, C. A., (1985). Characterization of lacto-N-hexaose and two fucosylated derivatives from human milk by high-performance liquid chromatography and proton NMR spectroscopy. *Journal of Chromatography A, 328*, 259–269.
26. Eiwegger, T., Stahl, B., Schmitt, J., Boehm, G., & Gerstmayr, M., (2004). Human milk-derived oligosaccharides and plant-derived oligosaccharides stimulate cytokine production of cord blood T-cells *in vitro*. *Pediatric Research, 56*(4), 536–540.
27. Fergusson, D. M., Beautrais, A. L., & Silva, P. A., (1982). Breast-feeding and cognitive development in the first seven years of life. *Social Science & Medicine, 16*(19), 1705–1708.
28. Fievre, S., Wieruszeski, J. M., Michalski, J. C., Lemoine, J., Montreuil, J., & Strecker, G., (1991). Primary structure of a trisialylated oligosaccharide from human milk. *Biochemical and Biophysical Research Communications, 177*(2), 720–725.
29. Fong, B., Ma, K., & McJarrow, P., (2011). Quantification of bovine milk oligosaccharides using liquid chromatography-selected reaction monitoring-mass spectrometry. *Journal of Agricultural and Food Chemistry, 59*(18), 9788–9795.
30. Ginsburg, V., Zopf, D. A., Yamashita, K., & Kobata, A., (1976). Oligosaccharides of human milk: Isolation of a new pentasaccharide, lacto-N-fucopentaose V. *Archives of Biochemistry and Biophysics, 175*(2), 565–568.
31. Gopal, P. K., & Gill, H. S., (2000). Oligosaccharides and glycoconjugates in bovine milk and colostrum. *British Journal of Nutrition, 84*(S1), 69–74.
32. Grimmonprez, L., & Montreuil, J., (1968). Etude physico-chemique de six nouveaux oligosaccharides isole du lai de femme. *Bulletin de la Societe de Chimie Biologique (Paris), 50*(4), 843–855.
33. Grönberg, G., Lipniunas, P., Lundgren, T., Erlansson, K., Lindh, F., & Nilsson, B., (1989). Isolation of monosialylated oligosaccharides from human milk and structural analysis of three new compounds. *Carbohydrate Research, 191*(2), 261–278.
34. Grönberg, G., Lipniunas, P., Lundgren, T., Lindh, F., & Nilsson, B., (1990). Isolation and structural analysis of three new desialylated oligosaccharides from human milk. *Archives of Biochemistry and Biophysics, 278*(2), 297–311.
35. Grönberg, G., Lipniunas, P., Lundgren, T., Lindh, F., & Nilsson, B., (1992). Structural analysis of five new monosialylated oligosaccharides from human milk. *Archives of Biochemistry and Biophysics, 296*(2), 597–610.

36. Guérardel, Y., Morelle, W., Plancke, Y., Lemoine, J., & Strecker, G., (1999). Structural analysis of three sulfated oligosaccharides isolated from human milk. *Carbohydrate Research, 320*(3, 4), 230–238.
37. Haeuw-Fievre, S., Wieruszeski, J. M., Plancke, Y., Michalski, J. C., Montreuil, J., & Strecker, G., (1993). Primary structure of human milk octa-, dodeca-and tridecasaccharides determined by a combination of 1H-NMR spectroscopy and fast-atom-bombardment mass spectrometry: Evidence for a new core structure, the para-lacto-N-octaose. *European Journal of Biochemistry, 215*(2), 361–371.
38. Hakkarainen, J., Toivanen, M., Leinonen, A., Frängsmyr, L., Strömberg, N., Lapinjoki, S., Nasssif, X., & Tikkanen-Kaukanen, C., (2005). Human and bovine milk oligosaccharides inhibit *Neisseria* meningitidis pili attachment *in vitro. The Journal of Nutrition, 135*(10), 2445–2448.
39. Heuvel, V. D. E. G., Schoteraman, M. H., & Muijs, T., (2000). Transgalacto-oligosaccharides stimulate calcium absorption in postmenopausal women. *The Journal of Nutrition, 130*(12), 2938–2942.
40. Johansson, P., Nilsson, J., Ångström, J., & Miller-Podraza, H., (2005). Interaction of *Helicobacter pylori* with sialylated carbohydrates: The dependence on different parts of the binding trisaccharide Neu5Acα3Galβ4GlcNAc. *Glycobiology, 15*(6), 625–636.
41. Kimura, K., Watanabe, Y., Matsumoto, K., & Miyagi, A., (1997). *Yakult Reports, 17*, 1–7.
42. Kiskini, A., & Difilippo, E., (2013). Oligosaccharides in goat milk: Structure, health effects and isolation. *Cellular and Molecular Biology, 59*(1), 25–30.
43. Kitagawa, H., Nakada, H., Fukui, S., Funakoshi, I., Kawasaki, T., Yamashina, I., Tate, S. I., & Inagaki, F., (1993). Novel oligosaccharides with the sialyl-lea structure in human milk. *The Journal of Biochemistry, 114*(4), 504–508.
44. Kitagawa, H., Nakada, H., Fukui, S., Funakoshi, I., Kawasaki, T., Yamashina, I., Tate, S. I., & Inagaki, F., (1991). Novel oligosaccharides with the sialyl-lea structure in human milk. *Biochemistry, 30*(11), 2869–2876.
45. Kitagawa, H., Nakada, H., Kurosaka, A., Hiraiwa, N., Numata, Y., Fukui, S., Funakoshi, I., et al., (1989). Three novel oligosaccharides with the sialyl-lea structure in human milk: Isolation by immunoaffinity chromatography. *Biochemistry, 28*(22), 8891–8897.
46. Kitagawa, H., Takaoka, M., Nakada, H., Fukui, S., Funakoshi, I., Kawasaki, T., Tate, S. I., et al., (1991). Isolation and structural studies of human milk oligosaccharides that are reactive with a monoclonal antibody MSW 113. *The Journal of Biochemistry, 110*(4), 598–604.
47. Kobata, A., (2010). Structures and application of oligosaccharides in human milk. In: *Proceedings of the Japan Academy, Series B., 86*(7), 731–747.
48. Kobata, A., & Ginsburg, V., (1969). Oligosaccharides of human milk II. Isolation and characterization of a new pentasaccharide, lacto-N-fucopentaose III. *Journal of Biological Chemistry, 244*(20), 5496–5502.
49. Kobata, A., & Ginsburg, V., (1972). Oligosaccharides of human milk III. Isolation and characterization of a new hexasaccharide, lacto-N-hexaose. *Journal of Biological Chemistry, 247*(5), 1525–1529.
50. Kobata, A., & Ginsburg, V., (1972). Oligosaccharides of human milk: IV. Isolation and characterization of a new hexasaccharide, lacto-N-neohexaose. *Archives of Biochemistry and Biophysics, 150*(1), 273–281.

51. Kogelberg, H., Piskarev, V. E., Zhang, Y., Lawson, A. M., & Chai, W., (2004). Determination by electrospray mass spectrometry and 1H-NMR spectroscopy of primary structures of variously fucosylated neutral oligosaccharides based on the iso-lacto-N-octaose Core. *European Journal of Biochemistry, 271*(6), 1172–1186.
52. Kuhn, R., & Baer, H. H., (1956). Die konstitution der lacto-N-tetraose. *Chemische Berichte, 89*(2), 504–511.
53. Kuhn, R., Baer, H. H., & Gauhe, A., (1958). Die konstitution der lacto-N-fucopentaose II. Ein beitragzur spezifität der blutgruppensubstanz lea. *Chemische Berichte, 91*(2), 364–374.
54. Kuhn, R., Baer, H. H., & Gauhe, A., (1956). Kristallisation und konstitutionsermittlung der lacto-N-fucopentaose I. *Chemische Berichte, 89*(11), 2514–2523.
55. Kuhn, R., Baer, H. H., & Gauhe, A., (1956). Kristallisierte fucosido-lactose. *Chemische Berichte, 89*(11), 2513.
56. Kuhn, R., & Brossmer, R., (1959). Über das durch viren der influenza-gruppespaltbare trisaccharid der milch. *Chemische Berichte, 92*(7), 1667–1671.
57. Kuhn, R., & Gauhe, A., (1965). Bestimmung der bindungsstelle von sialinsäureresten in oligosaccharidenmit hilfe von perjodat. *Chemische Berichte, 98*(2), 395–413.
58. Kuhn, R., & Gauhe, A., (1962). Die konstitution der lacto-nneotetraose. *Chemische Berichte, 95*(2), 518–522.
59. Kuhn, R., & Gauhe, A., (1958). Über die lacto-difuco-tetraose der frauenmilch. Ein beitragzur strukturspezifität der blutgruppensubstanz le. *Justus Liebigs Annalen Der Chemie, 611*(1), 249–253.
60. Kuhn, R., & Gauhe, A., (1960). Über einkristallisiertes, lea-aktives hexasaccharidaus frauenmilch. *Chemische Berichte, 93*(3), 647–651.
61. Kunz, C., & Rudloff, S., (2006). Health-promoting aspects of milk oligosaccharides. *International Dairy Journal, 16*(11), 1341–1346.
62. Kunz, C., Rudloff, S., Baier, W., Klein, N., & Strobel, S., (2000). Oligosaccharides in human milk: Structural, functional, and metabolic aspects. *Annual Review of Nutrition, 20*(1), 699–722.
63. Kunz, C., Rudloff, S., & Michaelsen, K. F., (2001). Concentration of human milk oligosaccharides during late lactation. *FASEB Journal, 15*(5), A986.
64. Kurakevich, E., Hennet, T., Hausmann, M., Rogler, G., & Borsig, L., (2013). Milk oligosaccharide sialyl (α2,3) lactose activates intestinal CD11c+ cells through TLR4. *Proceedings of the National Academy of Sciences, 110*(43), 17444–17449.
65. Lagström, H., Rautava, S., Ollila, H., Kaljonen, A., Turta, O., Mäkelä, J., Yonemitsu, C., et al., (2020). Associations between human milk oligosaccharides and growth in infancy and early childhood. *The American Journal of Clinical Nutrition, 111*(4), 769–778.
66. Lawson, M. A., O'Neill, I. J., Kujawska, M., Javvadi, S. G., Wijeyesekera, A., Flegg, Z., Chalklen, L., & Hall, L. J., (2020). Breast milk-derived human milk oligosaccharides promote bifidobacterium interactions within a single ecosystem. *The ISME Journal, 14*(2), 635–648.
67. Lu, J., Zhang, Y., Song, B., Zhang, S., Pang, X., Sari, R. N., Liu, L., Wang, J., & Lv, J., (2020). Comparative analysis of oligosaccharides in Guanzhong and Saanen goat milk by using LC-MS/MS. *Carbohydrate Polymers, 235*, 115965.
68. Lucas, A., Morley, R., Cole, T. J., Lister, G., & Leeson-Payne, C., (1992). Breast milk and subsequent intelligence quotient in children born preterm. *The Lancet, 339*(8788), 261–264.

69. Martin, M. J., Martin-Sosa, S., García-Pardo, L. A., & Hueso, P., (2001). Distribution of bovine milk sialo glycoconjugates during lactation. *Journal of Dairy Science, 84*(5), 995–1000.
70. Martin, M. J., Martin-Sosa, S., & Hueso, P., (2002). Binding of milk oligosaccharides by several enterotoxigenic *Escherichia coli* strains isolated from calves. *Glycoconjugate Journal, 19*, 5–11.
71. Martinez-Ferez, A., Rudloff, S., Guadix, A., Henkel, C. A., Pohlentz, G., Boza, J. J., et al., (2006). Goats' milk as a natural source of lactose-derived oligosaccharides: Isolation by membrane technology. *International Dairy Journal, 16*(2), 173–181.
72. Mehra, R., & Kelly, P., (2006). Milk oligosaccharides: Structural and technological aspects. *International Dairy Journal, 16*(11), 1334–1340.
73. Mills, S., Ross, R. P., Hill, C., Fitzgerald, G. F., & Stanton, C., (2011). Milk intelligence: Mining milk for bioactive substances associated with human health. *International Dairy Journal, 21*(6), 377–401.
74. Montreuil, J., (1956). Structure de deux triholosides isolés du lait de femme. *Comptes Rendus Hebdomadaires des Seances de l'Academie des Sciences, 242*(1), 192, 193.
75. Montreuil, J., & Mullet, S., (1959). Evolution de la constitution glucidique du lait de femme au cours de la lactation. *Comptes Rendus des Seances de la Societe de Biologie et de SesFiliales, 153*(8, 9), 1364–1366.
76. Morrow, A. L., Ruiz-Palacios, G. M., Jiang, X., & Newburg, D. S., (2005). Human-milk glycans that inhibit pathogen binding protect breast-feeding infants against infectious diarrhea. *The Journal of Nutrition, 135*(5), 1304–1307.
77. Mussatto, S. I., & Mancilha, I. M., (2007). Non-digestible oligosaccharides: A review. *Carbohydrate Polymers, 68*(3), 587–597.
78. Newburg, D. S., & Neubauer, S. H., (1995). Carbohydrates in milks: Analysis, quantities, and significance. Chapter 4. In: Jensen, R. G., (ed.), *Handbook of Milk Composition* (pp. 273–349). San Diego-California; USA; Academic Press.
79. Ninonuevo, M. R., Park, Y., Yin, H., Zhang, J., Ward, R. E., Clowers, B. H., German, J. B., et al., (2006). A strategy for annotating the human milk glycome. *Journal of Agricultural and Food Chemistry, 54*(20), 7471–7480.
80. Parkkinen, J., & Finne, J., (1987). Isolation of sialyl oligosaccharides and sialyl oligosaccharide phosphates from bovine colostrums and human urine. In: Ginsberg, V., (ed.), *Methods in Enzymology* (Vol. 138, pp. 289–300). New York; Academic Press.
81. Qiang, X., Yonglie, C., & Qiabing, W., (2009). Health benefit application of functional oligosaccharides. *Carbohydrate Polymers, 77*(3), 435–441.
82. Rivero-Urgëll, M., & Santamaria-Orleans, A., (2001). Oligosaccharides: Application in infant food. *Early Human Development, 65*(suppl 2), S43–S52.
83. Rousseu, V., Lepargneur, J. P., Roques, C., Remaud-Simeon, M., & Paul, F., (2005). Prebiotic effects of oligosaccharides on selected vaginal lactobacilli and pathogenic microorganisms. *Anaerobe, 11*(3), 145–153.
84. Rudloff, S., Pohlentz, G., Diekmann, L., Egge, H., & Kunz, C., (1996). Urinary excretion of lactose and oligosaccharides in preterm infants fed human milk or infant formula. *Acta Paediatrica, 85*(5), 598–603.
85. Ruiz-Palacios, G. M., Cervantes, L. E., Ramos, P., Chavez-Munguia, B., & Newburg, D. S., (2003). *Campylobacter jejuni* binds intestinal H (O) antigen (Fucα1, 2Galβ1, 4GlcNAc), and fucosyl oligosaccharides of human milk inhibit its binding and infection. *Journal of Biological Chemistry, 278*(16), 14112–14120.

86. Sabharwal, H., Nilsson, B., Chester, M. A., Lindh, F., Grönberg, G., Sjöblad, S., & Lundblad, A., (1988). Oligosaccharides from faeces of a blood-group B, breast-fed infant. *Carbohydrate Research, 178*(1), 145–154.
87. Sabharwal, H., Nilsson, B., Grönberg, G., Chester, M. A., Dakour, J., Sjöblad, S., & Lundblad, A., (1988). Oligosaccharides from feces of preterm infants fed on breast milk. *Archives of Biochemistry and Biophysics, 265*(2), 390–406.
88. Sabharwal, H., Sjöblad, S., & Lundblad, A., (1991). Affinity chromatographic identification and quantitation of blood group A-active oligosaccharides in human milk and feces of breast-fed infants. *Journal of Pediatric Gastroenterology and Nutrition, 12*(4), 474–479.
89. Saito, T., Itoh, T., & Adachi, S., (1987). Chemical Structure of three neutral trisaccharides isolated in free form bovine colostrum. *Carbohydrate Research, 165*(1), 43–51.
90. Saito, T., Itoh, T., & Adachi, S., (1984). Presence of two neutral disaccharides containing N-acetyl hexosamine in bovine colostrum as free forms. *Biochimica et Biophysica Acta (BBA)-General Subjects, 801*(1), 147–150.
91. Schaur, R., & Kamerling, J. P., (1997). Chemistry, biochemistry, and biology of sialic acids. Chapter 11. In: Montreuil, J., Schachter, H. M., & Vilegenthart, J. F. G., (eds.), *Glycoproteins II* (1st edn., Vol. 29, pp. 243–402). Amsterdam-Netherland; Elsvier Science.
92. Schimdt, R., (1989). Glycoproteins involved in long-lasting plasticity in the teleost brain. In: Rahmann, H., & Lindauer, M., (eds.), *Progress in Zoology: Fundamentals of Memory Formation: Neuronal Plasticity and Brain Function* (pp. 327–339). Fischer-Verlag; Stuttgart.
93. Schneir, M. L., & Rafelson, Jr. M. E., (1966). Isolation and characterization of two structural isomers of N-acetylneuraminyllactose from bovine colostrum. *Biochimica et Biophysica Acta (BBA)-General Subjects, 130*(1), 1–11.
94. Scholz-Ahrens, K. E., Schaafsma, G., Heuvel, V. D. E. G., & Schrezenmeir, J., (2001). Effects of prebiotics on mineral metabolism. *The American Journal of Clinical Nutrition, 73*(2), 459s–464s.
95. Schumacher, G., Bendas, G., Stahl, B., & Beermann, C., (2006). Human milk oligosaccharides affect p-selectin binding capacities: *In vitro* investigation. *Nutrition, 22*(6), 620–627.
96. Smith, D. F., Prieto, P. A., McCrumb, D. K., & Wang, W. C., (1987). A novel sialylfucopentaose in human milk. Presence of this oligosaccharide is not dependent on expression of the secretor or Lewis fucosyltransferases. *Journal of Biological Chemistry, 262*(25), 12040–12047.
97. Smith, M. M., Durkin, M., Hinton, V. J., Bellinger, D., & Kuhn, L., (2003). Influence of breastfeeding on cognitive outcomes at age 6–8 years: Follow-up of very low birth weight infants. *American Journal of Epidemiology, 158*(11), 1075–1082.
98. Strecker, G., Fievre, S., Wieruszeski, J. M., Michalski, J. C., & Montreuil, J., (1992). Primary structure of four human milk octa-, Nona-, and undeca-saccharides established by 1H-and 13C-nuclear magnetic resonance spectroscopy. *Carbohydrate Research, 226*(1), 1–14.
99. Strecker, G., Wieruszeski, J. M., Michalski, J. C., & Montreuil, J., (1988). Structure of a new nonasaccharide isolated from human milk: VI 2 Fuc, V 4 Fuc, III 3 Fuc-p-Lacto-N-hexaose. *Glycoconjugate Journal, 5*(4), 385–396.

100. Strecker, G., Wieruszeski, J. M., Michalski, J. C., & Montreuil, J., (1989). Primary structure of human milk Nona-and decasaccharides determined by a combination of fast atom bombardment mass spectrometry and 1 H-/13 C-nuclear magnetic resonance spectroscopy. Evidence for a new core structure, iso-lacto-N-octaose. *Glycoconjugate Journal, 6*(2), 169–182.
101. Sugawara, M., & Idota, T. A., (1995). New oligosaccharide; 4'-galactosyl-lactose in human milk. In: *Proceedings of the Annual Meeting of Japan Society for Bioscience, Biotechnology, and Agrochemistry* (p. 132). Sapporo; Japan.
102. Svennerholm, L., Bostrom, K., Fredman, P., Mansson, J. E., Rosengren, B., & Rynmark, B. M., (1989). Human brain gangliosides: Developmental changes from early fetal stages to advanced ages. *Biochimica et Biophysica Acta, 1005*(2), 109–117.
103. Tachibana, Y., Yamashita, K., & Kobata, A., (1978). Oligosaccharides of human milk: Structural studies of di-and trifucosyl derivatives of lacto-N-octaose and lacto-N-neooctaose. *Archives of Biochemistry and Biophysics, 188*(1), 83–89.
104. Tao, N., DePeters, E. J., Freeman, S., German, J. B., Grimm, R., & Lebrilla, C. B., (2008). Bovine milk glycome. *Journal of Dairy Science, 91*(10), 3768–3778.
105. Tao, N., DePeters, E. J., German, J. B., Grimm, R., & Lebrilla, C. B., (2009). Variations in bovine milk oligosaccharides during early and middle lactation stages analyzed by high-performance liquid chromatography-chip/mass spectrometry. *Journal of Dairy Science, 92*(7), 2991–3001.
106. Tarrago, M. T., Tucker, K. H., Van, H. H., & Smith, D. F., (1988). A novel sialylhexasaccharide from human milk: Purification by affinity chromatography on immobilized wheat germ agglutinin. *Archives of Biochemistry and Biophysics, 267*(1), 353–362.
107. Thomas, P. G., Carter, M. R., Atochina, O., Da'Dara, A. A., Piskorska, D., McGuire, E., & Harn, D. A., (2003). Maturation of dendritic cell 2 phenotype by a helminth glycan uses a toll-like receptor 4-dependent mechanism. *The Journal of Immunology, 171*(11), 5837–5841.
108. Urashima, T., Asakuma, S., & Messer, M., (2007). Milk oligosaccharides. Chapter 8. In: McSweeney, P. L. H., & Fox, P. F., (eds.), *Advanced Dairy Chemistry* (3rd edn., Vol. 3, pp. 295–349). New York; Springer.
109. Urashima, T., Bubb, W. A., Messer, M., Tsuji, Y., & Taneda, Y., (1994). Studies of the neutral trisaccharides of goat (Capra Hircus) colostrum and of the one-and two-dimensional 1H and 13C NMR spectra of 6′-N-acetylglucosaminyllactose. *Carbohydrate Research, 262*(2), 173–184.
110. Urashima, T., Murata, S., & Nakamura, T., (1997). Structural determination of monosialyl trisaccharides obtained from caprine colostrum. *Comparative Biochemistry and Physiology Part B: Biochemistry and Molecular Biology, 116*(4), 431–435.
111. Urashima, T., Saito, T., Nakamura, T., & Messer, M., (2001). Oligosaccharides of milk and colostrum in non-human mammals. *Glycoconjugate Journal, 18*(5), 357–371.
112. Urashima, T., Saito, T., Ohmisya, K., & Shimazaki, K., (1991). Structural determination of three neutral oligosaccharides in bovine (Holstein-Friesian) colostrum, including the novel trisaccharide; GalNAcα1-3Galβ1-4Glc. *Biochimica et Biophysica Acta, 1073*(1), 225–229.
113. Urashima, T., & Taufik, E., (2010). Oligosaccharides in milk: Their benefits and future utilization. *Media Peternakan, 33*(3), 189–197.

114. Veh, R. W., Michalski, J. C., Corfield, A. P., Sander-Wewer, M., Gies, D., & Schauer, R., (1981). New chromatographic system for the rapid analysis and preparation of colostrum sialyloligosaccharides. *Journal of Chromatography A, 212*(3), 313–322.
115. Velupillai, P., & Harn, D. A., (1994). Oligosaccharide-specific induction of interleukin 10 production by B220+ cells from schistosome-infected mice: A mechanism for regulation of CD4+ T-cell subsets. *Proceedings of the National Academy of Sciences, 91*(1), 18–22.
116. Viverge, D., Grimmonprez, L., Cassanas, G., Bardet, L., Bonnet, H., & Solère, M., (1985). Variations of lactose and oligosaccharides in milk from women of blood types secretor A or H, secretor Lewis, and secretor H/nonsecretor Lewis during the course of lactation. *Annals of Nutrition and Metabolism, 29*(1), 1–11.
117. Viverge, D., Grimmonprez, L., & Solere, M., (1997). Chemical characterization of sialyl oligosaccharides isolated from goat (Capra Hircus) milk. *Biochimica et Biophysica Acta, 1336*(2), 157–164.
118. Von-Itzstein, M., & Thomson, R. J., (1997). Sialic Acids and sialic acid-recognizing proteins: Drug discovery targets and potential glycopharmaceuticals. *Current Medicinal Chemistry, 4*(3), 185–210.
119. Wang, B., & Brand-Miller, J., (2003). The role and potential of sialic acid in human nutrition. *European Journal of Clinical Nutrition, 57*(11), 1351–1369.
120. Wang, B., Yu, B., Karim, M., Hu, H., Sun, Y., McGreevy, P., Petocz, P., et al., (2007). Dietary sialic acid supplementation improves learning and memory in piglets. *The American Journal of Clinical Nutrition, 85*(2), 561–569.
121. Weinborn, V., Li, Y., Shah, I. M., Yu, H., Dallas, D. C., German, J. B., Mills, D. A., et al., (2020). Production of functional mimics of human milk oligosaccharides by enzymatic glycosylation of bovine milk oligosaccharides. *International Dairy Journal, 102*, 104583.
122. Wieruszeski, J. M., Chekkor, A., Bouquelet, S., Montreuil, J., Strecker, G., Peter-Katalinic, J., & Egge, H., (1985). Structure of two new oligosaccharides isolated from human milk: Sialylated lacto-N-fucopentaoses I and II. *Carbohydrate Research, 137*, 127–138.
123. Wu, S., Tao, N., German, J. B., Grimm, R., & Lebrilla, C. B., (2010). Development of an annotated library of neutral human milk oligosaccharides. *Journal of Proteome Research, 9*(8), 4138–4151.
124. Yamashita, K., Tachibana, Y., & Kobata, A., (1977). Oligosaccharides of human milk. Structural studies of two new octasaccharides, difucosyl derivatives of para-Lacto-N-hexaose and para-Lacto-N-neohexaose. *Journal of Biological Chemistry, 252*(15), 5408–5411.
125. Yamashita, K., & Kobata, A., (1974). Oligosaccharides of human milk: V. Isolation and characterization of a new trisaccharide, 6′-Galactosyllactose. *Archives of Biochemistry and Biophysics, 161*(1), 164–170.
126. Yamashita, K., Tachibana, Y., & Kobata, A., (1977). Oligosaccharides of human milk: Structures of three lacto-N-hexaose derivatives with H-haptenic structure. *Archives of Biochemistry and Biophysics, 182*(2), 546–555.
127. Yamashita, K., Tachibana, Y., & Kobata, A., (1976). Oligosaccharides of human milk: Isolation and characterization of three new disialylfucosyl hexasaccharide. *Archives of Biochemistry and Biophysics, 174*(2), 582–591.

128. Yamashita, K., Tachibana, Y., & Kobata, A., (1976). Oligosaccharides of Human milk: Isolation and characterization of two new monosaccharides, monofucosyllacto-N-octaose, and monofucosyllacto-N-neooctaose. *Biochemistry, 15*(18), 3950–3955.
129. Zehra, S., Khambati, I., Vierhout, M., Mian, M. F., Buck, R., & Forsythe, P., (2018). Human milk oligosaccharides attenuate antigen-antibody complex induced chemokine release from human intestinal epithelial cell lines. *Journal of Food Science, 83*(2), 499–508.
130. Zivkovic, A. M., & Barile, D., (2011). Bovine milk as a source of functional oligosaccharides for improving human health. *Advances in Nutrition, 2*(3), 284–289.

CHAPTER 6

CONJUGATED LINOLEIC ACID: SYNTHESIS, PHYSIOLOGICAL, AND FUNCTIONAL ASPECTS

PRIYAE B. GAUTAM, KAMAL GANDHI, RAJAN SHARMA, NEELIMA SHARMA, and BIMLESH MANN

ABSTRACT

Conjugated linoleic acids (CLA) belong to class of sterio-isomers (*cis* or *trans*) of the octadecadienoic acid, with the double bond present at alternate position in either cis or trans configuration. Lamb, beef, and milk products are a good source of CLA. It is synthesized in the mammary gland and rumen of the lactating animal from linolenic and linoleic acid by a series of isomerization, saturation, and desaturation reactions. About 28 isomers of CLA exist naturally, of which the cis-9 trans-11 (c9t11) form is most predominant among them and is called rumenic acid, as it gets synthesized in the rumen. The Δ^9-desaturase system has a crucial part in the synthesis of c9t11 isomer in milk. CLA isomer mixture (1:1) in combination or the c9t11 and t10c12 isomers alone, are available commercially and are known to confer several health benefits. Both of these isomers have the biological potential to improve various cellular functions, body composition, and metabolism and also, proved to have impending anti-adipogenic, anti-diabetic, anti-inflammatory, anticarcinogenic, and osteogenic effects. CLA also improves the immune functions of the body. These physiological effects of CLA have been studied in various *in vitro* and *in vivo* models and also, many clinical experiments. Some data is also reported from human studies. All these evidences clearly prove that either the individual isomers of CLA, i.e., c9t11 and t10c12 or

Functional Dairy Ingredients and Nutraceuticals: Physicochemical, Technological, and Therapeutic Aspects. Megh R. Goyal, N. Veena, PhD & Santosh K. Mishra (Eds.)

a combination thereof, can be used as a protective and effective remedy in improving human health.

6.1 INTRODUCTION

Nowadays, the consumer plays emphasis on the consumption of foods which can address food security, nutritional value and health. These requirements can be fulfilled by animal products, as they can have a vital role in an individual's well-being, which may be attributed to the presence of various types of fatty acids. One such fatty acid which has gained ample amount of interest is CLA. CLAs are the inherent fatty acids in lamb, beef, and different milk products. The rumen microflora in the rumen synthesizes CLA by converting the dietary linoleic acid (LA) through the process of biohydrogenation. It is synthesized mainly during the biohydrogenation process of dietary LAs by the microflora present in the rumen of animals [65, 69].

The health benefits of CLA like effect on body composition, bone health, cardiovascular diseases, atherosclerosis, immune modulation, insulin resistance, different types of cancer and their action as mediators of inflammatory responses are well known and documented (Figure 6.1). All these health-related aspects of CLA have been proved using various *in vitro*, animal, and clinical models. In this chapter, we will discuss about the synthesis, biological effects of CLA and its mechanism.

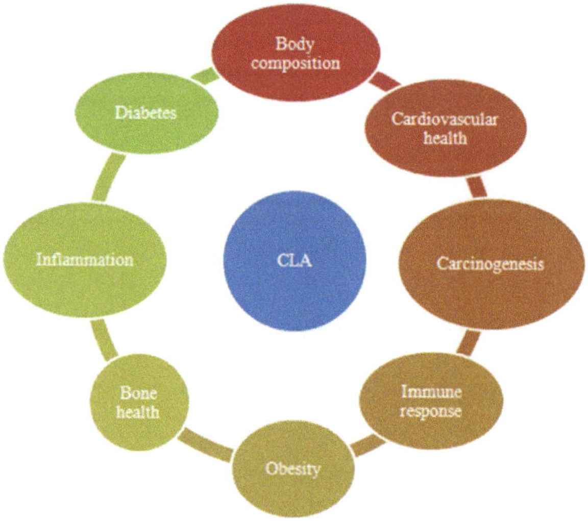

FIGURE 6.1 Role of CLA in different health complications.

This book chapter provides information on the synthesis of CLA and its physiological role. It will help them in getting a better understanding about the health-related aspects of CLA and its mechanism.

6.2 SYNTHESIS OF CLA

Pariza and Hargraves [64] discovered CLA for the first time from the grilled beef. The extract obtained from beef had certain components which exhibited anti-tumor activity in the epithelial cells of mice. Such anti-mutagenic components in the lipid fractions of the meat were isolated and characterized using spectrophotometric and chromatographic techniques [27]. They found four different isomers of LA, containing double bond at alternate position, thus being called conjugated linoleic acids (CLAs). Therefore, CLA may be defined as a group of positional [9] and geometric isomers (cis or trans) of the octadecadienoic acid (LA, 18:2), with the double bond present at alternate position in either cis or trans configuration (Figure 6.2).

The CLA isomers are a class of unsaturated fatty acids, having conjugated double bonds (i.e., a single bond separates 2 double bonds), which are present mainly at positions 11 and 13, 10 and 12, 9 and 11 or 8 and 10, in either cis (c) or trans (t) configuration. Till date, 28 different types of CLA isomers have been reported. Out of which, the most prominent (occurrence around 80%) is the c9t11 CLA isomer of the total dietary CLA [7]. Two isomers of CLA namely c9t11 and t10c12 endow to the health-related benefits. Most of the commonly used preparation of CLA's contains approximately equivalent amount of the c9t11 and t10c12 isomers.

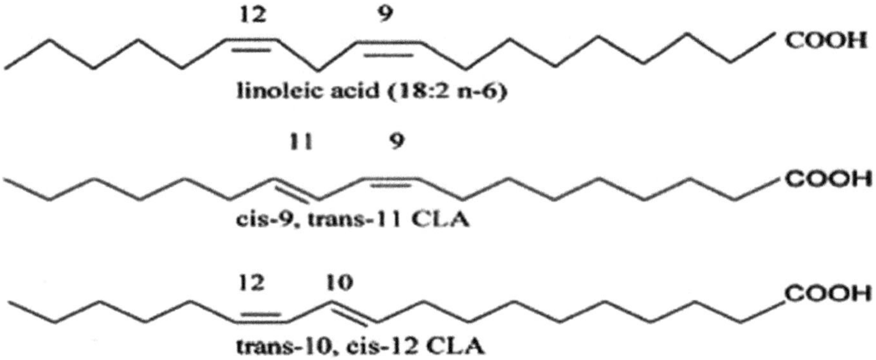

FIGURE 6.2 Structures and position of double bonds in LA and CLA [9].

CLA is either formed in the rumen by the action of rumen microflora on the dietary polyunsaturated fatty acids (PUFAs) or by the desaturation of vaccenic acid $C_{18:1}$ (t11) under the influence of stearoyl-CoA desaturase. The best-known microorganism for the biohydrogenation is *Butyrivibrio fibrosolvens* [23]. LA, when biohydrogenated in the rumen, leads to the formation of the c9t11 C18:2 isomer. This CLA isomer is called rumenic acid (c9t11-octadecadienoic acid) [49]. The enzyme Δ^{12} cis, Δ^{11} trans isomerase catalyzes this reaction. The rumenic acid thus produced is further partially or completely hydrogenated to vaccenic acid and stearic acid, respectively [94]. This process of biohydrogenation occurs completely, but a part of the rumenic acid passes to the mammary gland via the blood stream and thus is reflected in milk.

Another pathway for the synthesis of CLA is, the vaccenic acid produced in the rumen, enters into the mammary gland, which is then further desaturated to c9t11 isomer of CLA by the action of enzyme Δ^9 desaturase (Figure 6.3). For commercial purposes, it can be prepared by subjecting the LA to heat under the alkaline conditions, or by partially hydrogenating the LA [22], or endogenously by the action of Δ^9-desaturase enzyme on the dietary vaccenic acid to produce c9t11 CLA isomer [108].

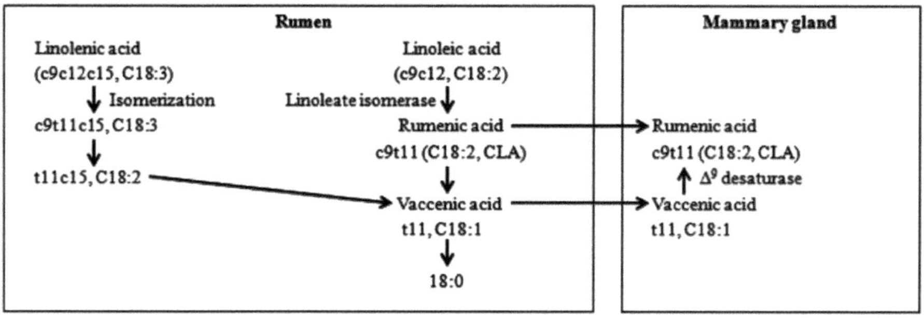

FIGURE 6.3 Pathways for CLA synthesis in mammary gland and rumen.

6.3 FUNCTIONAL BENEFITS OF CLA

6.3.1 EFFECT ON BODY COMPOSITION

It has been reported that intake of 0.5% CLA (1:1 mixture of c9t11 and t10c12) ensued in a substantial reduction of the body fat deposits with subsequently increasing the lean body mass of rats. Such changes are chiefly associated with the elevated fatty acid oxidation, increased lipolysis or reduction in the fatty acid uptake efficiency by the adipocytes [90]. It has been observed that the administration of CLA causes an increase in the activity of an enzyme called carnitine palmitoyltransferase (CPT). It is also hypothesized that CLA (t10c12 isomer) curtails the action of an enzyme (i.e., lipoprotein lipase (LPL)) [68], which maintains (reducing) the intracellular concentrations of glycerol and triacylglycerol by regulating uptake of the fatty acids from the blood into the adipose tissues [66].

Dietary CLA supplementation leads to stimulation of β-oxidation of fatty acid and inhibiting the triglyceride accumulation in the adipose tissue. Pigs which were fed on CLA showed lesser development of body fat with improved feed conversion efficiency [15]. It has been found that mice fed on CLA had reduced carcass fat content [100]. Previously, many studies have reported that this effect is primarily induced by the t10c12 isomer only and not by the c9t11 isomer. Similar results were obtained in both *in vitro* 3T3-L1 and *in vivo* experiments for these isomers [68]. Hamsters when fed a hyper-cholesterolemic diet comprising of 1% CLA (mixture of c9t11 and t10c12), 0.2% c9t11 CLA or 1% LA, the lowest weight gain was observed in the CLA isomer mix fed group [25].

Reduction in certain obesity-related complications/parameters like increase in the body mass, epididymal body fat and white corpus adiposum mass, etc., has been observed when t10c12 CLA isomer was fed to the genetically obese mice for 3 weeks [29, 95]. CLA down-regulates the expression of the leptin mRNA in the peritoneal fat pads while exercise stimulates or fastens the oxygen consumption and energy expenditure [10]. Both of these effects ultimately control the whole metabolic activity.

Another study in mice reported that incorporation of the mixture of CLA isomer for a period of 10 weeks; led to a significant decrease in the fat mass, prohibited the aggregated loss of muscle mass and preserved the elevated gastrocnemius and extensor muscle mass. This is because CLA intake decreased the adipocyte size, energy intake along with inhibiting the enzymes implicated with the metabolism of fatty acid and lipogenesis [86, 99]. This hypothesis has been further strengthened by various studies, suggesting that the isomer t10c12 increases the oxygen utilization and energy dissipation, to a greater extent than the c9t11 isomer [67]. A description of different types of effects in different models is described in Table 6.1.

TABLE 6.1 Impact of c9t11 and t10c12 CLA on the Body Composition

Animal Model	Results	References
Pigs/ rats/ mice	• Lean mass increases • Body fat decreases	[31, 66, 91, 99, 100]
Mice	• Consumption of diet enriched with t10c12; fat mass decreases *vis-a-vis* to the diet fortified with c9t11 • Body weight and epididymal fat mass decreases, on intake of the diet lacking in essential fatty acid.	[28, 68]
Zucker diabetic fatty rats	• Weight gain and fat mass decreases when fed with CLA	[77]
Atherogenic diet-fed hamsters	• CLA fed group decreases weight gain, *vis-à-vis* to c9t11 fed group	[25]
	• Fat mass decreases in c9t11 and t10c12 group but no impact on body weight	[57]
Stearoyl-CoA desaturase 1 null mice	• Decreased over weight with t10c12, free of stearoyl-CoA desaturase 1 expression	[34]
High-lipid-diet-supplemented mice (exercised)	• Body weight and body fat decreases but lean mass increases	[10]
Rats (exercised)	• Body mass, body composition, adipose weight remains unchanged	[20, 53]
Healthy exercising humans	• Supplementation of CLA for a period of 12 weeks, body fat decreases with a increases in lean mass while the overall body weight remain unchanged	[87]
Healthy volunteers	• Fat mass decreases	[55]

CLA causes a significant reduction in the food intake, as it can exert an inhibitory effect on certain genes which regulate the appetite, retards certain orexigenic neuropeptides like NPY and agouti-related proteins, which can increase the food intake [3]. Another study carried on the healthy adult women was to visualize the effect of CLA (3 g daily) intake on body fat, percent fat mass and fat-free mass. But no significant effect was observed in sedentary young women [71]. However, in the case of healthy exercising humans (CLA, when fed @ 1.8 g daily) and in heavy and obese person (fed at 1.7, 3.4, 5.1, and 6.8 g of CLA daily) for 3 months significantly decreased fat mass [87]. CLA supplementation for 4 weeks reduced abdominal fat significantly (i.e., sagittal abdominal diameter in men having abdominal visceral obesity) [76]. Furthermore, another study in athletes reported that CLA when fed even at a lower level (1.8–3.4 g/day) resulted in reducing the body fat mass. CLA (4.2 g/day) intake for 1 month had a significant effect in reducing adipose tissue and also, abdominal fat mass in cardiovascular disease's patients. This effect is mainly attributed to increase in the lean body weight without an increase in the total body weight, along with an increase in the metabolic rate [75].

6.3.2 EFFECT ON CARDIOVASCULAR HEALTH

CLA exhibits a positive impact on cardiovascular health. In the initial studies, it was observed that aortas of CLA-fed (0.5 g/day) rabbits along with 0.1% cholesterol as atherogenic diet had lower atherosclerosis with marked fallout in LDL and total cholesterol, LDL to HDL ratio. Also CLA fed rabbits developed fewer lesions in their abdominal aorta than control rabbits [44]. Nicolosi et al. [58] proclaimed significant fallout in plasma concentration of triglycerides, total cholesterol and low-density lipoproteins (LDL) when the hamsters were fed with CLA. CLA at even 0.05% was sufficient enough to trim down the extremity of abrasions caused due to atherosclerosis in rabbits [38] while a 30% regression in the atherosclerotic lesions was found at 1% CLA intake [39]. Another study reported that treatment of the hamsters fed with hyper-cholesterolemic diet (0.12% cholesterol and 10% saturated lipid) with CLA (0.06%, 0.11% or 1.1%) for 11 weeks showed a downward slide of the atherosclerosis in hamsters [63]. The c9t11 isomer was more effective in arresting the atherosclerotic lesions [9].

Feeding 20% butterfat (high in saturated fat) for 12 weeks to the hamsters substantially minimized the deposition of lipid molecules in the arteries *vis-a-vis* to the diet fortified with 1% rumenic acid [92]. Many studies in mice reported that intake of 0.6% CLA or 2:8 mixture of t10c12:c9t11 CLA isomer decreased aortic fatty streaks formation and prevented progression of atherosclerosis [89].

In CLA fed hypercholesterolemic hamsters and rabbits, the serum levels of non-high-density lipoproteins, triglycerides, and total cholesterol significantly declined without affecting the high-density lipoprotein cholesterol (HDL-C) [59]. It has been depicted in hamsters that t10c12 isomer is actively involved in modifying the lipid profiles while the c9t11 isomer significantly reduces the non-HDL-C/HDL-C ratio when administered at 1% level of the diet [92]. The serum levels of HDL-C improved while that of total cholesterol and triglycerides levels reduced when the atherosclerotic mice were supplied with CLA or modified structured lipids having 0.6% CLA [43].

The main mechanism targeting above mentioned effects comprises CLA's activating action on peroxisome proliferator-activated receptors (PPARs). PPARs are a special type of nuclear receptors which get activated by certain ligands; they help in controlling the lipid and glucose homeostasis by regulating the gene expression. PPAR-α and PPAR-γ have been found to hinder the advancement of atherosclerosis in mouse deficient with LDL receptor. Peroxisome proliferator (PP) has anti-atherosclerotic properties which require PPAR-α. CLA is one of the members of the PP family and acts as a PPAR-α activator by induction of the peroxisomal enzyme acyl-CoA oxidase (ACO) at the protein and mRNA level in the mice liver when fed with CLA. CLA isomer, in particular the c9t11 isomer is a high affinity ligand and a strong catalyst for the activation of PPARα [108]. Hence, the anti-inflammatory and lipid-modifying effect on the arterial wall helps in reducing atherosclerosis complications [6].

CLA, especially c9t11 promotes the storage of the lipids along with LPL, a vital enzyme linked with the elimination of the triacylglycerol-rich lipoproteins. In another study by Ha et al. [26], on incubation of CLA at 40°C for 15 days, it was less prone to peroxide formation than LA. Also, CLA was found to be more protective than ascorbic acid, α-tocopherol or butylated hydroxyl-toluene (BHT). CLA isomers have the ability to donate hydrogen atoms and stop the oxidation. Spectrophotometric and photoemission methods indicated the free-radical quenching property of CLA [19].

Another hypothesis supporting the anti-atherogenic property of CLA is: it decreases apoB100 production and secretion from the liver. Cholesterol level in the liver is an important regulating factor for the production of apoB100, a component mandatory for the generation of the atherogenic very low and LDLs. This study also certified that when human hepatocyte HepG2 cells were enriched with CLA, a lower cholesterol level was observed. This reduction in cholesterol level leads to the increased degradation of apoB100 cells [63].

White et al. [103] showed that the generation and the release of very LDL is dependent on the rate at which apoB100 is degraded in the hepatic cells. Cholesterol/cholesterol content (the substrate) and the intracellular triglycerides in the hepatic cells regulate the degradation of apoB100. Thus, more the amount of intracellular lipids, more resistant apoB100 is to the proteolytic enzymes, as a result it does not loses its identity, finally assembling into very low density lipoprotein and secreted. CLA also limits the production of foam cells. Foam cells are important for atheroma development, is formed when the oxidized LDL is consumed by macrophages, through CD36 and SR-A1, which are a type of scavenger receptors [41].

CLA acts as an anti-atherogenic agent, as they limit the formation of the foam cells and regulate the RCT expressing genes, which subsequently increase the elimination of cholesterol from the plasma [17]. CLA also controls atherosclerosis by regulating a gene called PGC-1α. This gene can lead to the co-activation of numerous nuclear receptors and is also proved as an atheroprotective mediator through various *in vivo* studies. The increase in the expression of this gene is inversely related with development of lesions. PGC-1α also reduces the uptake of oxidized LDL and foam cell formation, signifying that the gene expression is inversely related to the atheroma formation or progression [51]. Thus, the above-mentioned evidence suggests CLA as drug molecules for patients suffering from different health-related disorders.

6.3.3 EFFECT ON CARCINOGENESIS

6.3.3.1 EFFECT ON GASTROINTESTINAL AND COLON CANCER

The most important effect of CLA is inhibition of HT-29 colon cancer cell growth as reported in *in-vitro* studies. This is mainly attributed to the repression of procreation of cell while stimulation of apoptosis through the down-regulation of ErbB3 signaling and PI3-kinase/Akt pathway. Another study in Caco2 colon cancer cells proved that out of c9t11 and t10c12, the most potent inhibitor is t10c12 [37]. In mice, CLA prevented the initiation of tumorigenesis in their forestomach as induced by benzo (a)pyrene (BAP), thus confirming better antioxidant activity of CLA than that of α-tocopherol or BHT. This tumor incidences are reduced mainly through induction of apoptosis. The main mechanism of apoptosis is reduction in the mucosal levels of arachidonic acid, thromboxane B2 and prostaglandin E2 (PGE_2) which is mainly responsible for decreasing the tumor incidence [26].

Another study in rats reported that oncogenesis in the mammary and colon cells, as induced by 7, 12-dimethylbenzanthracene and 1, 2-dimethylhydrazine was inhibited when supplemented with CLA [13]. In one of the studies, Colo320 (human colon cancer cells) and MKN28 (human gastric cancer cells) when injected peritoneally in BALB/c *nu/nu* mice, the metastatic foci developed in the peritoneum decreased with ingestion of CLA. This study revealed that CLA can efficiently act at the metastatic stage in human enteric and colon malignant cells [40]. In the Swedish mammography cohort study, an increase in high-fat dairy foods (enriched in CLA) for two servings daily, decreased the danger of bowel and distal colon cancer by 13% and 34%, respectively [42].

6.3.3.2 EFFECT ON BREAST CANCER

CLA showed growth-inhibitory and cytotoxic effect on MCF-7 breast cancer cells. It commands cell cycle inhibition via affecting estrogen receptor in MCF-7 cells [16]. Another mechanism is mediated through lipo-oxygenase inhibition which induces lipid peroxidation hence cytotoxic for cells [61]. CLA especially the c9t11 isomer, bound to the milk-fat triacylglycerol possesses the growth suppressing effects, related mainly to the modification of the arachidonic acid disposition and altering or modifying the prostaglandin profile. The mode of action for both the isomers is different on the MCF-7 cells. The t10c12 isomer inhibits cell propagation influenced by hormones like insulin and estrogen and down-regulates the estrogen receptor at both protein and mRNA levels. So, this anti-estrogenic property contributes to the antitumor property of t10c12 CLA isomer in the breast cancer cells [84].

Pretreatment with CLA diets primarily with t10c12 isomer appeared to be more effective, hence being the more effective isomer in inhibiting the occurrence of the tumor and their total weight. CLA even prevented growth and metastasis of human breast adenocarcinoma cells in severe combined immune deficient (SCID) rats [93].

6.3.3.3 EFFECT ON PROSTATE CANCER

Regular intake of CLA in diet @1% level for 14 weeks by the SCID mice, showed a marked reduction in local tumor load and metastasis to other organs when the DU-145 human prostatic carcinoma cells were injected in them. The anti-proliferative activity of both the isomers is different and is described below [12]:

1. Mechanisms of t10c12 isomer:
   i. caspase-dependent apoptosis;
   ii. Decrease in Bcl-2 expression;
   iii. Increase in p21 (WAF/Cip1) expression.

2. Mechanism of c9t11 isomer:
   i. Altering the eicosanoid production by manipulating the 5-lipo-oxygenase and cyclooxygenase expression;
   ii. Alternation in amino acid metabolism.

6.3.3.4 EFFECT ON ANGIOGENESIS

Angiogenesis has been found to be inhibited by CLA under *in vivo* conditions in implanted rat breast tumors. The mechanism involved in lowering the level of vascular endothelial growth factor (VEGF) and its receptor Flk-1 in serum and mammary gland. Endothelial cell propagation and synthesis of DNA decreased due to CLA in an *in vitro* study [18]. A study reported that t10c12 CLA inhibits breast cancer by decreasing leptin, the pro-angiogenic hormone [10].

6.3.4 EFFECT OF CLA ON OBESITY AND DIABETES

Supplementation of CLA in the diet may lead to a better management of diabetes, generally the type II diabetes [8]. Intake of CLA isomer mix (50:50) at 1.5% of the diet-controlled attenuated hyperinsulinemia and impaired glucose tolerance; in contrast to the intake of 91% enriched c9t11 doesn't have any such result [31]. So, these beneficial effects are mainly contributed by the t10c12 isomer. The mechanism of action involves decreasing the liver weight and mRNA expression in white adipose tissues. In mice, it was reported that intake of a diet having 4% fat/100 g diet with 1% CLA, was efficient in significantly reducing the body fat mass accompanied with lipoatrophic diabetes mellitus syndrome, i.e., insulin resistance and hepatomegaly [91].

Norris et al. [60] conducted a study on postmenopausal obese women and published that supplementing CLA @ 6.4 g/day, substantially lowered the body weight and the total adipose mass. So, the above studies clearly indicate that supplementation of CLA along with fat content in the diet plays a vital role in making CLA as an anti-diabetic agent. The action of CLA on diabetes is due to binding to PPAR-γ and activation of PPAR receptor which are involved in the insulin metabolism. It has similarity to thiazolidinedione drugs, a type of drug similar to the insulin-sensitizing agents which have the ability to improve glucose utilization without stimulating the release of insulin. It also serves as an agent against diabetes through enhancement of the adipocyte differentiation [108].

Leukotriene B4 (LTB4), a lipoxygenase product of arachidonic acid which can promote insulin resistance in obese mice is reduced by CLA [48, 82]. The t10c12 isomer of CLA restricts the uptake of glucose and fatty acid and their metabolism by controlling the expression or activity of PPAR-γ and its downstream targets, which significantly regulate the metabolism of the fatty acid and glucose [11]. Another possible mechanism of CLA to act as an anti-diabetic agent is that it alleviates hyperinsulinemia (observed in ZDF rats), via increasing the secretion of adiponectin which enhances insulin sensitivity [56].

6.3.5 ANTI-INFLAMMATORY ACTION OF CLA

Initially, the studies carried aimed at scrutinizing the anti-inflammatory effects of CLA on chicks and rats to which endotoxin (lipopolysaccharide (LPS)) was injected [14, 52]. The results from these studies were inconclusive. Some studies revealed that CLA fed to the mice had higher levels of interleukin-2 (IL-2) [30, 104] with a concomitant decrease in interleukin-6 (IL-6) and transforming necrosis factor-α (TNF-α) [73] as compared to the control group. One human trial indicated that feeding CLA @ 3 g/day for 12 weeks enhanced plasma concentration of immunoglobulin (IgA and IgM) and IL-10 (anti-inflammatory cytokine), with subsequent drop in the amount of IgE and pro-inflammatory cytokines (PIC) (IL-1β and TNF-α) [81].

Another study showed that on feeding the mixture of CLA isomers (c9t11 and t10c12 @ 50:50 or 80:20) to the healthy human subjects increased the effect of the vaccine for hepatitis B while no change was noted in the activity of natural killer (NK) cells, IL (–2, –4, –6), TNF-α and PGE_2 [1]. A similar result was observed when CLA was fed to moderate overweight subjects [74]. The generation of certain PGE_2, TNF-α, IL (–1β and –6) and nitric oxide in macrophages, activated with gamma interferon (IFN-γ) *in vitro*, decreased when treated with CLA [107]. Similar effects were also obtained with human aortic endothelial cells *in vitro*. In co-cultured human bronchial epithelial cells and eosinophils, c9t11 isomer effectively inhibited the formation of inflammatory mediator, i.e., eosinophil cationic protein formation than t10c12. But another study reported that both isomers (c9t11 and t10c12) suppressed the generation and activity of TNF-α, IL (–6 and –1β) and also amplified the stimulation of PPAR-γ. It was also shown by some studies that CLA and medium workout when combined with high-fat diet lowered TNF-α, leptin levels, fat mass and IL-6 in the serum of sedentary mice. CLA also acts as a protective agent against anorexia and LPS-induced body weight wasting. In addition, CLA reduces lipid-derived inflammatory mediators [102].

CLA reduced the inflammation via inhibiting antigen-induced histamine. Additionally, it was also concluded that no significant difference existed in the activity between c9t11 and t10c12 isomers w.r.t. their activity or effect towards the immune function [101]. CLA also enhanced peripheral bone marrow cells (PBMC) propagation in response to inhibitor, concanavalin A and inflated the generation of the antibody against sheep red blood cells. Moreover, 1% CLA in the diet decreased both systemic and peripheral blood lymphocytic PGE_2 synthesis. CLA is also protective against auto-immunity. A trial conducted on a healthy human depicted that ingestion of both 80% c9t11 and t10c12 isomers for two months had a positive impact on immune functions. CLA also decreased delayed-type hypersensitivity. But t10c12 isomer is found to be a more potent anti-inflammatory agent than the c9t11 isomer [4].

The neuroprotective (anti-oxidative or anti-inflammatory) effect for the combo of CLA isomers (c9t11 and t10c12) was studied in mice having neuro-psychiatric disease and was found that CLA can improve the antioxidant and detoxifying defenses by activating the nuclear factor-E2-relate factor [54].

The possible reason behind the anti-inflammatory properties of CLA is attributed to:

- CLA regulates the production of arachidonic acid derived eicosanoids. The eicosanoids derived from CLA exhibit anti-inflammatory activity [101]. Also, CLA has the ability to alter eicosanoid signaling. This altered signaling could affect cytokine synthesis [32, 65]. As CLA inhibits the synthesis of PGE_2 affecting a number of biological functions like modification of cell membrane by changing the action of several membrane proteins which act as enzymes, transporters, ion channels, and receptors signal transducers [21].
- CLA isomers upsurge or activate PPAR-γ factor and a decline in the IFN-γ or LPS-induced PIC [50, 107]. The macrophages express the PPAR-γ factor, which is a critical factor in regulating the level of inflammation. PPARs being fatty acids receptors regulate the expression of the genes which are responsible for maintaining energy balance and immune function. PPARs are involved in the expression

of various genes which are related with procreation of lymphocytes and monocytes, apoptosis, and inflammation. Synthetic PPAR-γ agonists act in a similar way as CLA does, by obstructing the generation of the PIC, including TNF-α in monocytes [33] and also affects the differentiation of monocytes and macrophages [92].

- CLA isomers have been found to reduce LPS-induced nuclear transcription factor activity in mouse macrophage cell lines [13]. It also inhibits COX2 (cyclooxygenase 2) and PGE_2 production [45].
- CLA can also retard the rate of DNA damage, protein oxidation, and lipid peroxidation by enhancing the activity of certain anti-oxidative enzymes present in the tissue homogenate, erythrocyte lysate, and plasma obtained from rats [78, 80]. Two isomers of CLA, i.e., punicic acid and eleostearic acid, were found to reduce the lipid and lipoprotein peroxidation, oxidative stress by reviving the activities and levels of certain anti-oxidative and pro-inflammatory enzymes like catalase, glutathione, superoxide dismutase in the blood, pancreas, and erythrocyte lysate obtained from streptozotocin-induced diabetes in rats [79].

Administration of CLA significantly reduced the oxidative stress markers (carbonyl proteins and thiobarbituric acid reactive compounds), with an increment in glutathione levels in the spleen, polymorphonuclear leukocytes, and thymus obtained from CLA and acrolein fed rats [5].

6.3.6 ROLE OF CLA ON BONE HEALTH

CLA has a significant impact on the bone health. The mobilization of calcium in the cells increased when the intestinal-like CaCo-2 cells, were treated with CLA. It monitors bone development by regulating insulin-like growth factors (IGF-I and IGFBP). Another potential mechanism is enhanced calcium absorption [36]. Supplementation of CLA for 8 weeks to an ovariectomized 12-month-old Fisher rat observed a decrease in the bone resorption rate [35]. A study in postmenopausal women showed that CLA intake and bone mineral density values are directly co-related. So, CLA can be used as a hormone replacement therapy in osteoporotic patients. This is because CLA supplementation is capable of reducing osteoclastogenesis [107].

Watkins et al. [97] reported that bone formation occurred at a higher rate in chicks given CLA rich butterfat vis-à-vis those fed on n-6 fatty acids rich diet. A study carried on 52 patients suffering from rheumatoid arthritis indicated that in the various bone markers between CLA fed and control group was observed to be significantly different. Osteocalcin, IGF-1 increased significantly while telopeptides c decreased significantly. Osteocalcin is considered as one of the bone formation marker, which attenuates the bone formation by increasing the mineralization of bones while telopeptides-C are considered as markers for bone resorption [2].

Consumption of CLA decreases the amount of body fat with a subsequent increase in the ash, protein, and water content of the body. This increase in the ash content of the body indicates that CLA effects in the prevention or treatment of bone degeneration disorders like osteoporosis [72].

CLA improves the bone health by altering the following physiological actions of the body:

- Altered prostaglandin E_2-dependent bone resorption in the presence of PUFAs;
- Decreased osteoclastogenesis
- Dietary CLA can regulate the action and expression of COX enzymes, hence it reduces the prostaglandin-dependent bone resorption [85, 96].
- Downregulation of PGE_2, thus influencing IGFs, IGFBPs, and bone fabrication [108];
- Increase synthesis of collagen;
- Increases the expression and circulation of leptin. Leptin affects the central nervous system, thus stimulates the bone marrow stromal cells with a subsequent reduction in bone resorption by inhibiting the osteoclast cells and thus increase in the bone formation [2];
- It regulates the process of osteoclastogenesis and adipogenesis in mice thus preventing the age-induced bone loss [48];
- Limiting the secretion of LTB4 (an important and strong bone resorption factor) [24, 82], a lipoxygenase product of arachidonic acid [24];
- Promotes osteogenesis;
- Reduce the release of parathyroid hormone [98];
- Reduced PIC, thus reduced bone resorption [9];
- Reduction in the level of arachidonic acid and a subsequent reduction in arachidonate derived eicosanoids [47].

6.3.7 EFFECT OF CLA ON IMMUNE RESPONSE

When healthy human beings were given 80:20 mixture of c9t11: t10c12-CLA isomers for 2 months, there was a decrease in T-cell lymphocyte activation, which had a positive impact on the immune function [90]. In human trial research by Penedo et al. [70] showed that c9t11 isomer of CLA in butter exhibited immune-modifying properties on the healthy adolescents with a subsequent fall out in the generation of the pro-inflammatory biomarkers like the extracellular matrix metalloproteinases MMP-9 and MMP-2; the nuclear transcription factor; and the cytokines TNFα, IL-2 and IL-8, which are related to overweight and obesity. Cook et al. [14] in their study found that, CLA influences the immune response of the aged mammals. Chicks which were fed on CLA and injected with the endotoxin LPS showed significant growth, although those kept deprived of CLA, were unable to grow or kept on losing their weight following LPS injection. The possible role of CLA in modulating immune response may be related to a decrease in the adverse effect of immune-mediated catabolism, thus enhancing immune response. It increases protective Ig production [106], IgA, IgM, IL-10 while decreasing IgE, TNF-α and IL-1β production [7]. CLA can prevent the catabolic losses without affecting the generation of adaptive immunity could improve growth and development.

Song et al. [81] conducted an experiment on healthy men and women, to study the effect of this fatty acid supplementation on the immune system. The subjects were divided and were given oleic acid rich sunflower oil or CLA (equal ratios of c9t11: t10c12 isomers). It was found that CLA decreased the TNF-α, IFN-γ, IgE, and increased the levels of IgM and IL-10. Another study carried by Sugano et al. [83] found a higher concentration of Igs (A, G, and M) in the mesenteric lymph cells from CLA fed rats than that of control rats with a subsequent reduction in the level of IgE. Such studies clearly indicate that CLA can regulate class specific production of immunoglobulins [62].

IgE is generally related with the allergic reactions in humans, the reduction in IgE by CLA shows its anti-allergic activity. The possible hypothesis supporting the anti-allergic potential of CLA is due to the fact that CLA helps in maintaining a balance between T helper cells, i.e., Th1 and Th2 responses. The Th1 cells trigger the fabrication of IgG_2, a isotype antibodies, while Th2 stimulates the production of other types of antibodies like IgG_1 and IgE by B cells. This balance maintained by CLA can be correlated by a decrease in IgE was observed with a subsequent increase in IgM, IgA, and IgG [62]. Such immune specific response is exhibited by t10c12-isomer rather than c9t11-isomer, which has no effect [105].

6.4 SUMMARY

Nowadays, hyperlipidemia, cancer, osteoporosis, obesity, and diabetes are some of the major health problems encountered in the western society especially during the aging process. The major cause for such problems is mainly due the dietary habits and sedentary lifestyle. Some of the latest research reports have concentrated on the effects of supplementation of CLA with respect to body composition, cardiovascular health, diabetes, and cancer, etc. Many research studies carried out in sedentary, overweight, healthful, and exercised persons have illustrated that CLA has a significant role in lowering body fat and atherogenic lipids without altering the body-weight significantly. Consumption of CLA has been found to either decrease or prevent the growth of preformed atherosclerotic lesions in animal studies. Intake of CLA isomers (c9t11 and t10c12) in equal proportion or t10c12 isomer alone prevented the gastrointestinal tract cancer and colon cancer. CLA also has a protective effect against auto-immunity. It also has an inhibitory effects on the pro-inflammatory cytokines was mainly attributable to the t10c12 form while the other major isomer, i.e., c9t11 CLA contributes in improving the cardiovascular health effectively. CLAs also have beneficial effects on bone metabolism. However, more research needs to be done worldwide involving patients all over the world, so as to draw conclusive results. As both the isomers of CLA have different health improving properties, most of the research is done by administration of the crude mixture of CLA. Thus, various factors such as nature of CLA administered, procedure or form of administration (in free or triglyceride form), age, sex, gender, conditions during trials are needed to be taken into account so as to recommend CLA as a true nutraceuticals/functional food. Still in short, with the present literature available, it can be concluded that CLA supplementation along with diet can improve the deteriorating status of health in western countries and can prevent different types of disorders responsible for enhanced mortality rate in modern time.

KEYWORDS

- **anti-inflammatory**
- **anticarcinogenic**
- **atherosclerosis**
- **bone health**
- **CLA isomers**
- **conjugated linolenic acid**
- **eicosanoids**

REFERENCES

1. Albers, R., Vander-Wielen, R. P. J., Brink, E. J., Hendriks, H. F. J., Dorovska-Taran, V. N., & Mohede, I. C. M., (2003). Effects of cis-9, trans-11 and trans-10, cis-12 conjugated linoleic acid (CLA) isomers on immune function in healthy men. *European Journal of Clinical Nutrition, 57*(4), 595–603.
2. Aryaeian, N., Shahram, F., & Djalali, M., (2016). CLA has a useful effect on bone markers in patients with rheumatoid arthritis. *Lipids, 51*(12), 1397–1405.
3. Asmaa, G. A., Mohamed, A. B., Abdou, S. E., & Ghada, S., (2018). A study on the immunological changes in adult male obese rats and the possible immune modulator effect of conjugated linoleic acid supplementation. *The Medical Journal of Cairo University, 86*, 4115–4124.
4. Avenell, A., Richmond, P. R., Lean, M. E., & Reid, D. M., (1994). Bone loss associated with a high fiber weight-reduction diet in postmenopausal women. *European Journal of Clinical Nutrition, 48*(8), 561–566.
5. Aydın, B., Atlı, S. Z., & Sekeroglu, V., (2018). Acrolein-induced oxidative stress and genotoxicity in rats: Protective effects of whey protein and conjugated linoleic acid. *Drug and Chemical Toxicology, 41*(2), 225–231.
6. Barbier, O., Torra, I. P., & Duguay, Y., (2002). Pleiotropic actions of peroxisome proliferator-activated receptors in lipid metabolism and atherosclerosis. *Arteriosclerosis, Thrombosis, and Vascular Biology, 22*(5), 717–726.
7. Belda, B. J., Lee, Y., & Heuvel, J. P. V., (2010). Conjugated linoleic acids and inflammation: Isomer-and tissue-specific responses. *Clinical Lipidology, 5*(5), 699–717.
8. Benjamin, S., & Spener, F., (2009). Conjugated linoleic acids as functional food: An insight into their health benefits. *Nutrition & Metabolism, 6*(1), 36.
9. Bhattacharya, A., Banu, J., Rahman, M., Causey, J., & Fernandes, G., (2006). Biological effects of conjugated linoleic acids in health and disease. *The Journal of Nutritional Biochemistry, 17*(12), 789–810.
10. Bhattacharya, A., Rahman, M. M., & Sun, D., (2005). The combination of dietary conjugated linoleic acid and treadmill exercise lowers gain in body fat mass and enhances lean body mass in high fat-fed male BALB/C mice. *The Journal of Nutrition, 135*(5), 1124–1130.
11. Brown, J. M., Boysen, M. S., Chung, S., Fabiyi, O., Morrison, R. F., Mandrup, S., & McIntosh, M. K., (2004). Conjugated linoleic acid induces human adipocyte delipidation: Autocrine/paracrine regulation of MEK/ERK signaling by adipocytokines. *Journal of Biological Chemistry, 279*(25), 26735–26747.
12. Cesano, A., Visonneau, S., Scimeca, J. A., Kritchevsky, D., & Santoli, D., (1998). Opposite effects of linoleic acid and conjugated linoleic acid on human prostatic cancer in SCID mice. *Anticancer Research, 18*(3A), 1429–1434.
13. Cheng, J. L., Futakuchi, M., & Ogawa, K., (2003). Dose-response study of conjugated fatty acid derived from safflower oil on mammary and colon carcinogenesis pretreated with 7,12-dimethylbenz[a]anthracene (DMBA) and 1,2-dimethylhydrazine (DMH) in female Sprague-Dawley rats. *Cancer Letters, 196*(2), 161–168.
14. Cook, M. E., Miller, C. C., Park, Y., & Pariza, M., (1993). Immune modulation by altered nutrient metabolism: Nutritional control of immune-induced growth depression. *Poultry Science, 72*(7), 1301–1305.

15. Dugan, M. E. R., Aalhus, J. L., Schaefer, A. L., & Kramer, J. K. G., (1997). The effect of conjugated linoleic acid on fat to lean repartitioning and feed conversion in pigs. *Canadian Journal of Animal Science, 77*(4), 723–725.
16. Durgam, V. R., & Fernandes, G., (1997). The growth inhibitory effect of conjugated linoleic acid on MCF-7 cells is related to estrogen response system. *Cancer Letters, 116*(2), 121–30.
17. Ecker, J., Langmann, T., Moehle, C., & Schmitz, G., (2007). Isomer specific effects of conjugated linoleic acid on macrophage ABCG1 transcription by a SREBP-1c dependent mechanism. *Biochemical and Biophysical Research Communications, 352*(3), 805–811.
18. Eder, K., Schleser, S., Becker, K., & Korting, R., (2003). Conjugated linoleic acids lower the release of eicosanoids and nitric oxide from human aortic endothelial cells. *The Journal of Nutrition, 133*(12), 4083–4089.
19. Fagali, N., & Catala, A., (2008). Antioxidant activity of conjugated linoleic acid isomers, linoleic acid and its methyl ester determined by photoemission and DPPH techniques. *Biophysical Chemistry, 137*(1), 56–62.
20. Faulconnier, Y., Arnal, M. A., Patureau, M. P., Chardigny, J. M., & Chilliard, Y., (2004). Isomers of conjugated linoleic acid decrease plasma lipids and stimulate adipose tissue lipogenesis without changing adipose weight in post-prandial adult sedentary or trained Wistar rat. *The Journal of Nutritional Biochemistry, 15*(12), 741–748.
21. Field, C. J., & Schley, P. D., (2004). Evidence for potential mechanisms for the effect of conjugated linoleic acid on tumor metabolism and immune function: Lessons from n-3 fatty acids. *The American Journal of Clinical Nutrition, 79*(6), 1190S–1198S.
22. Fritsche, J., Rickert, R., & Steinhart, H., (1999). Formation, contents, and estimation of daily intake of conjugated linoleic acid isomers and trans-fatty acids in foods. Chapter 30. In: Yurawecz, M. P., Mossoba, M. M., Kramer, J. K. G., Pariza, W. M., & Nelson, G. J., (eds.), *Advances in Conjugated Linoleic Acid Research* (Vol. 1, pp. 378–396). Champaign Illinois-USA: AOCS Press.
23. Fuke, G., & Nornberg, J. L., (2017). Systematic evaluation on the effectiveness of conjugated linoleic acid in human health. *Critical Reviews in Food Science and Nutrition, 57*(1), 1–7.
24. Garcia, C., Boyce, B. F., Gilles, J., Dallas, M., Qiao, M., Mundy, G. R., & Bonewald, L. F., (1996). Leukotriene B4 stimulates osteoclastic bone resorption both *in vitro* and *in vivo. Journal of Bone and Mineral Research, 1*(11), 1619–1627.
25. Gavino, V. C., Gavino, G., Leblanc, M. J., & Tuchweber, B., (2000). An isomeric mixture of conjugated linoleic acids but not pure cis-9, trans-11-octadecadienoic acid affects body weight gain and plasma lipids in hamsters. *The Journal of Nutrition, 130*(1), 27–29.
26. Ha, Y. L., Storkson, J., & Pariza, M. W., (1990). Inhibition of benzo(a)pyrene-induced mouse forestomach neoplasia by conjugated dienoic derivatives of linoleic acid. *Cancer Research, 50*(4), 1097–1101.
27. Ha, Y. L., Grimm, N. K., & Pariza, M., (1987). Anticarcionogens from fried ground beef: Heat-altered derivatives of linoleic acid. *Carcinogenesis, 8*(12), 1881–1887.
28. Hargrave, K. M., Meyer, B. J., Li, C., Azain, M. J., Baile, C. A., & Miner, J. L., (2004). Influence of dietary conjugated linoleic acid and fat source on body fat and apoptosis in mice. *Obesity Research, 12*(9), 1435–1444.
29. Hargrave, K. M., Azain, M. J., & Miner, J. L., (2005). Dietary coconut oil increases conjugated linoleic acid-induced body fat loss in mice independent of essential fatty

acid deficiency. *Biochimica et Biophysica Acta (BBA)-Molecular and Cell Biology of Lipids, 1737*(1), 52–60.
30. Hayek, M. G., Han, S. N., Wu, D., Watkins, B. A., Meydani, M., Dorsey, J. L., Smith, D. E., & Meydani, S. N., (1999). Dietary conjugated linoleic acid influences the immune response of young and old C57BL/6NCrlBR mice. *The Journal of Nutrition, 129*(1), 32–38.
31. Houseknecht, K. L., VandenHeuvel, J. P., & Moya-Camarena, S. Y., (1998). Dietary conjugated linoleic acid normalizes impaired glucose tolerance in the Zucker diabetic fatty fa/fa rat. *Biochemical and Biophysical Research Communications, 244*(3), 678–682.
32. Hur, S. J., & Park, Y., (2007). Effect of conjugated linoleic acid on bone formation and rheumatoid arthritis. *European Journal of Pharmacology, 568*(1–3), 16–24.
33. Jiang, C., Ting, A. T., & Seed, B., (1998). PPAR-γ agonists inhibit production of monocyte inflammatory cytokines. *Nature, 391*(6662), 82–86.
34. Kang, K., Miyazaki, M., Ntambi, J. M., & Pariza, M. W., (2004). Evidence that the anti-obesity effect of conjugated linoleic acid is independent of effects on stearoyl-CoA desaturase 1 expression and enzyme activity. *Biochemical and Biophysical Research Communications, 315*(3), 532–537.
35. Kelly, O., & Cashman, K. D., (2004). The effect of conjugated linoleic acid on calcium absorption and bone metabolism and composition in adult ovariectomized rats. *Prostaglandins, Leukotrienes, and Essential Fatty Acids (PLEFA), 71*(5), 295–301.
36. Kelly, O., Cusack, S., Jewell, C., & Cashman, K. D., (2003). The effect of polyunsaturated fatty acids, including conjugated linoleic acid, on calcium absorption and bone metabolism and composition in young growing rats. *British Journal of Nutrition, 90*(4), 743–750.
37. Kim, E. J., Kang, I. J., Cho, H. J., Kim, W. K., Ha, Y. L., & Park, J. H., (2003). Conjugated linoleic acid downregulates insulin-like growth factor-I receptor levels in HT-29 human colon cancer cells. *The Journal of Nutrition, 133*(8), 2675–2681.
38. Kritchevsky, D., Wright, S., & Czarnecki, S. K., (2002). Influence of graded levels of conjugated linoleic acid (CLA) on experimental atherosclerosis in rabbits. *Nutrition Research, 22*(11), 1275–1279.
39. Kritchevsky, D., Tepper, S. A., Wright, S., Tso, P., & Czarnecki, S. K., (2000). Influence of conjugated linoleic acid (CLA) on establishment and progression of atherosclerosis in rabbits. *Journal of the American College of Nutrition, 19*(4), 472S–477S.
40. Kuniyasu, H., Yoshida, K., Sasaki, T., Sasahira, T., Fujii, K., & Ohmori, H., (2005). Conjugated linoleic acid inhibits peritoneal metastasis in human gastrointestinal cancer cells. *International Journal of Cancer, 118*(3), 571–576.
41. Kzhyshkowska, J., Neyen, C., & Gordon, S., (2012). Role of macrophage scavenger receptors in atherosclerosis. *Immunobiology, 217*(5), 492–502.
42. Larsson, S. C., Bergkvist, L., & Wolk, A., (2005). High-fat dairy food and conjugated linoleic acid intakes in relation to colorectal cancer incidence in the Swedish mammography cohort. *The American Journal of Clinical Nutrition, 82*(4), 894–900.
43. Lee, J. H., Cho, K. H., Lee, K. T., & Kim, M. R., (2005). Antiatherogenic effects of structured lipid containing conjugated linoleic acid in C57BL/6J mice. *Journal of Agricultural and Food Chemistry, 53*(18), 7295–301.
44. Lee, K. N., Kritchevsky, D., & Parizaa, M. W., (1994). Conjugated linoleic acid and atherosclerosis in rabbits. *Atherosclerosis, 108*(1), 19–25.

45. Li, G., Barnes, D., Butz, D., Bjorling, D., & Cook, M. E., (2005). 10t, 12c-conjugated linoleic acid inhibits lipopolysaccharide-induced cyclooxygenase expression *in vitro* and *in vivo*. *Journal of Lipid Research, 46*(10), 2134–2142.
46. Li, P., Bandyopadhyay, G., Lagakos, W. S., Talukdar, S., Osborn, O., Johnson, A., Chung, H., et al., (2015). LTB4 promotes insulin resistance in obese mice by acting on macrophages, hepatocytes, and myocytes. *Nature Medicine, 21*(3), 239–247.
47. Li, Y., Seifert, M. F., Ney, D. M., Grahn, M., Grant, A. L., Allen, K. G., & Watkins, B. A., (1999). Dietary conjugated linoleic acids alter serum IGF-I and IGF binding protein concentrations and reduce bone formation in rats fed (n-6) or (n-3) fatty acids. *Journal of Bone and Mineral Research, 14*(7), 1153–1162.
48. Lin, G., Wang, H., Dai, J., Li, X., Guan, M., Gao, S., Ding, Q., Wang, H., & Fang, H., (2017). Conjugated linoleic acid prevents age-induced bone loss in mice by regulating both osteoblastogenesis and adipogenesis. *Biochemical and Biophysical Research Communications, 490*(3), 813–820.
49. Lock, A. L., & Bauman, D. E., (2004). Modifying milk fat composition of dairy cows to enhance fatty acids beneficial to human health. *Lipids, 39*(12), 1197–1206.
50. Luongo, D., Bergamo, P., & Rossi, M., (2003). Effects of conjugated linoleic acid on growth and cytokine expression in Jurkat T cells. *Immunology Letters, 90*(2, 3), 195–201.
51. McCarthy, C., Lieggi, N. T., Barry, D., Mooney, D., De Gaetano, M., James, W. G., McClelland, S., et al., (2013). Macrophage PPAR gamma co-activator-1 alpha participates in repressing foam cell formation and atherosclerosis in response to conjugated linoleic acid. *EMBO Molecular Medicine, 5*(9), 1443–1457.
52. Miller, C. C., Park, Y., Pariza, M. W., & Cook, M. E., (1994). Feeding conjugated linoleic acid to animals partially overcomes catabolic responses due to endotoxin injection. *Biochemical and Biophysical Research Communications, 198*(3), 1107–1112.
53. Mirand, P. P., Arnal-Bagnard, M. A., Mosoni, L., Faulconnier, Y., Chardigny, J. M., & Chilliard, Y., (2004). Cis-9, trans-11 and trans-10, cis-12 conjugated linoleic acid isomers do not modify body composition in adult sedentary or exercised rats. *The Journal of Nutrition, 134*(9), 2263–2269.
54. Monaco, A., Ferrandino, I., Boscaino, F., Cocca, E., Cigliano, L., Maurano, F., & Bergamo, P., (2018). Conjugated linoleic acid prevents age-dependent neurodegeneration in a mouse model of neuropsychiatric lupus via the activation of an adaptive response. *Journal of Lipid Research, 59*(1), 48–57.
55. Mougios, V., Matsakas, A., Petridou, A., Ring, S., Sagredos, A., Melissopoulou, A., Tsigilis, N., & Nikolaidis, M., (2001). Effect of supplementation with conjugated linoleic acid on human serum lipids and body fat. *The Journal of Nutritional Biochemistry, 12*(10), 585–594.
56. Nagao, K., Inoue, N., Wang, Y. M., & Yanagita, T., (2003). Conjugated linoleic acid enhances plasma adiponectin level and alleviates hyperinsulinemia and hypertension in Zucker diabetic fatty (fa/fa) rats. *Biochemical and Biophysical Research Communications, 310*(2), 562–566.
57. Navarro, V., Zabala, A., Macarulla, M. T., Fernández-Quintela, A., Rodríguez, V. M., Simón, E., & Portillo, M. P., (2003). Effects of conjugated linoleic acid on body fat accumulation and serum lipids in hamsters fed an atherogenic diet. *Journal of Physiology and Biochemistry, 59*(3), 193–199.

58. Nicolosi, R. J., Courtemanche, K. V., Laitinen, L., Scimeca, J. A., & Huth, P. J., (1993). Effect of feeding diets enriched in conjugated linoleic acid on lipoproteins and aortic atherogenesis in hamsters. *Circulation, 88*(1), 2458.
59. Nicolosi, R. J., Rogers, E. J., Kritchevsky, D., Scimeca, J. A., & Huth, P. J., (1997). Dietary conjugated linoleic acid reduces plasma lipoproteins and early aortic atherosclerosis in hypercholesterolemic hamsters. *Artery, 22*(5), 266–277.
60. Norris, L. E., Collene, A. L., Asp, M. L., Hsu, J. C., Liu, L. F., Richardson, J. R., Li, D., et al., (2009). Comparison of dietary conjugated linoleic acid with safflower oil on body composition in obese postmenopausal women with type 2 diabetes mellitus. *The American Journal of Clinical Nutrition, 90*(3), 468–476.
61. O'Shea, M., Stanton, C., & Devery, R., (1999). Antioxidant enzyme defense responses of human mcf-7 and sw480 cancer cells to conjugated linoleic acid. *Anticancer Research, 19*(3A), 1953–1959.
62. O'Shea, M., Bassaganya-Riera, J., & Mohede, I. C., (2004). Immunomodulatory properties of conjugated linoleic acid. *The American Journal of Clinical Nutrition, 79*(6), 1199S–1206S.
63. Pal, S., Takechi, R., & Ho, S. S., (2005). Conjugated linoleic acid suppresses the secretion of atherogenic lipoproteins from human HepG2 liver cells. *Clinical Chemistry and Laboratory Medicine, 43*(3), 269–274.
64. Pariza, M. W., & Hargraves, W. A., (1985). A beef-derived mutagenesis modulator inhibits initiation of mouse epidermal tumors by 7,12-dimethylbenz[a]anthracene. *Carcinogenesis, 6*(4), 591–593.
65. Pariza, M. W., Park, Y., & Cook, M., (2000). Mechanisms of action of conjugated linoleic acid: Evidence and speculation (44457). *Proceedings of the Society for Experimental Biology and Medicine, 223*(1), 8–13.
66. Park, Y., Albright, K. J., Lin, W., Storkson, J. M., Cook, M. E., & Pariza, M. W., (1997). Effect of conjugated linoleic acid on body composition in mice. *Lipids, 32*(8), 853–858.
67. Park, Y., & Pariza, M. W., (2001). The effects of dietary conjugated nonadecadienoic acid on body composition in mice. *Biochimica et Biophysica Acta (BBA)-Molecular and Cell Biology of Lipids, 1533*(3), 171–174.
68. Park, Y., Storkson, J. M., Albright, K. J., Liu, W., & Pariza, M. W., (1999). Evidence that the trans-10, cis-12 isomer of conjugated linoleic acid induces body composition changes in mice. *Lipids, 34*(3), 235–241.
69. Parodi, P. W., (1997). Cows' milk fat components as potential anticarcinogenic agents. *The Journal of Nutrition, 127*(6), 1055–1060.
70. Penedo, L. A., Nunes, J. C., Gama, M. A. S., Leite, P. E. C., Quirico-Santos, T. F., & Torres, A. G., (2013). Intake of butter naturally enriched with cis9, trans11 conjugated linoleic acid reduces systemic inflammatory mediators in healthy young adults. *The Journal of Nutritional Biochemistry, 24*(12), 2144–2151.
71. Petridou, A., Mougios, V., & Sagredos, A., (2003). Supplementation with CLA: Isomer incorporation into serum lipids and effect on body fat of women. *Lipids, 38*(8), 805–811.
72. Platt, I. D., & El-Sohemy, A., (2009). Regulation of osteoblast and adipocyte differentiation from human mesenchymal stem cells by conjugated linoleic acid. *The Journal of Nutritional Biochemistry, 20*(12), 956–964.
73. Rahman, M. M., Bhattacharya, A., Banu, J., & Fernandes, G., (2007). Conjugated linoleic acid protects against age-associated bone loss in C57BL/6 female mice. *The Journal of Nutritional Biochemistry, 18*(7), 467–474.

74. Ramakers, J. D., Plat, J., Sebedio, J. L., & Mensink, R. P., (2005). Effects of the individual isomers cis-9, trans-11 vs. Trans-10, cis-12 of conjugated linoleic acid (CLA) on inflammation parameters in moderately overweight subjects with LDL-phenotype B. *Lipids, 40*(9), 909–918.
75. Riserus, U., Arner, P., Brismar, K., & Vessby, B., (2002). Treatment with dietary trans-10 cis-12 conjugated linoleic acid causes isomer-specific insulin resistance in obese men with the metabolic syndrome. *Diabetes Care, 25*(9), 1516–1521.
76. Riserus, U., Smedman, B., Basu, S., & Vessby, B., (2003). CLA and body weight regulation in humans. *Lipids, 38*(2), 133–137.
77. Ryder, J. W., Portocarrero, C. P., Song, X. M., Cui, L., Yu, M., Combatsiaris, T., Galuska, D., et al., (2001). Isomer-specific antidiabetic properties of conjugated linoleic acid: Improved glucose tolerance, skeletal muscle insulin action, and UCP-2 gene expression. *Diabetes, 50*(5), 1149–1157.
78. Saha, S. S., Dasgupta, P., Sengupta, S., & Ghosh, M., (2012). Synergistic effect of conjugated linolenic acid isomers against induced oxidative stress, inflammation, and erythrocyte membrane disintegrity in rat model. *Biochimica et Biophysica Acta (BBA)-General Subjects, 1820*(12), 1951–1970.
79. Saha, S. S., & Ghosh, M., (2012). Antioxidant and anti-inflammatory effect of conjugated linolenic acid isomers against streptozotocin-induced diabetes. *British Journal of Nutrition, 108*(6), 974–983.
80. Saha, S. S., & Ghosh, M., (2013). Protective effect of conjugated linolenic acid isomers present in vegetable oils against arsenite-induced renal toxicity in rat model. *Nutrition, 29*(6), 903–910.
81. Song, H. J., Grant, I., Rotondo, D., Mohede, I., Sattar, N., Heys, S. D., & Wahle, K. W. J., (2005). Effect of CLA supplementation on immune function in young healthy volunteers. *European Journal of Clinical Nutrition, 59*(4), 508–517.
82. Sugano, M., Tsujita, A., Yamasaki, M., Noguchi, M., & Yamada, K., (1998). Conjugated linoleic acid modulates tissue levels of chemical mediators and immunoglobulins in rats. *Lipids, 33*(5), 521–527.
83. Sugano, M., Yamasaki, M., Yamada, K., & Huang, Y. S., (1999). Effect of conjugated linoleic acid on polyunsaturated fatty acid metabolism and immune function. Chapter 25. In: Yurawecz, M. P., Mossoba, M. M., Kramer, J. K. G., Pariza, W. M., & Nelson, G. J., (eds.), *Advances in Conjugated Linoleic Acid Research* (Vol. 1, pp. 327–339). Champaign Illinois-USA: AOCS Press.
84. Tanmahasamut, P., Liu, J., Hendry, L. B., & Sidell, N., (2004). Conjugated linoleic acid blocks estrogen signaling in human breast cancer cells. *The Journal of Nutrition, 134*(3), 674–680.
85. Tarnopolsky, M. A., & Safdar, A., (2008). The potential benefits of creatine and conjugated linoleic acid as adjuncts to resistance training in older adults. *Applied Physiology, Nutrition, and Metabolism, 33*(1), 213–227.
86. Terpstra, A. H., Beynen, A. C., Everts, H., Kocsis, S., Katan, M. B., & Zock, P. L., (2002). The decrease in body fat in mice fed conjugated linoleic acid is due to increases in energy expenditure and energy loss in the excreta. *The Journal of Nutrition, 132*(5), 940–945.
87. Thom, E., Wadstein, J., & Gudmundsen, O., (2001). Conjugated linoleic acid reduces body fat in healthy exercising humans. *Journal of International Medical Research, 29*(5), 392–396.

88. Tontonoz, P., Nagy, L., Alvarez, J. G., Thomazy, V. A., & Evans, R. M., (1998). PPARγ promotes monocyte/macrophage differentiation and uptake of oxidized LDL. *Cell, 93*(2), 241–252.
89. Toomey, S., Harhen, B., Roche, H. M., Fitzgerald, D., & Belton, O., (2006). Profound resolution of early atherosclerosis with conjugated linoleic acid. *Atherosclerosis, 187*(1), 40–49.
90. Tricon, S., Burdge, G. C., Kew, S., Banerjee, T., Russell, J. J., Grimble, R. F., Williams, C. M., et al., (2004). Effects of cis-9, trans-11 and trans-10, cis-12 conjugated linoleic acid on immune cell function in healthy humans. *The American Journal of Clinical Nutrition, 80*(6), 1626–1633.
91. Tsuboyama-Kasaoka, N., Takahashi, M., & Tanemura, K., (2000). Conjugated linoleic acid supplementation reduces adipose tissue by apoptosis and develops lipodystrophy in mice. *Diabetes, 49*(9), 1534–1542.
92. Valeille, K., Ferezou, J., & Amsler, G., (2005). A cis-9, trans-11-conjugated linoleic acid-rich oil reduces the outcome of the atherogenic process in hyperlipidemic hamster. *American Journal of Physiology-Heart and Circulatory Physiology, 289*(2), H652–H659.
93. Visonneau, S., Cesano, A., Tepper, S. A., Scimeca, J. A., Santoli, D., & Kritchevsky, D., (1997). Conjugated linoleic acid suppresses the growth of human breast adenocarcinoma cells in SCID mice. *Anticancer Research, 17*(2A), 969–973.
94. Wahle, K. W., Heys, S. D., & Rotondo, D., (2004). Conjugated linoleic acids: Are they beneficial or detrimental to health? *Progress in Lipid Research, 43*(6), 553–587.
95. Wargent, E., Sennitt, M. V., & Stocker, C., (2005). Prolonged treatment of genetically obese mice with conjugated linoleic acid improves glucose tolerance and lowers plasma insulin concentration: Possible involvement of PPAR activation. *Lipids in Health and Disease, 4*(1), 3.
96. Watkins, B. A., Li, Y., Lippman, H. E., Reinwald, S., & Seifert, M. F., (2004). A test of ockham's razor: Implications of conjugated linoleic acid in bone biology. *The American Journal of Clinical Nutrition, 79*(6), 1175S–1185S.
97. Watkins, B. A., Shen, C. L., McMurtry, J. P., Xu, H., Bain, S. D., Allen, K. G., & Seifert, M. F., (1997). Dietary lipids modulate bone prostaglandin E2 production, insulin-like growth factor-I concentration and formation rate in chicks. *The Journal of Nutrition, 127*(6), 1084–1091.
98. Weiler, H., Austin, S., Wong, S. F., Nitschmann, E., Calic, N. B., Mollard, R., Aukema, H., & Ogborn, M., (2004). Conjugated linoleic acid reduces parathyroid hormone in health and in polycystic kidney disease in rats. *The American Journal of Clinical Nutrition, 79*(6), 1186S–1189S.
99. West, D. B., Blohm, F. Y., Truett, A. A., & DeLany, J. P., (2000). Conjugated linoleic acid persistently increases total energy expenditure in AKR/J mice without increasing uncoupling protein gene expression. *The Journal of Nutrition, 130*(10), 2471–2477.
100. West, D. B., DeLany, J. P., Camet, P. M., Blohm, F., Truett, A. A., & Scimeca, J., (1998). Effects of conjugated linoleic acid on body fat and energy metabolism in the mouse. *American Journal of Physiology-Regulatory, Integrative, and Comparative Physiology, 275*(3), R667–R672.
101. Whigham, L. D., Cook, E. B., & Stahl, J. L., (2001). CLA reduces antigen induced histamine and PGE2 release from sensitized guinea pig tracheae. *American Journal of Physiology-Regulatory, Integrative, and Comparative Physiology, 280*(3), R908–R912.

102. Whigham, L. D., Higbee, A., Bjorling, D. E., Park, Y., Pariza, M. W., & Cook, M. E., (2002). Decreased antigen-induced eicosanoid release in conjugated linoleic acid-fed guinea pigs. *American Journal of Physiology-Regulatory, Integrative, and Comparative Physiology, 282*(4), R1104–R1112.
103. White, D. A., Bennett, A. J., Billett, M. A., & Salter, A. M., (1998). The assembly of triacylglycerol-rich lipoproteins: An essential role for the microsomal triacylglycerol transfer protein. *British Journal of Nutrition, 80*(3), 219–229.
104. Wong, M. W., Chew, B. P., Wong, T. S., Hosick, H. L., Boylston, T. D., & Shultz, T. D., (1997). Effects of dietary conjugated linoleic acid on lymphocyte function and growth of mammary tumors in mice. *Anticancer Research, 17*(2A), 987–993.
105. Yamasaki, M., Chujo, H., Hirao, A., Koyanagi, N., Okamoto, T., Tojo, N., Oishi, A., et al., (2003). Immunoglobulin and cytokine production from spleen lymphocytes is modulated in C57BL/6J mice by dietary cis-9, trans-11 and trans-10, cis-12 conjugated linoleic acid. *The Journal of nutrition, 133*(3), 784–788.
106. Yamasaki, M., Kishihara, K., Mansho, K., Ogino, Y., Kasai, M., Sugano, M., Tachibana, H., & Yamada, K., (2000). Dietary effect of conjugated linoleic acid increases immunoglobulin productivity of Sprague-Dawley rat spleen lymphocytes. *Bioscience, Biotechnology, and Biochemistry, 64*(10), 2159–2164.
107. Yu, Y., Correll, P. H., & Heuvel, J. V., (2002). Conjugated linoleic acid decreases production of pro-inflammatory products in macrophages: Evidence for a PPARγ-dependent mechanism. *Biochimica et Biophysica Acta (BBA)-Molecular and Cell Biology of Lipids, 1581*(3), 89–99.
108. Yurawecz, M. P., Mossoba, M. M., Kramer, J. K. G., Pariza, M. W., & Nelson, G. J., (1999). *Advances in Conjugated Linolenic Acid Research* (Vol. 1, p. 480). Illinois; AOCS Press.

CHAPTER 7

LACTOFERRIN: ISOLATION AND FUNCTIONAL ASPECTS

SUDHEENDRA V. K. CHAGANTI, TANMAY HAZRA, AKASHKUMAR SOLANKI, and VIMAL M. RAMANI

ABSTRACT

Milk is considered a wholesome food, due to the fact that it contains micro as well as macronutrients. Among these micronutrients, lactoferrin (LF) is one such compound that occupies a significant place. LF is a globular protein belonging to the transferrin family which possess higher affinity towards iron in comparison to other transferrin proteins. This distinctive property plays a significant role in the humans as well as animals for iron uptake into the system. Highest concentration of LF occurs in human milk while its concentration is low in ruminant's milk. Sorensen and Sorensen in 1939 first isolated crude form of LF from whey. In later years method was standardized to extract the purest form of LF from milk and its products. It is found in different isomeric forms and is known to influence various physiological events in the body. LF is part of the immune mechanism of animal and humans and hence have a profound impact on immunity. It also possesses antimicrobial, antiviral, anti-allergic, and anti-cancerous activities. Owing to these beneficial aspects of LF, it has been supplemented in foods especially infant and baby foods.

7.1 INTRODUCTION

Milk is a primary food source for newborn which comprises many antimicrobial agents, i.e., lysozyme, lactoperoxidase (LPO), lactoferrin (LF), etc.

Functional Dairy Ingredients and Nutraceuticals: Physicochemical, Technological, and Therapeutic Aspects. Megh R. Goyal, N. Veena, PhD & Santosh K. Mishra (Eds.)

Other than the nutritional components, milk also comprises many biologically active compounds in trace amounts. These bioactive agents present in milk protect the consumer from several microbial infections. The highest concentration of bioactive components is present in colostrum than mature milk, and these compounds exhibits antimicrobial, antiviral, antifungal, anti-carcinogenic, anti-inflammatory, antioxidant, immune-enhancing and mineral binding properties [24, 59, 79, 115]. Generally, bioactive compounds are usually in dormant condition and are liberated upon hydrolysis of proteins. Such proteolysis occurs either by specific microbial enzymes or gastrointestinal enzymes or during food processing [49].

LF is the member of the transferrin family [94] with molecular weight (MW) of 78–80 kDa and 690–702 amino acids residues. It is a glycoprotein, which possesses iron-binding ability. LF's crude form was first discovered by Sorensen and Sorensen in 1939, but it's pure form was isolated in the early 1960s from bovine milk and human milk [52, 62, 95]. Rachman et al. [123] reported that in goat milk, concentration of LF varies with lactation stage and the highest concentration was found on the day one of lactation, i.e., 5–7 mg/ml and in the later lactation stage it was reduced to 2–3 mg/ml.

LF is present in body fluids, such as saliva, nasal mucosa, mucous membrane of bronchus, seminal fluids, tears, vaginal fluids and in some white blood cells [59]. LF was isolated from milk in the late fifties [62], however it was characterized in 1971 [101]. Later, it was noted that LF is a glycoprotein and the concentration of this biomolecule varies from species to species [33, 85, 180] as shown in Table 7.1.

TABLE 7.1 LF Concentration in Different Mammals

Species	Concentration (g/L)
Bovine colostrum	0.34–1.96
Bovine milk	0.02–0.5
Camel milk	0.02–7.28
Goat milk	0.2–0.3
Human colostrum	5–6
Human milk	1.5–2.0
Sheep milk	0.6–0.8

The present chapter gives an overview of the LF isolation process, its different forms, functional aspects and applications in foods.

7.2 STRUCTURE AND VARIOUS FORMS OF LF

LF is a glycoprotein with a MW of around 80 kDa. Human and bovine milk LF contains almost 696 and 691 amino acids residues, respectively [14, 96] with 70% similarity in amino acid sequences [182]. It is a single-chain protein with 33–34% α-helix has been observed in secondary structure [165] containing 17 disulfide bonds [96]. Outer surface of LF has highly positive charged and these charges help to bind carbohydrate indeed [165]. Its structure has two lobes, namely N-lobe consisting of 1–333 amino acids residues while C-lobe consisting of 345–691 amino acid residues and each lobe can bind with a ferric ion and a carbohydrate molecule (Figure 7.1). Isoelectric point (pI) of LF is very high which is around 8–9.

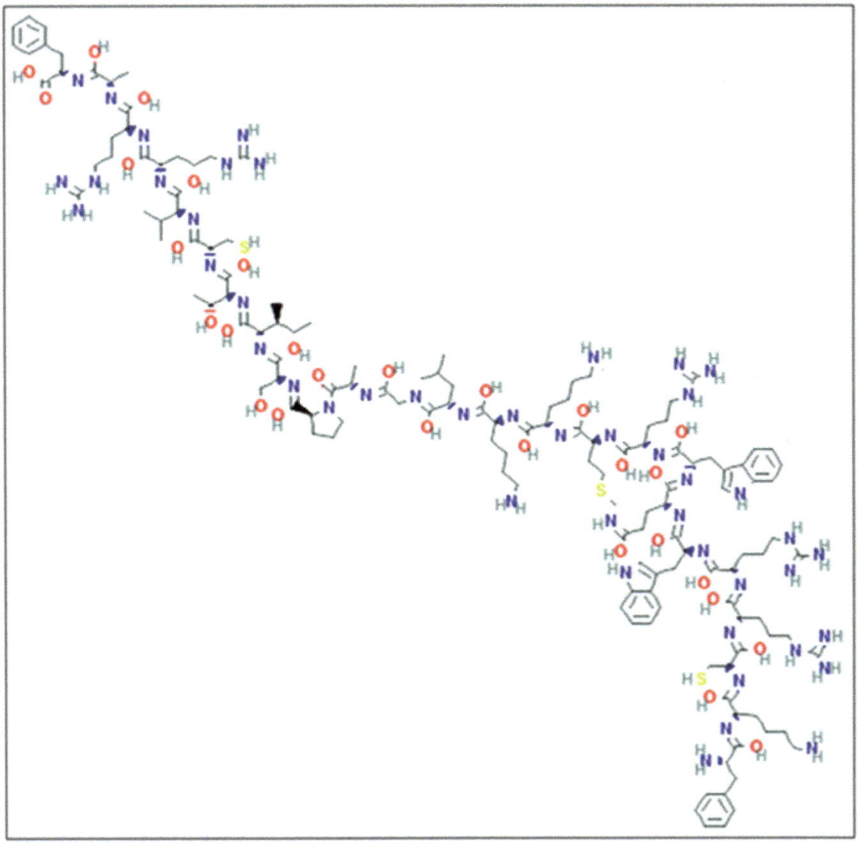

FIGURE 7.1 Structure of LF.

Source: URL: https://pubchem.ncbi.nlm.nih.gov

LF exists mainly in dual states, apo-LF which does not contain metal (Fe) and has an open structure whereas halo-LF contains metal ions and has a closed structure. Apo-LF is highly stable when compared with halo form. Iron content in native, apo, and halo forms of LF is 12.9, 0.9, and 99.7%, respectively. Presence of iron in LF affect its properties like increase in iron content increases the thermal stability and reduction in iron content decreases the surface tension [21].

7.3 ISOLATION TECHNIQUES FOR LF

LF is a glycol protein, therefore, many researchers have tried different isolation and purification protocols based on chromatographic, filtration based and liquid-liquid extraction methods [2] for an efficient extraction from sources like milk or whey, while its characterization was usually carried out by electrophoresis or mass spectrometry (MS).

7.3.1 ION EXCHANGE CHROMATOGRAPHY (IEC)

It is a widely used method for isolation of LF from milk or whey [173]. Ye et al. [176] used this technique to isolate and purify LF from whey. The efficiency of this technique depended on the type of ion exchanger used for the isolation. Yoshida and Ye-Xiuyun [181] applied carboxymethyl cation exchanger system to separate LF A and LF B from both bovine milk and whey. Elagamy et al. [40] applied the carboxymethyl-cellex cation exchanger followed by electrophoresis to characterize bovine and camel LF whose MW were 79.5 and 76 kDa, respectively.

Purification of LF from milk of different species like goat, camel, sheep, elephant, alpaca, grey seal and humans was reported by using an IEC [27]. In a study, ultrafiltration process followed by strong cation exchange chromatography was used for isolation and purification of LF from bovine colostrum [88]. Zainab et al. [183] suggested that LF can be purified from goat colostrum by ion-exchange chromatography followed by gel filtration chromatography.

Fee and Chand [46] reported that cation exchange chromatography followed by enzyme linked immunosorbent assay (ELISA) is the most efficient technique for the isolation as well as purification of LF from raw milk. Liang et al. [80] applied cation-exchange resin (SPEC 70 SLS) for simultaneous isolation of LF and LPO from fat-free cow colostrum. This resin proved maximum static binding capacity with LF and hence, better yields of LF. Faraji et al. [44] employed cation exchange column chromatography with two-step optimization approach for isolation of LF and LPO. Heparin-sepharose [20] and carboxymethyl (CM) Sephadex C-50 [97] IEC was also applied to isolate the LF from human whey and bovine colostrums, respectively.

7.3.2 HYDROPHOBIC INTERACTION CHROMATOGRAPHY (HIC)

HIC is a method popularly employed to separate and purify proteins without any structural damages. Basically, it is a salting-out method for protein isolation and purification. Yoshida [179] successfully applied this technique to isolate LF from whey, followed by characterization of the LF using sodium dodecyl sulfate-polyacrylamide gel electrophoresis (SDS-PAGE). Rachman et al. [123] applied this technique to isolate and characterize LF from goat colostrum and confirm that molecular weight of LF varies with breeds of goat.

7.3.3 FAST PROTEIN LIQUID CHROMATOGRAPHY (FPLC)

FPLC technique is more reliable, sensitive, and faster when compared to IEC in isolating LF from milk [39]. FPLC method has been adopted for isolation of LF from camel milk using CM Sephadex C-25 resin [102]

7.3.4 METAL CHELATING CHROMATOGRAPHY

Metal-chelate affinity chromatography (MCAC) is also known as immobilized metal affinity chromatography (IMAC) is another popular and common technique employed to isolate LF either from whey or milk. Al-Mashikhi et al. [6] isolated LF from cheddar cheese by MCAC followed by SDS-PAGE for characterization. Lonnerdal et al. [84] used this technique to isolate LF from milk while immune-electrophoresis for the characterization instead of SDS-PAGE.

7.3.5 IMMUNOAFFINITY CHROMATOGRAPHY (IAC)

LF can be isolated from raw skim milk or cheese whey using the antibodies which were covalently attached with agarose beads without changing other protein constituents [81, 107]. Tu et al. [151] used anti-LF immunoglobulin in yolk-Sepharose 4B-IAC technique to isolate LF from colostral whey with a recovery rate of 82–99% of LF.

Noppe et al. [106] purified LF from skimmed milk using peptides expressed by cloned phages. These peptides were embedded onto the macro-porous poly(dimethyl acrylamide) monolithic column which possess high binding affinity towards LF. Bound LF was removed from the column using 1 M sodium chloride with more than 95% of purity. High quality LF could be isolated using this technique in an efficient manner. Magnetic affinity separation also can be used in the isolation of LF from milk [26].

7.3.6 GEL FILTRATION AND ADSORPTIVE MEMBRANE TECHNOLOGY

Gel filtration on heparin-sepharose was used to isolate LF from sweet whey, and the adsorbed LF was subsequently eluted from the column using veronal-hydrochloric acid (5 mM) containing sodium chloride (0.5 M) and pH 7.4. Later its purity was established by SDS-PAGE and immune electrophoresis [7].

Bovine LF and lactoferricin were isolated from sweet cheese whey using membrane adsorber technology where cation-exchange membrane (Sartobind S) was used. The identification of the LF was performed with matrix-assisted laser desorption ionization-mass spectrometry (MALDI-MS) [119]. It was suggested to use this procedure for industrial-scale purification of LF with recovery of more than 90%.

7.3.7 HYDROPHOBIC IONIC LIQUID EXTRACTION

A novel separation method that has been tried to isolate LF from bovine whey [9] in a U-shaped bulk membrane tube by using hydrophobic ionic liquid-like 1-butyl-3-methylimidazolium bis[(trifluoromethyl)[sulfonyl] imide with extraction efficiency of 20% was accomplished. This method results in better selectivity over chromatography for isolation of LF might be because of the interaction of ionic liquid with iron atom of the LF.

7.3.8 SIMULATED MOVING BED (SMB) TECHNOLOGY

In the 1960s, the continuous chromatographic technique was first introduced by Broughton and Gerhold and has many advantages over traditional batch chromatographic method. The 20 column SMB technique was applied for isolation of LF and LPO from whey protein concentrate [11]. This method results in increased productivity, higher protein extraction and less buffer utilization in comparison with the same column set up in non-moving bed.

7.3.9 REVERSE MICELLES (RM) EXTRACTION

It is a type of liquid-liquid extraction process and used as an alternative to chromatography, since it can be easily scaled up and permits uninterrupted separation process. In this technique, RM are used to extract protein from the aqueous phase based on hydrophobic interaction as well as protein size relative to the droplets. The protein like LF from whey was extracted using nano-scale RMs of non-ionic (triton-X-100), cationic (cetyltrimethylammonium bromide, CETB) and combination of surfactant (triton-X-100/ CETB) in organic solvent was employed [12]. Results show that best separation was achieved using CETB than combination mode with recovery of LF is more than 96%. However, most of these techniques are lab-based technique therefore for applications in commercial scale is still a big question [105].

7.3.10 IMMOBILIZED BEADS

LF in cow milk was extracted by interacting with α-casein-immobilized beads after treating the milk with poly-l-lysine-immobilized beads. The combination of poly-l-lysine- and α-casein-immobilized beads is the simplest technique for isolating LF from milk which based on protein-protein interaction [100].

7.4 FUNCTIONAL ASPECTS OF LF

LF is one of the immune-nutrient with an inexplicable display of functions that are still being elucidated even after so many years of its initial discovery in 1960 [41, 95]. Host's innate and adaptive defense mechanism is highly influenced by the LF [73, 74, 136] and also in iron metabolism [139]. In addition to these benefits, LF also plays an important role as an antioxidant [69], anti-cancerous [149] and antimicrobial properties.

In recent times its role in suppressing metabolic disorders like obesity, diabetes, and coronary heart diseases were well reported [93]. Studies also suggested that bovine LF can be used in treating conditions like inflammation, anemia, and oral diseases in athletes [127].

7.4.1 IRON UPTAKE

LF role in iron transport and intestinal absorption especially in infants is well established [4]. Reports suggest that infants fed with mother milk were found to possess better iron absorption capability than infants fed with infant formula [43]. The reason is that around 30% iron present in mother's milk is bound to LF that is absorbed in intact form [50, 58]. LF-bound iron absorption will be increased when specific receptors on the surface of enterocytes are released in response to lack of intracellular iron [139].

7.4.2 ANTIMICROBIAL PROPERTIES

This protein has shown good bacteriostatic and bactericidal activity, which has been well documented since its discovery. Reports suggest that it is highly effective against numerous bacteria, including gram-positive (G +ve) and gram-negative (G –ve) organisms. Antibacterial property is due to the LF's structure [1, 48, 75]. LF can scavenge and bind iron even at very low pH 3.5, which is the reason behind LF's bacteriostatic properties against G –ve bacteria (such as coliforms) as these organisms require high iron for their metabolism. It is also antagonistic against *Bacillus, S. aureus, L. monocytogenes, Bacteroides, Clostridium, E. coli, and Salmonella* which also require high iron content [169]. Various proposed mechanisms by which LF exhibit antibacterial property are as follows:

- Attachment of LF onto receptors like glycosaminoglycans (GAGs) on host cells which are used by pathogens and viruses to enter in to host system. This competitive inhibition of microorganisms like *E. coli* and *Staphylococcus aureus* by LF reduces microorganism's entry into the host cells [140, 159].
- Bacteria require iron for their growth, but LF binds to the free iron available. This may lead to bacteriostatic condition or bactericidal condition.
- LF in its glycosylated form binds onto adhesion sites of microorganisms and host cells. Thereby avoiding the adhesion of bacteria.

- LF's iron-binding activity causes depletion of iron causing bacteria to move towards iron rich environments which cause disruption of biofilms. [159].
- LF's proteolytic activity degrades virulence factors produced by many bacteria. *In vitro* antimicrobial activity was observed against pathogenic organisms like *Shigella* and *E. coli* and viruses like *H. influenza* [110].
- Microorganism's cell membrane is disrupted and this disruption mechanism differs from G+ve to G –ve bacteria. In G –ve bacteria, lipopolysaccharide (LPS) gets released from the cell wall due to attachment of LF on to porins and calcium chelation. This causes instability in the bacterial cell wall [47, 140]. In cell wall of G+ve organism, cationic residues and hydrophobic residues present at N-terminal gets disrupted leading to destruction of microorganisms. Extracellular cation concentration in fungal cell wall is important role in destruction of fungi.

Highest antimicrobial effect was reported in camel LF [134] followed by goat and sheep against *E. coli* O157:H7 [72, 134]. Human LF shows lower antibacterial activity in comparison with bovine LF [45], which is due to their structural differences. Changes in the glycans makes Human LF more impervious to proteolysis than bovine LF [160].

Apart from LF's antibacterial properties, it also reported that it shows antiviral activity against a broad spectrum of viruses [117, 155]. Few mechanisms of antiviral activity of LF have been proposed, but the universally accepted mechanism is the blockage of viral receptors, especially heparin sulfate (HS). Thus, it prevents viral adhesion on to the cell and avoiding infection [158]. Many studies on LF have been reported the primary defense action against viral attack [10, 16, 18, 54, 61, 82, 118, 128, 162].

LF was found to inhibit Respiratory Syncytial Virus at very low concentration. Its activity against viruses such as adenoviruses and enteroviruses was also reported [130]. Scientific reports suggest that LF has a strong inhibitory effect against human immunodeficiency virus (HIV), by inhibiting virus replication in the host [92, 121, 141, 166]. LF's antiviral effect was observed even in animals such as rodents [87] and cats [3, 128] against viruses like friend virus complex, feline calicivirus and immunodeficiency virus. LF could be used for treatment of COVID-19 as it can prevent this virus from attaching itself to host cells [64].

LF shows a wide spectrum of antifungal activity against many yeasts and molds [135, 157]. Antifungal activity of LF was first studied by Kirkpatrick and coworkers [66] on *Candida* species. This effect was attributed to the Fe^{3+} sequestering capability of the LF. Various mechanisms were proposed for antifungal activity of LF viz., by iron deprivation [156], damage to the cell wall [103, 104, 174] and by stimulating the host immune mechanism [161]. Bovine LF reported to show fungicidal activity against *Candida* species such as *C. krusei, C. albicans, C. guilliermondii,* and *C. tropicalis* [174]. Concentrations as low as 10 to 100 μg/mL of LF inhibits *C. albicans* and *C. glabrata* [67].

LF also exhibits antiparasitic effect whose mechanism of action needs further study. Parasites like *Babesia* is inhibited by LF by obstructing the iron intake [60]. There are reports suggested that peptides of LF, i.e., lactoferricin also exhibits antiparasitic effect on *Giardia lamblia* [152]. *Entamoeba histolytica* is inhibited by both iron deprivation as well as by the peptides of LF [78]. LF and lactoferricin prevent the penetration of parasites into host cell [112]. A study carried out on Peruvian infants reported that reduction of Giardia's colonization on administration of LF [109]. On contrary to the above reports, there are few studies which suggest that LF may also promote the growth of some parasites like *Tritrichomonas foetus* [143] and *Entamoeba histolytica* [77].

7.4.3 ANTI-INFLAMMATORY AND IMMUNO-MODULATORY EFFECTS

LF is known to alter both innate as well as adaptive immunity by changing the immune system components [62, 124, 129] and immune related gene expressions [163]. LF is a positively charged protein that can bind with negatively charged surface cells of the immune system [14, 38] can activate different signaling pathways leading to the various cellular responses like activation, differentiation, and proliferation. It is also observed that LF is taken into the nucleus to bind with genetic material. [17, 76]. LF interaction with receptors leads to the activation of antigen presenting cells (dendritic cells, lymphocytes, macrophages, etc.), which in turn syntheses interleukins, chemokine CXCL10 and chemokine CCL20. This stimulates the development of Th1 cell responses which is a powerful inducer of interferon-gamma production by T and natural killer (NK) cells. [1, 120, 136, 138].

LF is found to influence leucocytes and regulate increased or decreased production of pro-inflammatory cytokine [70, 90, 137, 186–188]. LF also increases NK cells and adaptative T cells number [56]. It boosts polymorphonuclear leukocytes in blood [133] and induces phagocytosis [71]. It influences myeloid progenitor cell which produces neutrophils, monocytes, and dendritic cells, which are part of the innate immunity [142]. LF exerts a major role in communication between congenital and adaptive immune responses. Anti-inflammatory activity of LF can help in preventing sepsis by controlling a cytokine (tumor necrosis factor-α) which directly associated with septic shock [37].

LF is known to stimulate skin immunity and inhibit allergic responses [28, 31, 51]. Reports revealed that LF reduces the harmful effect of reactive oxygen species (ROS) at the inflammation sites [168]. It also play an important role in anti-inflammation in non-infectious conditions like rheumatoid arthritis, inflammatory bowel disease (IBD's), neurodegenerative diseases and drug-induced or bacterial infectious intestinal injury [31, 35, 36]. LF could be used in the treatment of disease conditions like anemia and inflammation [127].

7.4.4 ANTI-TUMOR ACTIVITY

Anti-cancer properties of LF have been reported in numerous studies since its discovery in rodents. Oral delivery of LF inhibited the tumorigenesis induced by chemicals in rats [19, 167, 172]. Bovine LF was found to inhibit various types of cancers such as lung, intestinal, colonic, throat cancer in rats [150]. Apoptosis is induced by Bovine LF in human leukemic cells [178]. Ingestion of LF has shown positive results in humans against cancer development [57]. A double-blind, controlled trial reported reduction of adenomatous colorectal polyp's size [68, 149]. Recombinant human LF was reported to inhibit tumors in mice by activating anti-carcinogenic cytokines [132, 167]

LF exerts antitumor property by different mechanisms based on type of cancer, including altering the plasma membrane, immunomodulation, inducing cell death and apoptosis and inhibition of angiogenesis and metastasis [98]. Much of the studies on tumor inhibition by LF are concentrated on iron-chelating activity as iron concentration in tissues increases the oxidative stress that leads to carcinogenesis [124] but in another study iron-free apo-form of LF expresses apoptotic property in human epithelial cancer cells

[89]. Mechanisms of anti-tumor action of LF at molecular level need further elucidation [184].

7.4.5 BRAIN HEALTH

LF is known to attenuate some aging-related changes by reducing the deposition of iron, malondialdehyde (MDA) and ROS levels in the hippocampus region of the brain. The potential of LF to protect cerebral functions can be used in treating neurodegenerative disorders associated with aging [185]. Dietary LF can be used in treating re-oxygenation injury to brain [175]. It improves the health condition of patients suffering from Alzheimer's disease [53, 55, 170]. On contrary to the above statement, a study reported that in neurodegenerative disorders like Alzheimer's disease, LF accumulates in the brain along with iron causing such disorders [65].

7.4.6 INTESTINAL HEALTH

LF has a major role in protecting the GI tract from infections and also promotes immune system [63, 91, 93]. Bovine LF's administration orally in rats reduces the chemically induced colitis [148]. Necrotizing enterocolitis in children can be prevented by intake of LF [83]. One of the studies reported that LF induces cell proliferation and reduces inflammation. On the contrary LF administration in calves did not show any beneficial effect in treating diarrhea [116]. LF intervention in children aged 12–18 months suffering from diarrhea was found to reduce the severity and its longitudinal prevalence [108]. This reduction in severity is due to change in gut microbiome in infants and children's who consumed LF fortified infant formula. Hence infant formulas are fortified with LF up to 200 mg/100 g. Apart from antimicrobial activity LF also improves the GI health.

7.4.7 BONE HEALTH

LF is known to stimulate the production of osteoblast and osteocytes and improve the absorption of thymidine into osteocytes. Osteoblasts and chondrocytes apoptosis is reduced to 50–70% by LF intake [29]. It was also reported that LF inhibits the resorption of calcium from bones [86].

It has a direct effect on bones by inhibiting osteolytic cytokines and stabilizes the osseous tissue. Due to this reason, LF may be used in treating diseases like osteoporosis [29]. A study reported that dietary supplementation of LF has improved bone density in postmenopausal women [40].

7.4.8 REGENERATIVE PROPERTIES

LF was found to possess regenerative properties in different tissues of the body. It helps in activating osteoblasts and proliferation of bone matrix secretions [171]. It improves bone density and ameliorates osteoporosis [30]. It can be effectively used in treating injuries and wounds as LF promotes skin granulation and epithelialization [144, 146]. LF interacts with fibroblasts thus improving tissue regeneration by synthesizing collagen [42]. LF is also helpful in treating some skin problems like psoriasis and ulcerations caused by diabetes [55]. Low cost and its easy availability make LF as a potential and effective product which can be used as therapeutic and/ or nutraceutical [98].

7.5 FOOD APPLICATIONS

LF's global commercial market was valued at USD 206.8 million in the year 2018 and it is projected to grow at 7.7% annually. Expanding applications of LF in the food sector is driving LF production and sales. It has got GRAS status from the European Food Safety Authority (EFSA) and hence its utilization in beverage industry especially non-alcoholic drinks, diet food and milk based products has seen a raise. Addition of LF into food products is mainly due to its antimicrobial property [5]. Addition of activated LF, a new form of a naturally occurring milk protein enhanced the shelf life as well as visual characteristics of beef and poultry by inhibiting the pathogens like *E. coli* [99].

7.5.1 BIO-PRESERVATION

Application of LF in cheese at different levels results in increased shelf life with increased hardness, resiliency, springiness, and chewiness [131]. LF extracted from skim milk using pepsin digestion was used for the inhibition of *E. coli* in carrot juice [25]. Incorporation of LF into meat samples has reduced microbial load and increased the glucose content, considered as freshness index of meat [154]. Encapsulated and un-encapsulated LF addition during sausage manufacture has a profound effect on pathogenic *E. coli* O157: H7 strains while its effect on other pathogens is limited [8]. Incorporation of LF into fermented products like yogurt and other food products as a functional ingredient is commercialized in countries like Japan [164].

LF Immobilized onto food grade galactose-rich polysaccharide or carrageenan blocks affected adhesion/ colonization of microorganisms as it avoids microbial film formation and can even neutralize endotoxin formation [99]. Soybean oil powder's shelf life was found to be enhanced by the addition of LF as it prevents oxidation of unsaturated fatty acids present in the product. Quality of wines was enhanced by LF addition which inhibits the yeast *Dekkera bruxellesis* [177]. LF and its derivatives under high hydrostatic pressure improved the chicken fillet's quality by suppressing the growth of *E. coli* O157:H7 and *P. fluorescens* [34].

7.5.2 FOOD PACKAGING

Food packaging plays a profound position in food industry. Techniques like antimicrobial packaging and active packaging have been in use for quite some time. Research on various novel polymers and antimicrobial substances is going on at present and would develop in leaps and bounds in the future [13]. The use of antimicrobials from natural sources like LF as a food preservative is attracting consumer's interest. An edible film of chitosan incorporated with LF was found to inhibit pathogens like *E. coli* and *L. monocytogenes* growth [22]. Studies revealed that various edible films added with LF and lysozyme could prevent the growth of pathogens like *E. coli, L. monocytogenes* and hydrogen sulfide producing bacteria [23, 111, 126]. Another study reported that addition of lysozyme and LF onto the Carboxyl-methyl cellulose-containing paper was effective against *Listeria species* [15]. Active packaging containing a coating of lactoferricin B controlled the growth of psychrotrophic organisms like *Pseudomonas* and extended the shelf life of cheeses [122].

LF antimicrobial activity is influenced by the levels of metal ions like calcium and phosphorus [13]. Microencapsulated LF had better antimicrobial activity than un-encapsulated LF [8]. Apo-form of LF is the most unstable form of protein, is encapsulated to protect its antibacterial properties by spray drying using whey proteins [32]. Encapsulated LF also showed antimicrobial activity against *Carnobacterium viridans* [8]. Electrostatic spraying of bovine LF onto red meat helped the detachment of microorganisms on the surface of meat and improved its quality [147].

7.5.3 LF AS A DRUG

LF was loaded onto chitosan microcapsules for controlled delivery, and it was found that LF was released gradually for over 7 hours [113]. LF enriched stirred yogurt stored at refrigerated temperature (4°C) for 21 days showed enhanced osteogenic activity *in vitro* [114]. Application of biomaterials tools and biomedical instruments are used to improve the transportation and persistent release of LF at the required region or part [145].

7.6 SUMMARY

LF is a glycoprotein with iron in its structure and exerts the capability to chelate iron ions from the surroundings. This distinctive property plays a significantly beneficial role in the body. Its physiological function in the body is still being studied and further in-depth research studies are essential to thoroughly understand the metabolism of LF in human beings. Studies on the antiparasitic activity of LF are scarce and more studies are required in this direction. Further profound research in human beings is required to ascertain beneficial properties of LF and their mechanism at the molecular level.

KEYWORDS

- **food applications**
- **health benefits**
- **iron-binding**
- **lactoferrin**
- **purification**

REFERENCES

1. Actor, J. K., Hwang, S. A., & Kruzel, M. L., (2009). Lactoferrin as a natural immune modulator. *Current Pharmaceutical Design, 15*(17), 1956–1973.
2. Adam, V., Zitka, O., Dolezal, P., Zeman, L., Horna, A., Hubalek, J., Sileny, J., et al., (2008). Lactoferrin isolation using monolithic column coupled with spectrometric or micro-amperometric detector. *Sensors, 8*(1), 464–487.
3. Addie, D. D., Radford, A., Yam, P. S., & Taylor, D. J., (2003). Cessation of feline calicivirus shedding coincident with resolution of chronic gingivostomatitis in a cat. *Journal of Small Animal Practice, 44*(4), 172–176.
4. Adlerova, L., Bartoskova, A., & Faldyna, M., (2008). Lactoferrin: A review. *Veterinarni Medicina, 53*(9), 457–468.
5. Alkõn, E., (2008). Lactoferrin and its use in foods. *Journal of Food and Feed Science Technology, 10,* 31–38.
6. Al-Mashikhi, S. A., Li-Chan, E., & Nakai, S., (1988). Separation of immunoglobulins and lactoferrin from cheese whey by chelating chromatography. *Journal of Dairy Science, 71*(7), 1747–1755.
7. Al-Mashikhi, S. A., & Nakai, S., (1987). Isolation of bovine immunoglobulins and lactoferrin from whey proteins by gel filtration techniques. *Journal of Dairy Science, 70*(12), 2486–2492.
8. Al-Nabulsi, A. A., & Holley, R. A., (2007). Effects on *Escherichia coli* O157: H7 and meat starter cultures of bovine lactoferrin in broth and microencapsulated lactoferrin in dry sausage batters. *International Journal of Food Microbiology, 113*(1), 84–91.
9. Alvarez-Guerra, E., & Irabien, A., (2012). Extraction of lactoferrin with hydrophobic ionic liquids. *Separation and Purification Technology, 98*, 432–440.
10. Andersen, J. H., Jenssen, H., Sandvik, K., & Gutteberg, T. J., (2004). Anti-HSV activity of lactoferrin and lactoferricin is dependent on the presence of heparan sulphate at the cell surface. *Journal of Medical Virology, 74*(2), 262–271.
11. Andersson, J., & Mattiasson, B., (2006). Simulated moving bed technology with a simplified approach for protein purification-separation of lactoperoxidase and lactoferrin from whey protein concentrate. *Journal of Chromatography A, 110*(1, 2), 88–95.
12. Anjana, P., Kumar, P. A., Srinivasan, P., Suryaprakash, P., & Kumar, S. R., (2010). Reverse micelles extraction of lactoferrin using cationic surfactant from whey. *International Journal of Chemical Sciences, 8*(5), 49–56.
13. Appendini, P., & Hotchkiss, J. H., (2002). Review of antimicrobial food packaging. *Innovative Food Science and Emerging Technologies, 3*(2), 113–126.
14. Baker, E. N., & Baker, H. M., (2005). Molecular structure, binding properties and dynamics of lactoferrin. *Cellular and Molecular Life Sciences, 62*(22), 2531–2539.
15. Barbiroli, A., Bonomi, F., Capretti, Iematti, S., Manzoni, M., Piergiovanni, L., & Rollini, M., (2012). Antimicrobial activity of lysozyme and lactoferrin incorporated in cellulose-based food packaging. *Food Control, 26*(2), 387–392.
16. Beaumont, S. L., Maggs, D. J., & Clarke, H. E., (2003). Effects of bovine lactoferrin on *in vitro* replication of feline herpesvirus. *Veterinary Ophthalmology, 6*(3), 245–250.
17. Bennett, R. M., & Davis, J., (1982). Lactoferrin interacts with deoxyribonucleic acid: A preferential reactivity with double-stranded DNA and dissociation of DNA-anti-DNA complexes. *The Journal of Laboratory and Clinical Medicine, 99*(1), 127–138.

18. Berkhout, B., Van, W. J. L., Beljaars, L., Meijer, D. K., Visser, S., & Floris, R., (2002). Characterization of the anti-HIV effects of native lactoferrin and other milk proteins and protein-derived peptides. *Antiviral Research, 55*(2), 341–355.
19. Bezault, J., Bhimani, R., Wiprovnick, J., & Furmanski, P., (1994). Human lactoferrin inhibits growth of solid tumors and development of experimental metastases in mice. *Cancer Research, 54*(9), 2310–2312.
20. Blackberg, L., & Hernell, O., (1980). Isolation of lactoferrin from human whey by a single chromatographic step. *FEBS Letters, 109*(2), 180–184.
21. Bokkhim, H., Bansal, N., Grondahl, L., & Bhandari, B., (2013). Physico-chemical properties of different forms of bovine lactoferrin. *Food Chemistry, 141*(3), 3007–3013.
22. Brown, C. A., Wang, B., & Oh, J. H., (2008). Antimicrobial activity of lactoferrin against foodborne pathogenic bacteria incorporated into edible chitosan film. *Journal of Food Protection, 71*(2), 319–324.
23. Brown, C., Wang, B., Jun, S., & OH, J., (2004). Antimicrobial efficacy of lactoferrin in chitosan-based edible film against foodborne pathogenic bacteria. In: *IFT's 2004 Annual Meeting + Food Expo®*. Las Vegas, NV.
24. Bruni, N., Capucchio, M. T., Biasibetti, E., Pessione, E., Cirrincione, S., Giraudo, L., Corona, A., & Dosio, F., (2016). Antimicrobial activity of lactoferrin-related peptides and applications in human and veterinary medicine. *Molecules, 21*(6), 752–777.
25. Chantaysakorn, P., & Richter, R. L., (2000). Antimicrobial properties of pepsin-digested lactoferrin added to carrot juice and filtrates of carrot juice. *Journal of Food Protection, 63*(3), 376–380.
26. Chen, L., Guo, C., Guan, Y. P., & Liu, H. Z., (2007). Isolation of lactoferrin from acid whey by magnetic affinity separation. *Separation and Purification Technology, 56*(2), 168–174.
27. Conesa, C., Sánchez, L., Rota, C., Pérez, M. D., Calvo, M., Farnaud, S., & Evans, R. W., (2008). Isolation of lactoferrin from milk of different species: Calorimetric and antimicrobial studies. *Comparative Biochemistry and Physiology Part B: Biochemistry and Molecular Biology, 150*(1), 131–139.
28. Conneely, O. M., (2001). Anti-inflammatory activities of lactoferrin. *Journal of the American College of Nutrition, 20*(Sup 5), 389S–397S.
29. Cornish, J., Callon, K. E., Naot, D., Palmano, K. P., Banovic, T., Bava, U., Watson, M., et al., (2004). Lactoferrin is a potent regulator of bone cell activity and increases bone formation *in vivo*. *Endocrinology, 145*(9), 4366–4374.
30. Cornish, J., & Naot, D., (2010). Lactoferrin as an effector molecule in the skeleton. *Biometals, 23*(3), 425–430.
31. Cumberbatch, M., Bhushan, M., Dearman, R. J., Kimber, I., & Griffiths, C. E. M., (2003). IL-1β-induced Langerhans' cell migration and TNF-α production in human skin: Regulation by lactoferrin. *Clinical and Experimental Immunology, 132*(2), 352–359.
32. Darmawan, K. K., Karagiannis, T. C., Hughes, J. G., Small, D. M., & Hung, A., (2020). High temperature-induced structural changes of apo-lactoferrin and interactions with β-lactoglobulin and α-lactalbumin for potential encapsulation strategies. *Food Hydrocolloids, 105*, 105817.
33. Deeth, H., & Bansal, N., (2018). Whey proteins: An overview. Chapter 1. In: Deeth, H. C., & Bansal, N., (eds.), *Whey Proteins: From Milk to Medicine* (pp. 1–50). London-UK: Academic Press.

34. Del, O. A., Calzada, J., & Nuñez, M., (2012). Effect of lactoferrin and its derivatives, high hydrostatic pressure, and their combinations, on *Escherichia coli* O157:H7 and *Pseudomonas fluorescens* in chicken filets. *Innovative Food Science & Emerging Technologies, 13*, 51–56.
35. Dial, E. J., Dohrman, A. J., Romero, J. J., & Lichtenberger, L. M., (2005). Recombinant human lactoferrin prevents NSAID-induced intestinal bleeding in rodents. *Journal of Pharmacy and Pharmacology, 57*(1), 93–99.
36. Dial, E. J., & Lichtenberger, L. M., (2002). Effect of lactoferrin on *Helicobacter felis* induced gastritis. *Biochemistry and Cell Biology, 80*(1), 113–117.
37. Drago-Serrano, M. E., Campos-Rodríguez, R., Carrero, J. C., & De La Garza, M., (2017). Lactoferrin: Balancing ups and downs of inflammation due to microbial infections. *International Journal of Molecular Sciences, 18*(3), 501.
38. Duarte, D. C., Nicolau, A., Teixeira, J. A., & Rodrigues, L. R., (2011). The effect of bovine milk lactoferrin on human breast cancer cell lines. *Journal of Dairy Science, 94*(1), 66–76.
39. Ekstrand, B., & Björck, L., (1986). Fast protein liquid chromatography of antibacterial components in milk lactoperoxidase, lactoferrin, and lysozyme. *Journal of Chromatography, 358*(2), 429–433.
40. Elagamy, E. I., Ruppanner, R., Ismail, A., Champagne, C. P., & Assaf, R., (1996). Purification and characterization of lactoferrin, lactoperoxidase, lysozyme, and immunoglobulins from camel's milk. *International Dairy Journal, 6*(2), 129–145.
41. Embleton, N. D., Berrington, J. E., McGuire, W., Stewart, C. J., & Cummings, S. P., (2013). Lactoferrin: Antimicrobial activity and therapeutic potential. *Seminars in Fetal & Neonatal Medicine, 18*(3), 143–149.
42. Engelmayer, J., & Varadhachary, A., (2003). Properties and application of recombinant human lactoferrin to enhance healing of diabetic wounds. *Wounds: A Compendium of Clinical Research and Practice, 15*(9), 294–301.
43. Fairweather-Tait, S. J., Balmer, S. E., Scott, P. H., & Minski, M. J., (1987). Lactoferrin and iron absorption in newborn infants. *Pediatric Research, 22*(6), 651–654.
44. Faraji, N., Zhang, Y., & Ray, A. K., (2017). Optimization of lactoperoxidase and lactoferrin separation on an ion-exchange chromatography step. *Separations, 4*(2), 10.
45. Farnaud, S., & Evans, R. W., (2003). Lactoferrin: A multifunctional protein with antimicrobial properties. *Molecular Immunology, 40*(7), 395–405.
46. Fee, C. J., & Chand, A., (2006). Capture of lactoferrin and lactoperoxidase from raw whole milk by cation exchange chromatography. *Separation and Purification Technology, 48*(2), 143–149.
47. Gado, I., Erdie, J., Laszlo, V. G., Paszti, J., Czirok, E., Kontrohr, T., Toth, I., et al., (1991). Correlation between human lactoferrin binding and colicin susceptibility in *Escherichia coli. Antimicrobial Agents and Chemotherapy, 35*(12), 2538–2543.
48. García-Montoya, I. A., Cendón, T. S., Arévalo-Gallegos, S., & Rascón-Cruz, Q., (2012). Lactoferrin a multiple bioactive protein: An overview. *Biochimicaet Biophysica Acta (BBA)-General Subjects, 1820*(3), 226–236.
49. Gobbetti, M., Minervini, F., & Rizzello, C. G., (2007). Bioactive peptides in dairy products. Chapter 70. In: Hui Y. H., (ed.), *Handbook of Food Products Manufacturing* (Vol. 2, pp. 489–517). Hoboken, New Jersey-USA: John Wiley and Sons Inc.

50. Goldsmith, S. J., Eitenmiller, R. R., Barnhart, H. M., Toledo, R. T., & Rao, V. N. M., (1982). Unsaturated iron-binding capacity of human milk. *Journal of Food Science, 47*(4), 1298–1300.
51. Griffiths, C. E. M., Cumberbatch, M., Tucker, S. C., Dearman, R. J., Andrew, S., Headon, D. R., & Kimber, I., (2001). Exogenous topical lactoferrin inhibits allergen-induced Langerhans cell migration and cutaneous inflammation in humans. *British Journal of Dermatology, 144*(4), 715–725.
52. Groves, M. L., (1960). The isolation of a red protein from milk. *Journal of the American Chemical Society, 82*(13), 3345–3350.
53. Hao, L., Shan, Q., Wei, J., Ma, F., & Sun, P., (2019). Lactoferrin: Major physiological functions and applications. *Current Protein and Peptide Science, 20*(2), 139–144.
54. Hara, K., Ikeda, M., Saito, S., Matsumoto, S., Numata, K., Kato, N., Tanaka, K., & Sekihara, H., (2002). Lactoferrin inhibits hepatitis B virus infection in cultured human hepatocytes. *Hepatology Research, 24*(3), 228–235.
55. Hassoun, L. A., & Sivamani, R. K. A., (2017). Systematic review of lactoferrin use in dermatology. *Critical Reviews in Food Science and Nutrition, 57*(17), 3632–3639.
56. Håversen, L., Ohlsson, B. G., Hahn-Zoric, M., Hanson, L. Å., & Mattsby-Baltzer, I., (2002). Lactoferrin down-regulates the LPS-induced cytokine production in monocytic cells via NF-κB. *Cellular Immunology, 220*(2), 83–95.
57. Hayes, T. G., Falchook, G. F., Varadhachary, G. R., Smith, D. P., Davis, L. D., Dhingra, H. M., Hayes, B. P., & Varadhachary, A., (2006). Phase I trial of oral talactoferrin alfa in refractory solid tumors. *Investigational New Drugs, 24*(3), 233–240.
58. Hutchens, T. W., Henry, J. F., Yip, T. T., Hachey, D. L., Schanler, R. J., Motil, K. J., & Garza, C., (1991). Origin of intact lactoferrin and its DNA binding fragments found in the urine of human milk-fed preterm infants. Evaluation by stable isotopic enrichment. *Pediatric Research, 29*(3), 243–250.
59. Iigo, M., Alexander, D. B., Long, N., Xu, J., Fukamachi, K., Futakuchi, M., Takase, M., & Tsuda, H., (2009). Anti-carcinogenesis pathways activated by bovine lactoferrin in the murine small intestine. *Biochimie, 91*(1), 86–101.
60. Ikadai, H., Tanaka, T., Shibahara, N., Tanaka, H., Matsuu, A., Kudo, N., Shimazaki, K., et al., (2005). Inhibitory effect of lactoferrin on *in vitro* growth of *Babesia caballi*. *The American Journal of Tropical Medicine and Hygiene, 73*(4), 710–712.
61. Ikeda, M., Sugiyama, K., Tanaka, T., Tanaka, K., Sekihara, H., Shimotohno, K., & Kato, N., (1998). Lactoferrin markedly inhibits hepatitis C virus infection in cultured human hepatocytes. *Biochemical and Biophysical Research Communications, 245*(2), 549–553.
62. Johansson, B., (1960). Isolation of an iron-containing red protein from human milk. *Acta Chemica Scandinavica, 14*(2), 510–512.
63. Joslin, R. S., Erickson, P. S., Santoro, H. M., Whitehouse, N. L., Schwab, C. G., & Rejman, J. J., (2002). Lactoferrin supplementation to dairy calves. *Journal of Dairy Science, 85*(5), 1237–1242.
64. Kell, D. B., Heyden, E. L., & Pretorius, E., (2020). The Biology of lactoferrin, an iron-binding protein that can help defend against viruses and bacteria. *Frontiers in Immunology, 11*, 1221.
65. Khan, A. I., Liu, J., & Dutta, P., (2020). Bayesian inference for parameter estimation in lactoferrin-mediated iron transport across blood-brain barrier. *Biochimica et Biophysica Acta (BBA)-General Subjects, 1864*(3), 129459.

66. Kirkpatrick, C. H., Green, I., Rich, R. R., & Schadeet, L. A., (1971). Inhibition of growth of *Candida albicans* by iron-unsaturated lactoferrin: Relation to host-defense mechanisms in chronic mucocutaneous candidiasis. *Journal of Infectious Diseases, 124*(6), 539–544.
67. Komatsu, A., Satoh, T., Wakabayashi, H., & Ikeda, F., (2015). Effects of bovine lactoferrin to oral *Candida albicans* and *Candida glabrata* isolates recovered from the saliva in elderly people. *Odontology, 103*(1), 50–55.
68. Kozu, T., Iinuma, G., Ohashi, Y., Saito, Y., Akasu, T., Saito, D., Alexander, D. B., et al., (2009). Effect of orally administered bovine lactoferrin on the growth of adenomatous colorectal polyps in a randomized, placebo-controlled clinical trial. *Cancer Prevention Research, 2*(11), 975–983.
69. Kruzel, M. L., Actor, J. K., Zimecki, M., Wise, J., Płoszaj, P., Mirza, S., Kruzel, M., et al., (2013). Novel recombinant human lactoferrin: Differential activation of oxidative stress related gene expression. *Journal of Biotechnology, 168*(4), 666–675.
70. Kruzel, M. L., Harari, Y., Mailman, D., Actor, J. K., & Zimecki, M., (2002). Differential effects of prophylactic, concurrent, and therapeutic lactoferrin treatment on LPS-induced inflammatory responses in mice. *Clinical & Experimental Immunology, 130*(1), 25–31.
71. Kurose, I., Yamada, T., Wolf, R., & Granger, D. N., (1994). P-selectin-dependent leukocyte recruitment and intestinal mucosal injury induced by lactoferrin. *Journal of Leukocyte Biology, 55*(6), 771–777.
72. Lee, T. H., Yu, S. L., Nam, M. S., Kim, S. J., Lee, K. K., Yu, D. Y., & Shimazaki, K., (1997). Polymorphic sequence of Korean native goat lactoferrin exhibiting greater antibacterial activity. *Animal Genetics, 28*(5), 367–369.
73. Legrand, D., & Mazurier, J., (2010). A critical review of the roles of host lactoferrin in immunity. *Biometals, 23*(3), 365–376.
74. Legrand, D., Elass, E., Carpentier, M., & Mazurier, J., (2005). Lactoferrin: A modulator of immune and inflammatory responses. *Cellular and Molecular Life Sciences, 62*(22), 2549–2559.
75. Legrand, D., Pierce, A., Elass, E., Carpentier, M., Mariller, C., & Mazurier, J., (2008). Lactoferrin structure and functions. In: Bösze, Z., (ed.), *Bioactive Components of Milk* (pp. 163–194). New York: Springer.
76. Legrand, D., Vigié, K., Said, E. A., Elass, E., Masson, M., Slomianny, M. C., Carpentier, M., et al., (2004). Surface nucleolin participates in both the binding and endocytosis of lactoferrin in target cells. *European Journal of Biochemistry, 271*(2), 303–317.
77. León-Sicairos, N., Reyes-López, M., Canizalez-Román, A., Bermúdez-Cruz, R. M., Serrano-Luna, J., Arroyo, R., & De La Garza, M., (2005). Human holo lactoferrin: Endocytosis and use as an iron source by the parasite *Entamoeba histolytica. Microbiology, 151*(12), 3859–3871.
78. León-Sicairos, N., Reyes-López, M., Ordaz-Pichardo, C., & De La Garza, M., (2006). Microbicidal action of lactoferrin and lactoferricin and their synergistic effect with metronidazole in *Entamoeba histolytica. Biochemistry and Cell Biology, 84*(3), 327–336.
79. Levay, P. F., & Viljoen, M., (1995). Lactoferrin: A general review. *Haematologica, 80*(3), 252–267.
80. Liang, Y., Wang, X., Wu, M., & Zhu, W., (2011). Simultaneous isolation of lactoferrin and lactoperoxidase from bovine colostrum by SPEC 70 SLS cation exchange resin. *International Journal of Environmental Research and Public Health, 8*(9), 3764–3776.

81. Li-Chan, E. C. Y., Ler, S. S., Kummer, A., & Akita, E. M., (1998). Isolation of lactoferrin by immunoaffinity chromatography using yolk antibodies. *Journal of Food Biochemistry, 22*(3), 179–195.
82. Lin, T. Y., Chu, C., & Chiu, C. H., (2002). Lactoferrin inhibits enterovirus 71 infection of human embryonal rhabdomyosarcoma cells *in vitro. The Journal of Infectious Diseases, 186*(8), 1161–1164.
83. Liu, J., Li, B., Lee, C., Zhu, H., Zheng, S., & Pierro, A., (2019). Protective effects of lactoferrin on injured intestinal epithelial cells. *Journal of Pediatric Surgery, 54*(12), 2509–2513.
84. Lönnerdal, B., Carlsson, J., & Porath, J., (1977). Isolation of lactoferrin from human milk by metal chelate affinity chromatography. *FEBS Letter, 75*(1), 89–92.
85. Lönnerdal, B., & Iyer, S., (1995). Lactoferrin: Molecular structure and biological function. *Annual Review of Nutrition, 15*, 93–110.
86. Lorget, F., Clough, J., Oliveira, M., Daury, M. C., Sabokbar, A., & Offord, E., (2002). Lactoferrin reduces *in vitro* osteoclast differentiation and resorbing activity. *Biochemical and Biophysical Research Communications, 296*(2), 261–266.
87. Lu, L., Hangoc, G., Oliff, A., Chen, L. T., Shen, R. N., & Broxmeyer, H. E., (1987). Protective influence of lactoferrin on mice infected with the polycythemia-inducing strain of friend virus complex. *Cancer Research, 47*(15), 4184–4188.
88. Lu, R. R., Xu, S. Y., Wang, Z., & Yang, R. J., (2007). Isolation of lactoferrin from bovine colostrum by ultrafiltration coupled with strong cation exchange chromatography on a production scale. *Journal of Membrane Science, 297*(1, 2), 152–161.
89. Luzi, C., Brisdelli, F., Iorio, R., Bozzi, A., Carnicelli, V., Di Giulio, A., & Lizzi, A. R., (2017). Apoptotic effects of bovine apo-lactoferrin on Hela tumor cells. *Cell Biochemistry and Function, 35*(1), 33–41.
90. Machnicki, M., Zimecki, M., & Zagulski, T., (1993). Lactoferrin regulates the release of tumour necrosis factor-alpha and interleukin-6 *in vivo. International Journal of Experimental Pathology, 74*(5), 433–439.
91. Manzoni, P., (2019). Clinical studies of lactoferrin in neonates and infants: An update. *Breastfeeding Medicine, 14*(S1), S25–S27.
92. Marchetti, M., Longhi, C., Conte, M. P., Pisani, S., Valenti, P., & Seganti, L., (1996). Lactoferrin inhibits herpes simplex virus type 1 adsorption to Vero cells. *Antiviral Research, 29*(2, 3), 221–231.
93. Mayeur, S., Spahis, S., Pouliot, Y., & Levy, E., (2016). Lactoferrin, a pleiotropic protein in health and disease. *Antioxidants and Redox Signaling, 24*(14), 813–836.
94. Metz-Boutique, M. H., Jolles, J., Mazuries, J., Schoentgen, F., Legrand, D., Spik, G., Montreuil, J., & Jolles, P., (1984). Human lactotransferrin amino sequence and structural comparisons with other transferrins. *European Journal of Biochemistry, 145*(3), 659–676.
95. Montreuil, J., Tonnelat, J., & Mullet, S., (1960). Preparation and properties of lactosiderophilin (lactotransferrin) of human milk. *Biochimica et Biophysica Acta, 45*(3), 413–421.
96. Moore, S. A., Anderson, B. F., Groom, C. R., Haridas, M., & Baker, E. N., (1997). Three-dimensional structure of Di ferric bovine lactoferrin at 2.8 å resolution. *Journal of Molecular Biology, 274*(2), 222–236.
97. Moradian, F., Sharbafi, R., & Rafiei, A., (2014). Lactoferrin, isolation, purification, and antimicrobial effects. *Journal of Medical and Bioengineering, 3*(3), 203–206.

98. Moreno-Expósito, L., Illescas-Montes, R., Melguizo-Rodríguez, L., Ruiz, C., Ramos-Torrecillas, J., & De Luna-Bertos, E., (2018). Multifunctional capacity and therapeutic potential of lactoferrin. *Life Sciences, 195*, 61–64.
99. Naidu, A. S., (2002). Activated lactoferrin: A new approach to meat safety. *Food Technology, 56*(3), 40–46.
100. Naito, K., Iio, T., Katagi, M., Yoshikawa, Y., Ohtsuka, H., & Orino, K., (2020). Binding analysis of bovine milk proteins, especially casein interactions and the interaction between α-casein and lactoferrin, using beads immobilized with zinc ion, poly-l-lysine or α-casein. *International Dairy Journal, 105*, 104690.
101. Niaz, B., Saeed, F., Ahmed, A., Imran, M., Maan, A. A., Khan, M. K. I., Tufail, T., et al., (2019). Lactoferrin (LF): A natural antimicrobial protein. *International Journal of Food Properties, 22*(1), 1626–1641.
102. Niaz, B., Zahoor, T., Randhawa, M. A., & Jamil, A., (2017). Isolation of lactoferrin from camel milk through fast protein liquid chromatography and its antagonistic activity against *Escherichia coli* and *Staphylococcus aureus. Pakistan Journal of Zoology, 49*(4), 1307–1313.
103. Nikawa, H., Samaranayake, L. P., Tenovuo, J., Pang, K. M., & Hamada, T., (1993). The fungicidal effect of human lactoferrin on *Candida albicans* and *Candida krusei. Archives of Oral Biology, 38*(12), 1057–1063.
104. Nikawa, H., Samarayanake, L. P., & Hamada, T., (1995). Modulation of the anti-candida activity of apo-lactoferrin by dietary sucrose and tunicamycin *in vitro. Archives of Oral Biology, 40*(6), 581–584.
105. Noh, K. H., Rhee, M. S., & Imm, J. Y., (2005). Separation of lactoferrin from model whey protein mixture by reverse micelles formed by cationic surfactant. *Food Science and Biotechnology, 14*(1), 131–136.
106. Noppe, W., Plieva, F. M., Galaev, I. Y., Vanhoorelbeke, K., Mattiasson, B., & Deckmyn, H., (2006). Immobilized peptide displaying phages as affinity ligands purification of lactoferrin from defatted milk. *Journal of Chromatography-A, 1101*(1, 2), 79–85.
107. Nuijens, J. H., & Van, V. H. H., (1999). *Isolation of Lactoferrin from Milk* (p. 40). US Patent No. 5,861,491.
108. Ochoa, T. J., Chea-Woo, E., Baiocchi, N., Pecho, I., Campos, M., Prada, A., Valdiviezo, G., et al., (2013). Randomized double-blind controlled trial of bovine lactoferrin for prevention of diarrhea in children. *The Journal of Pediatrics, 162*(2), 349–356.
109. Ochoa, T. J., Chea-Woo, E., Campos, M., Pecho, I., Prada, A., McMahon, R. J., & Cleary, T. G., (2008). Impact of lactoferrin supplementation on growth and prevalence of giardia colonization in children. *Clinical Infectious Diseases, 46*(12), 1881–1883.
110. Ochoa, T. J., & Clearly, T. G., (2004). Lactoferrin Disruption of Bacterial Type III Secretion Systems. *Biometals, 17*(3), 257–260.
111. Oh, J., Wang, B., Dessai, A., & Aglan, H., (2003). The efficacy of natural anti-microbial agent lactoferrin in protein-based edible films against *E. Coli.* In: *Food Packaging: General IFT Annual Meeting.* Chicago.
112. Omata, Y., Satake, M., Maeda, R., Saito, A., Shimazaki, K., Uzuka, Y., Tanabe, S., et al., (2001). Reduction of the infectivity of toxoplasma gondii and *Eimeria stiedai* sporozoites by treatment with bovine lactoferricin. *Journal of Veterinary Medical Science, 63*(2), 187–190.

113. Onishi, H., Machida, Y., & Koyama, K., (2007). Preparation and *in vitro* characteristics of lactoferrin-loaded chitosan microparticles. *Drug Development and Industrial Pharmacy, 33*(6), 641–647.
114. Palmano, K., Ramos, R., Watson, M., Callon, K., & Cornish, J., (2011). Survival and bone-active properties of bovine lactoferrin supplemented into stirred yoghurt. *International Dairy Journal, 21*(7), 477–483.
115. Parhi, P., Mohanty, C., & Sahoo, S. K., (2012). Nanotechnology based combinational drug delivery: An emerging approach for cancer therapy. *Drug Discovery Today, 17*(17, 18), 1044–1052.
116. Pempek, J. A., Watkins, L. R., Bruner, C. E., & Habing, G. G., (2019). A multisite, randomized field trial to evaluate the influence of lactoferrin on the morbidity and mortality of dairy calves with diarrhea. *Journal of Dairy Science, 102*(10), 9259–9267.
117. Pierce, A., Legrand, D., & Mazurier, J., (2009). Lactoferrin: A multifunctional protein. *Médecine/Sciences, 25*(4), 361–369.
118. Pietrantoni, A., Di Biase, A. M., Tinari, A., Marchetti, M., Valenti, P., Seganti, L., & Superti, F., (2003). Bovine lactoferrin inhibits adenovirus infection by interacting with viral structural polypeptides. *Antimicrobial Agents and Chemotherapy, 47*(8), 2688–2691.
119. Plate, K., Beutel, S., Buccholz, H., Demmer, W., Fischer-Fruhholz, S., Reif, O., Ulber, R., & Scheper, T., (2006). Isolation of bovine lactoferrin, lactoperoxidase, and enzymatically prepared lactoferricin from proteolytic digestion of bovine lactoferrin using adsorptive membrane chromatography. *Journal of Chromatography-A, 1117*(1), 81–86.
120. Puddu, P., Valenti, P., & Gessani, S., (2009). Immunomodulatory effects of lactoferrin on antigen-presenting cells. *Biochimie, 91*(1), 11–18.
121. Qiu, J., Hendrixson, D. R., Baker, E. N., Murphy, T. F., Geme, J. W. S., & Plaut, A. G., (1998). Human milk lactoferrin inactivates two putative colonization factors expressed by *Haemophilus influenzae*. *Proceedings of the National Academy of Sciences, 95*(21), 12641–12646.
122. Quinteri, L., Pistillo, B. R., Caputo, L., Favia, P., & Baruzzi, F., (2013). Bovine lactoferrin and lactoferricin on plasma-deposited coating against spoilage *Pseudomonas* spp. *Innovative Food Science & Emerging Technologies, 20*, 215–222.
123. Rachman, A. B., Maheswari, R. R. A., & Bachroem, M. S., (2015). Composition and isolation of lactoferrin from colostrum and milk of various goat breeds. *Procedia Food Science, 3*, 200–210.
124. Rodrigues, L., Teixeira, J., Schmitt, F., Paulsson, M., & Månsson, H. L., (2009). Lactoferrin and cancer disease prevention. *Critical Reviews in Food Science and Nutrition, 49*(3), 203–217,
125. Rodríguez-Franco, D. A., Vázquez-Moreno, L., & Ramos-Clamont, M. G., (2005). Antimicrobial activity of lactoferrin: Mechanisms and potential clinical applications. *Latin American Journal of Microbiology, 47*(3, 4), 102–111.
126. Rollini, M., Nielsen, T., Musatti, A., Limbo, S., Piergiovanni, L., Munoz, P. H., & Gavara, R., (2016). Antimicrobial performance of two different packaging materials on the microbiological quality of fresh salmon. *Coatings, 6*(1), 6.
127. Rosa, L., Cutone, A., Lepanto, M. S., Paesano, R., & Valenti, P., (2017). Lactoferrin: A natural glycoprotein involved in iron and inflammatory homeostasis. *International Journal of Molecular Sciences, 18*(9), 1985.

128. Sato, R., Inanami, O., Tanaka, Y., Takase, M., & Naito, Y., (1996). Oral administration of bovine lactoferrin for treatment of intractable stomatitis in feline immunodeficiency virus (FIV)-positive and FIV-negative cats. *American Journal of Veterinary Research, 57*(10), 1443–1446.
129. Schanbacher, F. L., Goodman, R. E., & Talhouk, R. S., (1993). Bovine mammary lactoferrin: Implications from messenger ribonucleic acid (mRNA) sequence and regulation contrary to other milk proteins. *Journal of Dairy Science, 76*(12), 3812–3831.
130. Seganti, L., Di Biase, A. M., Marchetti, M., Pietrantoni, A., Tinari, A., & Superti, F., (2004). Antiviral activity of lactoferrin towards naked viruses. *Biometals, 17*(3), 295–299.
131. Shashikumar, C. S. S., & Puranik, D. B., (2012). Study on use of lactoferrin for the biopreservation of paneer. *Tropical Agricultural Research, 23*(1), 70–76.
132. Shimamura, M., Yamamoto, Y., Ashino, H., Oikawa, T., Hazato, T., Tsuda, H., & Ligo, M., (2004). Bovine lactoferrin inhibits tumor-induced angiogenesis. *International Journal of Cancer, 111*(1), 111–116.
133. Shimizu, K., Matsuzawa, H., Okada, K., Tazume, S., Dosako, S., Kawasaki, Y., Hashimoto, K., & Koga, Y., (1996). Lactoferrin-mediated protection of the host from murine cytomegalovirus infection by a T-cell-dependent augmentation of natural killer cell activity. *Archives of Virology, 141*(10), 1875–1889.
134. Shin, K., Yamauchi, K., Teraguchi, S., Hayasawa, H., Tomita, M., Otsuka, Y., & Yamazaki, S., (1998). Antibacterial activity of bovine lactoferrin and its peptides against enterohaemorrhagic *Escherichia coli* O157: H7. *Letters in Applied Microbiology, 26*(6), 407–411.
135. Silva, T., Adão, R., Nazmi, K., Bolscher, J. G. M., Funari, S. S., Uhríková, D., & Bastos, M., (2013). Structural diversity and mode of action on lipid membranes of three lactoferrin candidacidal peptides. *Biochimica et Biophysica Acta (BBA)-Biomembranes, 1828*(5), 1329–1339.
136. Siqueiros-Cendón, T., Arévalo-Gallegos, S., Iglesias-Figueroa, B. F., García-Montoya, I. A., Salazar-Martínez, J., & Rascón-Cruz, Q., (2014). Immunomodulatory effects of lactoferrin. *Acta Pharmacologica Sinica, 35*(5), 557–566.
137. Sorimachi, K., Akimoto, K., Hattori, Y., Ieiri, T., & Niwa, A., (1997). Activation of macrophages by lactoferrin: Secretion of TNF-alpha, IL-8 and NO. *Biochemistry and Molecular Biology International, 43*(1), 79–87.
138. Spadaro, M., Caorsi, C., Ceruti, P., Varadhachary, A., Forni, G., Pericle, F., & Giovarelli, M., (2008). Lactoferrin, a major defense protein of innate immunity, is a novel maturation factor for human dendritic cells. *FASEB Journal, 22*(8), 2747–2757.
139. Suzuki, Y. A., Lopez, V., & Lonnerdal, B., (2005). Mammalian lactoferrin receptors: Structure and function. *Cellular and Molecular Life Sciences, 62*(22), 2560–2575.
140. Suzuki, Y. A., Wong, H., Ashida, K. Y., Schryvers, A. B., & Lönnerdal, B., (2008). The N1 domain of human lactoferrin is required for internalization by caco-2 cells and targeting to the nucleus. *Biochemistry, 47*(41), 10915–10920.
141. Swart, P. J., Kuipers, M. E., Smit, C., Pauwels, R., De Béthune, M. P., De Clercq, E., Meijer, D. K. F., & Huisman, J. G., (1996). Antiviral effects of milk proteins: Acylation results in polyanionic compounds with potent activity against human immunodeficiency virus types 1 and 2 *in vitro. AIDS Research and Human Retroviruses, 12*(9), 769–775.

142. Szuster-Ciesielska, A., Kamińska, T., & Kandefer-Szerszeń, M., (1995). Phagocytosis-enhancing effect of lactoferrin on bovine peripheral blood monocytes *in vitro* and *in vivo. Archivum Veterinarium Polonicum, 35*(1, 2), 63–71.
143. Tachezy, J., Kulda, J., Bahníková, I., Suchan, P., Rázga, J., & Schrével, J., (1996). *Tritrichomonas foetus*: Iron acquisition from lactoferrin and transferrin. *Experimental Parasitology, 83*(2), 216–228.
144. Takayama, Y., & Aoki, R., (2012). Roles of lactoferrin on skin wound healing. *Biochemistry and Cell Biology, 90*(3), 497–503,
145. Takayama, Y., & Mizumachi, K., (2008). Effect of bovine lactoferrin on extracellular matrix calcification by human osteoblast-like cells. *Bioscience, Biotechnology, and Biochemistry, 72*(1), 226–230.
146. Tang, L., Wu, J. J., Ma, Q., Cui, T., Andreopoulos, F. M., Gil, J., Valdes, J., et al., (2010). Human lactoferrin stimulates skin keratinocyte function and wound re-epithelialization. *British Journal of Dermatology, 163*(1), 38–47.
147. Taylor, S., Brock, J., Kruger, C., Berner, T., & Murphy, M., (2004). Safety determination for the use of bovine milk-derived lactoferrin as a component of an antimicrobial beef carcass spray. *Regulatory Toxicology and Pharmacology, 39*(1), 12–24.
148. Togawa, J. I., Nagase, H., Tanaka, K., Inamori, M., Umezawa, T., Nakajima, A., Naito, M., et al., (2002). Lactoferrin reduces colitis in rats via modulation of the immune system and correction of cytokine imbalance. *American Journal of Physiology-Gastrointestinal and Liver Physiology, 283*(1), G187–G195.
149. Tsuda, H., Kozu, T., Iinuma, G., Ohashi, Y., Saito, Y., Saito, D., Akasu, T., et al., (2010). Cancer prevention by bovine lactoferrin: From animal studies to human trial. *Biometals, 23*(3), 399–409.
150. Tsuda, H., Sekine, K., Fujita, K., & Ligo, M., (2002). Cancer prevention by bovine lactoferrin and underlying mechanisms: A review of experimental and clinical studies. *Biochemistry and Cell Biology, 80*(1), 131–136.
151. Tu, Y. Y., Chen, C. C., Chang, J. H., & Chang, H. M., (2002). Characterization of lactoferrin (LF) from colostral whey using anti-LF antibody immunoaffinity chromatography. *Journal of Food Science, 67*(3), 996–1001.
152. Turchany, J. M., Aley, S. B., & Gillin, F. D., (1995). Giardicidal activity of lactoferrin and n-terminal peptides. *Infection and Immunity, 63*(11), 4550–4552.
153. Uchida, T., Dosako, S., Sato, K., & Kawakami, H., (2003). Sequential separation of lactoferrin, lactoperoxidase, and secretory component by lactoferrinate-linked ion-exchange chromatography. *Milchwissenschaft, 58*(9), 482–486.
154. Umuhumuza, L. C., Wei-min, N., & Sun, X., (2011). Effect of bovine lactoferrin and casein peptide powder on microbial growth and glucose utilization by microorganisms in pork meat during storage at 4°C. *Pakistan Journal of Nutrition, 10*(3), 208–213.
155. Valenti, P., & Antonini, G., (2005). Lactoferrin: An important host defense against microbial and viral attack. *Cellular and Molecular Life Sciences, 62*(22), 2576–2587.
156. Valenti, P., Visca, P., Antonini, G., & Orsi, N., (1986). Interaction between lactoferrin and ovotransferrin and candida cells. *FEMS Microbiology Letters, 33*(2, 3), 271–275.
157. Van, D. D. A. M., Joosten, S. A., Vroomans, E., Bogaards, S. J. P., Van, M. K. E., Ottenhoff, T. H. M., Van, D. J. T., & Nibbering, P. H., (2012). The antimicrobial peptide hLF1-11 drives monocyte-dendritic cell differentiation toward dendritic cells that promote antifungal responses and enhance Th17 polarization. *Journal of Innate Immunity, 4*(3), 284–292.

158. Van, D. S. B. W. A., Beljaars, L., Molema, G., Harmsen, M. C., & Meijer, D. K. F., (2001). Antiviral activities of lactoferrin. *Antiviral Research, 52*(3), 225–239.
159. Van, D. S. B. W. A., Harmsen, M. C., Schäfer, P., Swart, P. J., The, T. H., Jahn, G., Speer, C. P., et al., (2001). Viral load in breast milk correlates with transmission of human cytomegalovirus to preterm neonates, but lactoferrin concentrations do not. *Clinical and Diagnostic Laboratory Immunology, 8*(4), 818–821.
160. Van, V. H. A., Geerts, M. E. J., Van, B. P. H. C., & Nuijens, J. H., (2004). The role of N-linked glycosylation in the protection of human and bovine lactoferrin against tryptic proteolysis. *European Journal of Biochemistry, 271*(4), 678–684.
161. Viejo-Díaz, M., Andrés, M. T., & Fierro, J. F., (2004). Modulation of *in vitro* fungicidal activity of human lactoferrin against *Candida albicans* by extracellular cation concentration and target cell metabolic activity. *Antimicrobial Agents and Chemotherapy, 48*(4), 1242–1248.
162. Waarts, B. L., Aneke, O. J., Smit, J. M., Kimata, K., Bittman, R., Meijer, D. K., & Wilschut, J., (2005). Antiviral activity of human lactoferrin: Inhibition of alphavirus interaction with heparan sulfate. *Virology, 333*(2), 284–292.
163. Wakabayashi, H., Takakura, N., Yamauchi, K., & Tamura, Y., (2006). Modulation of immune related gene expression in small intestines of mice by oral administration of lactoferrin. *Clinical Vaccine Immunology, 13*(2), 239–245.
164. Wakabayashi, H., Yamauchi, K., & Takase, M., (2006). Lactoferrin research, technology, and applications. *International Dairy Journal, 16*(11), 1241–1251.
165. Wang, B., Timilsena, Y. P., Blanch, E., & Adhikari, B., (2017). Lactoferrin: Structure, function, denaturation, and digestion. *Critical Reviews in Food Science and Nutrition, 59*(4), 580–596.
166. Wang, H., Ye, X., & Ng, T. B., (2000). First demonstration of an inhibitory activity of milk proteins against human immunodeficiency virus-1 reverse transcriptase and the effect of succinylation. *Life Sciences, 67*(22), 2745–2752.
167. Wang, W. P., Ligo, M., Sato, J., Sekine, K., Adachi, I., & Tsuda, H., (2000). Activation of mucosal intestinal immunity in tumor-bearing mice by lactoferrin. *Japanese Journal of Cancer Research, 91*(10), 1022–1027.
168. Ward, P. P., Paz, E., & Conneely, O. M., (2005). Multifunctional roles of lactoferrin: A critical overview. *Cellular and Molecular Life Sciences, 62*(22), 2540–2548.
169. Weinberg, E. D., (2007). Antibiotic properties and applications of lactoferrin. *Current Pharmaceutical Design, 13*(8), 801–811.
170. Wirjantoro, T. I., Lewis, M. J., Grandison, A. S., Williams, G. C., & Delves-Broughton, J., (2001). The effect of nisin on the keeping quality of reduced heat-treated milks. *Journal of Food Protection, 64*(2), 213–219.
171. Włodarski, K. H., Galus, R., Brodzikowska, A., Włodarski, P. K., & Wojtowicz, A., (2014). The importance of lactoferrin in bone regeneration. *Polski Merkuriusz Lekarski: Organ Polskiego Towarzystwa Lekarskiego, 37*(217), 65–67.
172. Wolf, J. S., Li, D., Taylor, R. J., & O'Malley, Jr. B. W., (2003). Lactoferrin inhibits growth of malignant tumors of the head and neck. *ORL, 65*(5), 245–249.
173. Wu, M. B., & Xu, Y. J., (2009). Isolation and purification of lactoferrin and immunoglobulin g from bovine colostrum with serial cation-anion exchange chromatography. *Biotechnology and Bioprocess Engineering, 14*(2), 155–160.
174. Xu, Y. Y., Samaranayake, Y. H., Samaranayake, L. P., & Nikawa, H., (1999). *In vitro* susceptibility of candida species to lactoferrin. *Medical Mycology, 37*(1), 35–41.

175. Yang, H. G., Li, H. Y., Li, P., Bao, X. Y., Huang, G. X., Xing, L., Zheng, N., & Wang, J. Q., (2020). Modulation activity of heat-treated and untreated lactoferrin on the TLR-4 pathway in anoxia cell model and cerebral ischemia-reperfusion mouse model. *Journal of Dairy Science, 103*(2), 1151–1163.
176. Ye, X., Yoshida, S., & Ng, T. B., (2000). Isolation of lactoperoxidase, lactoferrin, α-lactalbumin, β-lactoglobulin B, and β-lactoglobulin a from bovine rennet whey using ion-exchange chromatography. *The International Journal of Biochemistry & Cell Biology, 32*(11–12), 1143–1150.
177. Yõlmaz, B., & Tosun, H., (2002). Sütte bulunan doğal antimikrobiyal sistemler ve bunlarõn gõda sanayinde kullanõmõ. *Celal Bayar Üniv. Fen Bilimleri Dergisi, 8*(1), 11–20.
178. Yoo, Y. C., Watanabe, R., Koike, Y., Mitobe, M., Shimazaki, K., Watanabe, S., & Azuma, I., (1997). Apoptosis in human leukemic cells induced by lactoferricin, a bovine milk protein-derived peptide: Involvement of reactive oxygen species. *Biochemical and Biophysical Research Communications, 237*(3), 624–628.
179. Yoshida, S., (1989). Preparation of lactoferrin by hydrophobic interaction chromatography from milk acid whey. *Journal of Dairy Science, 72*(6), 1446—1450.
180. Yoshida, S., Wei, Z., Shinmura, Y., & Fukunaga, N., (2000). Separation of lactoferrin-A and -B from bovine colostrum. *Journal of Dairy Science, 83*(10), 2211–2215.
181. Yoshida, S., & Ye-Xiuyun, (1991). Isolation of lactoperoxidase and lactoferrins from bovine milk acid whey by carboxymethyl cation exchange chromatography. *Journal of Dairy Science, 74*(5), 1439–1444.
182. Yount, Y., Andrés, M. T., Fierro, J. F., & Yeaman, M. R., (2007). The γ-core motif correlates with antimicrobial activity in cysteine-containing kaliocin-1 originating from transferrins. *Biochimica et Biophysica Acta (BBA)-Biomembranes, 1768*(11), 2862–2872.
183. Zainab, H. A., Kifah, S. D., & Nahi, Y. Y., (2015). Isolation, purification, and characterization of lactoferrin from goat colostrums whey. *Pakistan Journal of Nutrition, 14*(8), 517–523.
184. Zhang, Y., Lima, C. F., & Rodrigues, L. R., (2014). Anticancer effects of lactoferrin: Underlying mechanisms and future trends in cancer therapy. *Nutrition Reviews, 72*(12), 763–773.
185. Zheng, J., Xie, Y., Li, F., Zhou, Y., Qi, L., Liu, L., & Chen, Z., (2020). Lactoferrin improves cognitive function and attenuates brain senescence in aged mice. *Journal of Functional Foods, 65*, 103736.
186. Zimecki, M., Dawiskiba, J., Zawirska, B., Krawczyk, Z., & Kruzel, M., (2003). Bovine lactoferrin decreases histopathological changes in the liver and regulates cytokine production by splenocytes of obstructive jaundiced rats. *Inflammation Research, 52*(7), 305–310.
187. Zimecki, M., Spiegel, K., Właszczyk, A., Kübler, A., & Kruzel, M. L., (1999). Lactoferrin increases the output of neutrophil precursors and attenuates the spontaneous production of TNF-alpha and IL-6 by peripheral blood cells. *Archivum Immunologiae et Therapiae Experimentalis, 47*(2), 113–118.
188. Zimecki, M., Właszczyk, A., Wojciechowski, R., Dawiskiba, J., & Kruzel, M., (2001). Lactoferrin regulates the immune responses in post-surgical patients. *Archivum Immunologiae et Therapiae Experimentalis, 49*(4), 325–333.

PART III

Enhancing the Functionality of Dairy Foods

CHAPTER 8

FORTIFICATION OF VITAMINS AS DIETARY SUPPLEMENTS IN DAIRY FOODS

PRIYANKA S. RAO, HEENA SHARMA, SUMIT ARORA, VIVEK SHARMA, and RICHA SINGH

ABSTRACT

Vitamins are essential micronutrients which are required in very low concentrations (mg or μg) to perform various metabolic activities occurring in the human body system. Deficiency of vitamins results in various health-related disorders and certain complications. Fortifying vitamins in food seems to be a reliable and effective approach for their delivery among consumers in order to maintain recommended dietary allowances. Dairy products are important components in the global food chain; hence their fortification can play an critical role in meeting the daily dietary requirement of certain vitamins. The aim of the current chapter is to review technological interventions and nutritional aspects of vitamin (A and D) fortification in dairy products. Fortification with vitamins poses many technological challenges as these are sensitive to both light and heat. To overcome the losses of vitamins during processing and storage, overages should be calculated and further added to meet the regulatory standards related to vitamin fortification. Bioavailability of vitamins from fortified foods is another important aspect which has to be addressed since the form of vitamin (free or encapsulated) and type of carrier matrix significantly affects its bioavailability. The importance of legal issues regarding

Functional Dairy Ingredients and Nutraceuticals: Physicochemical, Technological, and Therapeutic Aspects. Megh R. Goyal, N. Veena, PhD & Santosh K. Mishra (Eds.)

fortification, risk of hypervitaminosis and the methods used for estimation of vitamins in fortified foods are also discussed in this chapter.

8.1 INTRODUCTION

Essential physiological functions of the human body require different classes of nutrients. Vitamins, an organic micronutrient, are required in small amount to perform normal physiological functions (development, growth, production, and maintenance). Vitamins available in food undergo activation process by conversion of biologically active to metabolically active form to perform physiological functions. Insufficient dietary intake, absence or underutilization of these vitamins is exhibited by the host in terms of deficiency symptoms as biochemical and physiological changes or disease signs. Deficiency of vitamins is usually noticed among consumers unless shown by particular symptoms; therefore, it is also referred to as 'hidden hunger.' One third population of the world has micronutrient deficiency, while in India, the majority of the population consume less than 50% of the daily required dose of micronutrients [60]. Micronutrient enriched foods such as fruits, vegetables, meat, and eggs are not the part of daily diet among Indians. The addition of vitamins to widely consumed foods, i.e., flour, milk, vegetable oil is common practice to overcome the deficiency and is known as fortification. The process by which concentration or quantity of a particular or an essential micronutrient is increased in food deliberately, mainly vitamins and minerals, is termed as fortification. It not only enhances nutritional food quality but also minimizes the risk to the health of consumers.

World Health Organization defines 'fortification' as deliberate integration of nutritive components singly or in combination, thus, increasing nutrient concentration/quantity, thereby, enhancing the nutritional food quality and preventing its deficiency [2]. However, certain considerations should be taken care of before its incorporation, viz., it should be non-toxic and non-competitive (no interference with assimilation and metabolism of other nutrients) in nature [20].

The importance of fortification of food items with vitamins has been realized over the past few years and is considered as useful technique to combat deficiencies related to vitamins. In western countries, food fortification has been used at commercial scale since time immemorial, for successfully combating the micronutrient deficiency including minerals viz., iron, and iodine, and vitamins viz., fat-soluble vitamins (A and D) and vitamin

B complex. On the contrary, in less developed countries, food fortification has become a recent trend owing to increased awareness among consumers and population realizing its health benefits. Several planned programs have been implemented successfully at a much rapid pace than ever thought. In this regard, dairy products have contributed a major role in successfully launching these goal-oriented programs as the former have emerged as efficient delivery vehicle and their popularity among majority of consumers. This has also ensured the uniform delivery of fortificants globally.

The concept of food fortification using dairy products as delivery vehicle is being given due importance and accepted worldwide; thereby expected to further augment the sales of dairy products manifolds.

The current chapter explores the basics of vitamin fortification (especially vitamin A and D) in dairy products, regulatory aspects regarding fortification, risk of hypervitaminosis and the estimation of vitamin (A and D) in fortified foods.

8.2 BASIC PRINCIPLE OF VITAMIN FORTIFICATION OF DAIRY PRODUCTS

Food based fortification is one of the key strategies among different policies and programs to improve vitamin deficiency [2, 75]. Different terms have been used to characterize for the fortification of micronutrients (Figure 8.1).

Dairy products are a useful vehicle for vitamin fortification point of view as they are part of the daily diet in major Indian population across the country. Certain other milk-based products such as yogurt, flavored milk, whey protein-based drink and low-fat milk is also desirable choice for vitamin fortification. Fortification of dairy products is preferred to target the vulnerable population, i.e., women, and infants. The main reason of preferring the dairy products for the fortification are as follows:

- Comparatively better stability and bioavailability of vitamins in milk;
- Globally consumed by various age group of consumers in anticipated quantity;
- Modest cost for targeted consumers;
- Organized sector processing leads to the uniform quality product in market.

Several countries have made it mandatory to fortify liquid milk with fat-soluble vitamins (mainly vitamin A and D) (Table 8.1). However, apart from liquid milk, there has been a recent trend to fortify several dairy products, including dried milk and ice-cream (vitamin A and D); butter (β-carotene as a natural coloring agent); milk-based weaning foods and infant formula (minerals and vitamins).

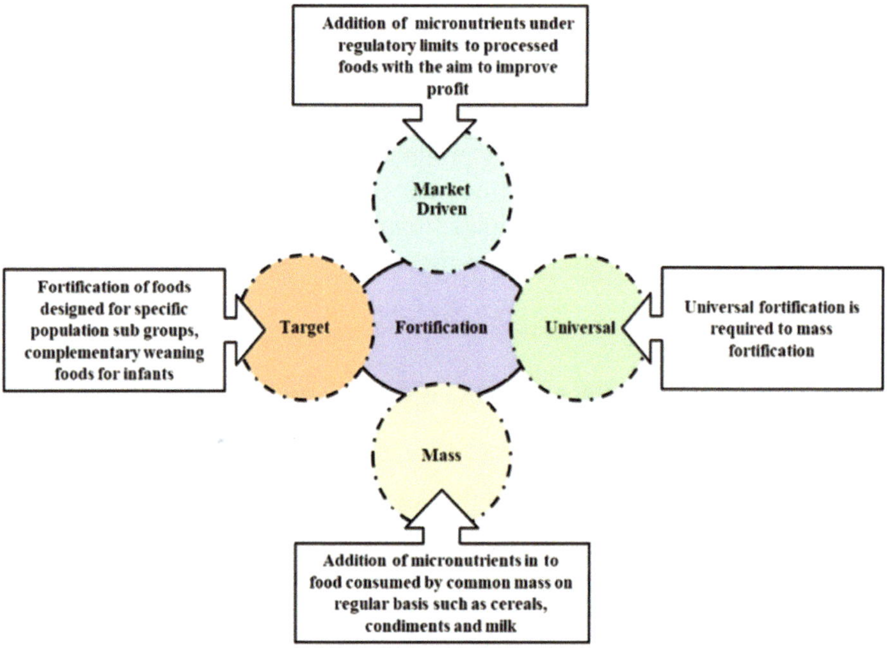

FIGURE 8.1 Types of food fortification to overcome the micronutrient deficiency.

Among two different classes of vitamins (i.e.: water and fat-soluble), globally fat-soluble vitamins (A and D) are majorly fortified in liquid milk and dairy products to combat their deficiency. These vitamins can be added to dairy products with limited additional technological changes in the raw milk processing line. Vitamins are commercially available in dry powder or oily form for fortification. Dry powder has the advantage of ease of mixing in milk, however; less stability makes it less desirable form for fortification. Oily vitamin preparation is mixed in bulk milk with prior dilution and homogenization in a small quantity of milk [61]. Vitamin is added in raw milk followed by pasteurization and homogenization step (Figure 8.2).

TABLE 8.1 Fortification of Dairy Products with Vitamins (A and D) in Different Countries*

Country	Type of Fortification	Milk and Milk Products	Vitamin A (IU)	Vitamin D (IU)
Canada	Mandatory	Milk	–	3000–4000/kg
		Skimmed milk	1200–1500/L	
		Partially skimmed milk; evaporated milk; milk powder	1200–2500/L	
USA	Mandatory	Fortified non-fat dried milk	2115/L	425/L
	Mandatory	Evaporated skim milk	4225/L	845/L
	Voluntary	Yogurt (Low fat, non-fat)	–	890/kg
	Voluntary	Cheese (excluding Ricotta, Cottage, and hard grating cheese)	–	2700/kg
Mexico	Mandatory	Sterilized, pasteurized milk (low fat), evaporated milk (both whole and low fat)	4000/L	400/L
Brazil	Mandatory	Dried skimmed milk for complimentary food	15000–25000/kg	2000–2400/kg
Guatemala	Mandatory	Skimmed milk	2000–3000/L	400–600/L
Venezuela	Mandatory	Dried milk powder	4000/L	400/L
Argentina	Mandatory	Fluid and dried milk (whole milk and skim milk)	2500/L	400/L
Honduras	Mandatory	Fluid milk	2000/L	400/L
Norway	Voluntary	Extra low-fat milk (also lactose free)	–	160/kg
Finland	Voluntary	Liquid milk	–	400/L
Sweden	Mandatory	Low fat milk	–	152–200/kg
Iceland	Voluntary	Low fat milk (1.5%)	–	200/kg
Philippines	Mandatory	Evaporated/ Condensed milk	4866/kg	973/L

TABLE 8.1 *(Continued)*

Country	Type of Fortification	Milk and Milk Products	Vitamin A (IU)	Vitamin D (IU)
Malaysia, Thailand, Indonesia	Mandatory	Evaporated milk, condensed milk, filled evaporated/ condensed milk	6700/kg	1111/kg
India	Voluntary	Processed milk	900–1500/L	200–300/L
China	Voluntary	Liquid milk-containing drinks	1000–3333/kg	400–1600/kg
		Modified milk powder	2000–3333/kg	400–1600 μg/kg
		Modified milk powder (for children)	4000–23333/kg	800–4480/kg
		Modified milk powder (for pregnant women)	6666–33333 /kg	800–4480/kg
		Ice cream	2000–4000/kg	400–800/kg

*Standards have been adopted from the regulations of the respective countries.

During processing of liquid milk, fortification is done before either pasteurization or ultra-heat treatment. However, oily preparation requires essentially homogenization step during processing. In this regard, two methods are being followed widely viz., batch process (small operations) and continuous process (metered additions). Ten parts of warm vegetable oil (45–5°C) is used to dilute oily preparations followed by homogenization with the appropriate quantity of skim milk/milk/cream. While, in case of water-soluble micronutrients, dilution is done using milk (45°C) [23]. Bector and Rani [5] have developed a batch process for vitamin A fortified skim milk without employing homogenizer; however, it is applicable only for low cost invested plants.

During yogurt preparation, lower pH makes it unsuitable for utilization as a delivery vehicle for vitamins mainly, fat-soluble vitamins. Losses may occur due to the utilization of vitamin by microorganisms [75]. It is worth noting that vitamins are extensively destroyed during manufacturing of ice cream. Majority of vitamin loss occurs in ice-cream mix during pasteurization; however, during whipping and further subsequent processing steps, minimal losses/oxidation of vitamins occur since all these are carried out at freezing temperature [14].

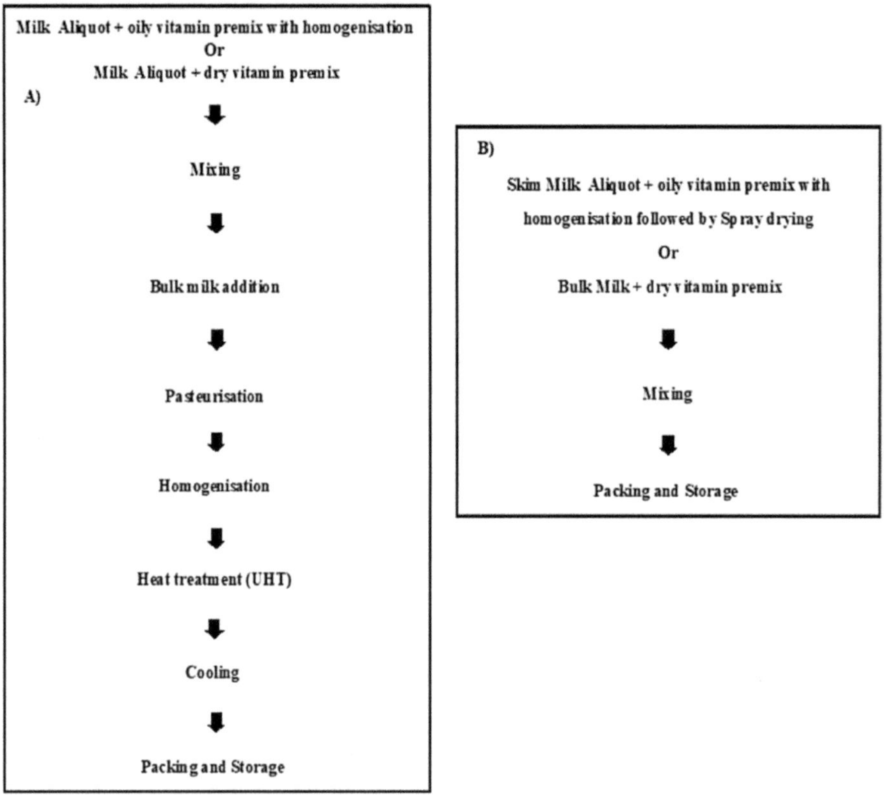

FIGURE 8.2 Fortification of milk (A) and dry milk powder; (B) with vitamins.

There are certain factors which need to be considered during the fortification of vitamins to dairy products, including liquid whole milk. Bioavailability of micronutrients plays a critical role in the absorption and metabolism of preparations/ products fortified with certain vitamins and minerals. Interactions between food components and vitamins could be antagonistic or synergistic. Absorption of iron was improved when taken along with vitamin C [47], while calcium interferes with absorption of iron/ zinc [89]. Therefore, it is imperative that fortified vitamin should not interact with any milk component as seen in the case of fat rancidity caused due to pro-oxidant iron. Fortified liquid milk and milk products also require special packaging requirements to avoid oxidation, whether it is oxygen or light-induced. Lastly, the levels of incorporation of vitamins in dairy products should be taken into consideration so that hypervitaminosis condition should be avoided to safeguard the health of consumers.

8.3 NUTRITIONAL AND BIOLOGICAL IMPORTANCE OF VITAMINS FORTIFIED MILK PRODUCTS

Vitamin A plays a significant role in different metabolic functions related to vision, reproduction, and growth, proliferation of cells and cohesion of immune system. The requirement of vitamin A is fulfilled from different animal and plant origin foods either in retinol form or pro-vitamin, i.e., β-carotene. Pro-vitamin form of vitamin A is believed to be less efficiently absorbed than retinol and subsequently, their conversion to vitamin A. Deficiency of vitamin A leads to xerophthalmia, night blindness, and keratinization [94]. Vitamin D plays a major role in maintaining the strength of bone via absorption of calcium and bone mineralization and its deficiency leads to osteoporosis (porosity in bones), osteomalacia (bone softening) in adults and rickets (skeletal disorder) in children [11].

Vitamin D has a direct link with the bone health of human population as it has critical role in calcium absorption and bone mineralization. Ergocalciferol (D_2) and cholecalciferol (D_3), biologically inactive vitamins from plant and animal sources, respectively, are converted to 25-dihydroxy cholecalciferol, which is a biologically active form, in visceral organs viz., liver, and kidney [36]. It is less likely to ascertain the predominance of vitamin D deficiency worldwide; however, it is possible that prevalence is common globally owing to less exposure of sunlight in combination with lower vitamin D dietary intake and places with limited duration of daylight, especially in winter months and at latitudes.

Prospective vitamin source in diet includes fat-rich dairy products specially butter, ghee, and cheese, however, the processing steps including skimming of milk, several thermal and pressure treatments affect the nutritional quality of the final product with respect to vitamins. Therefore, vitamin fortification in dairy products has been realized as an effective approach to manage and combat deficiencies related to fat-soluble vitamins (A and D). Care should be taken not to degrade or decrease the nutritional quality of milk products, while fortifying the dairy products with vitamins. Alteration in taste, shelf life of products or disturbance in the metabolism of other nutrients is also not accepted in the fortified products. Therefore, a sufficient number of trials should be performed by the manufacturers before product launch in order to investigate the stability and integrity of vitamins in dairy products along with any change in sensory evaluation of the product. In addition to this, higher temperature and humidity is required to store dried milk products, mainly in tropical countries [48].

8.4 TECHNOLOGICAL ASPECTS OF VITAMINS FORTIFICATION IN MILK PRODUCTS

Dairy products (including liquid milk) are not a significant source of vitamin A and D with concentration in milk is of 52 and 0.3 μg/kg, respectively [24], as they have limited stability during processing which make fortification mandatory to overcome the losses. Vitamin D is fortified in to fluid milk as D_2 or D_3 and vitamin A as retinyl acetate or retinyl palmitate. Fortification helps in eliminating nutritional deficiencies. However, it has been observed that vitamins added to the dairy products may also be destroyed during the routine milk processing and storage.

Various factors have been studied to affect the vitamin stability in a product, which includes temperature, time, exposure to light and humidity. Needless to mention, these factors must be determined prior to the storage of any novel fortified dairy product as the concentration of vitamins is significantly affected by storage conditions.

Literature shows contradictory reports on vitamin D stability during milk products processing and storage. Kreutler [50] reported that higher stability of vitamin D_3 to higher temperature (heat), light, and oxidizing conditions, while Kutsky [51] reported contradictory results indicating oxidation of vitamin D_3 upon exposure to light, but stability against acid and alkali. However, its stability to oxidation and alkali had also been reported [66]. Further, Cremin and Blanco reported unstable behavior of vitamin D under high temperature, acid, exposure to light and oxidizing conditions. During sterilization and pasteurization of milk, a considerable portion of natural vitamin D available in milk is reported to be lost [8, 16]. The same has been reported for milk in which fat has been separated off during standardization as the majority of the vitamins are fat-soluble [4]. Binding of vitamin D to casein micelles protects the vitamin against UV-light induced degradation [79].

Nonetheless, its stability is generally reported to be higher in several fortified products like homogenized whole milk [67], cheese [4, 17, 25, 45, 71, 88], UHT processed chocolate milk, yogurt [33, 37, 45] and ice cream [45]. Packaging material is also a key factor deciding the vitamin D stability in fortified products. Vitamin D_2 was found to be stable during processing (pasteurization, boiling, and sterilization), however packaging material (polyethylene pouches) has a significant effect on its stability [42]. Fortified milk stored at refrigeration temperature in glass as well as plastic bottles showed better stability than polyethylene pouches. Similarly, stability of vitamin D_3 in fortified yogurt was observed less when packed in translucent containers as compared with opaque containers [37].

Activity of vitamin A is significantly and rapidly reduced when subjected to sunlight exposure but gradually in the dark, when fortified food is packaged in glass bottles (transparent nature) or plastic containers (translucent nature). The mechanism behind this reduction is isomerization and degradation of retinoid compounds; other transformation reactions might also take place including dimerization reaction between two monomers of retinyl esters [59]. Comparatively lesser loss of vitamin A during thermal processing of milk and spray/roller drying or evaporation, however, it is highly unstable during processing which employs higher temperature process including prolonged heating of milk, butter, or butterfat [34]. However, the use of amber colored or pigmented packaging material incorporated with particular light barriers may prevent or delay the vitamin A degradation [5]. In fortified dairy foods, during processing, storage, and distribution chain, large loss of vitamin A activity occur [18]. Stability is also reported to be improved using oily form for fortification along other vitamins such as B_1, B_6, C, and D [18, 46, 54].

Forms of vitamins and food composition are the main factors which decide the stability of vitamins. Therefore, it is obvious that less stable vitamins are more prone to losses during processing, transportation, and storage of fortified food products.

8.5 BIOAVAILABILITY OF VITAMINS

Bioavailability of vitamins from food refers to the percentage of the vitamin being digested which are further absorbed and used by the human body [3]. Although vitamin-fortified foods are available in the global market, yet the deficiency is still existing, which indicates the issue of bioavailability from food (natural as well fortified food source) after digestion [9]. For sustainable fortification and compensate the losses, different strategies have been adopted, i.e., overage addition, co-fortification with other micronutrients, emulsification, and micro/nanoencapsulation.

During production, storage, and distribution of product, there occurs the loss of certain amount of micronutrient, the compensation of which is added to food vehicle well in advance and this addition of extra amount of micronutrient to the food product is known as overage [2]. The rate of losses of each vitamin fortified in the product has to be determined and to compensate for losses or variability due to processing, storage or analysis, excess amount needs to be added in the product during manufacturing as a means of maintaining at least the claimed amount of ingredients for the normal shelf life of the product and to allow for variation in assay performance. Overage also denotes the difference between these formulated and declared levels. Approximately, 20 and 30–50% overage addition is recommended for vitamin D and A, respectively, based on processing and storage conditions involved in manufacturing of different milk and milk products [2]. Vitamin losses vary according to storage conditions viz; humidity, time, temperature, and exposure to light, and should be determined accordingly by all the countries.

Combination of vitamins with other food components (lactose, casein, calcium, phosphorus, casein phosphor-peptides, and lactoferrin) improves the bioavailability of vitamins from milk [40, 41, 86, 93]. Vitamin D combination with calcium is apparently more effective when compared to individual fortification [21]. A similar effect was exhibited by vitamin A addition with iron due to the ability of vitamin A to solubilize iron over broad pH (2–6) range [27]. Fortification of vitamin D_2 with calcium and vitamin A with iron also showed increased bioavailability of both, indicating their synergistic interaction, when administered in combination [43, 74].

Irrespective of the type, concentration, and state of vitamin, delivery/carrier system is also essential, so that its stability and bioavailability could be enhanced in an aqueous medium and to maintain its integrity against the external damaging (oxidation/reduction) and environmental factors. One such technique is encapsulation (micro/nano), whose mechanism is based on the conversion of less stable oil form to more stable flowing powder, thus, making the latter available for blending with dry ingredients easily and uniformly [58]. Fortified vitamins are protected against chemical and biological deterioration and released at the particular site of interest [32, 77].

Several macromolecules viz., proteins, polysaccharides, and lipids have been evaluated for their efficiency as delivery systems of vitamins in food products [22, 31, 65, 81]. Among milk proteins, whey protein fraction, i.e., β-lactoglobulin has a strong tendency to bind with vitamin A due to the structural similarity with retinol-binding protein [84], results in improved stability [73] and bioavailability [19]. On the contrary, bovine serum albumin, α- and β-casein binds vitamin A by both hydrophilic and hydrophobic bonds [6, 10]. Natural nano vehicles, casein micelles are also being explored for the delivery of hydrophobic compounds like vitamins. Encapsulation or vitamin D complexed with reassembled casein micelles (rCM) [15, 32], lipids [76], casein micelles [56] and vitamin A with liposomes coated by N-trimethyl chitosan [35], reassembled succinylated-sodium caseinate [69] exhibited better vitamin absorption as compared to free vitamin after gastrointestinal digestion.

In rCM, soluble casein has hydrophobic domains which might entrap vitamin without changing size and morphology after interaction [79]. Encapsulation of vitamin D_3 in rCM with ultra-high-pressure homogenization treatment significantly improved temperature stability and bioavailability when consumed as fortified milk [32] and fortified yogurt [15] under clinical trials.

Casein micelles loading with hydrophobic vitamins are based on opening of structure at lower temperature (2°C) and pH (5.5) [56]. Vitamin A complex with native and modified sodium caseinate (succinylated and reassembled) showed higher binding capacity [31]. Further, they observed higher *in vitro* bio-accessibility [30] and *in vivo* bioavailability in Caco-2 cells [69] of these vitamin-protein complexes than a native form of vitamin A (free form). Chemical modification, i.e., succinylation of milk proteins improves protein solubility and application range as delivery vehicle for hydrophobic compounds [80]. The negative charge of succinic acid results in extensive opening of protein structure [85] and conformational changes expose the hydrophobic core for interaction with vitamins [38].

Emulsion based system were also developed to enhance the bioavailability as well as stability of lipid-soluble vitamins. The *in vitro* bio-accessibility and *in vivo* bioavailability of vitamin D_2 was assessed after emulsification in corn oil and showed contradictory results. The decreasing droplet size of emulsion showed better bio-accessibility of vitamin D_2, however, bioavailability decreased in rat models-based study. The difference in result might be due to the poor correlation between the simulated digestion conditions and animal digestion system. Winuprasith [90] has improved the stability of oil-in-water Pickering emulsion of vitamin D using a nanofiber cellulose (NFC) and observed limited impact of NFC on vitamin bio-accessibility when used at lower concentration (1:10, vitamin D: NFC). A pH based composite carbohydrate-protein gel system for encapsulation of vitamin D_3 also showed better storage stability. The compact structure of gel improved the *in vivo* stability of vitamin D_3 in intestine resulted in higher bioavailability [53]. The findings showed that encapsulation of vitamin A and D enhances the stability during processing and storage as well as their bioavailability.

8.6 RISK OF HYPERVITAMINOSIS AND VITAMIN FORTIFICATION

The maximum tolerable intake of vitamin is the amount one can ingest daily without harmful effect on health [92]. Vitamin A and D fall in the category of vitamins which have known risk and low safety [72]. The upper limit for vitamin A and D is 3300 and 50 µg while recommended daily intake is 600 and 5 µg, respectively [92]. Limited published reports are available on hypervitaminosis due to fortified dairy food consumption [13, 70, 72, 78]. The risk of hypervitaminosis is quite rare, although do occur due to over consumption of more than one staple fortified foods like sugar, milk, wheat flour, fortified food along with supplementation [26, 57] or vitamin-rich and fortified food consumption[82].

Excess intake of vitamin A/ hypervitaminosis A had been observed to cause calcification of aorta (calcium deposition in major blood vessel of heart), anaemia (lack of sufficient RBCs in blood), alopecia (hair loss), liver cirrhosis (liver damage) and ascites (excess abdominal fluid) when combined with alcohol consumption [68]. On the other side, excess dietary intake of vitamin D/ hypervitaminosis D leads to the condition of hypercalcemia, which refers to an increased level of calcium in the blood owing to its increased absorption through the gastrointestinal tract mediated by vitamin D [39]. The strategy to mitigate the hypervitaminosis A could be the use of β-carotene for the fortification of dairy food with natural yellow color, i.e., butter or ghee to mask the color of β-carotene or use of encapsulated form of β-carotene.

8.7 METHODS OF DETECTION

European Committee for Standardization, International Organization for Standardization (ISO) and Association of official analytical chemists (AOAC) have reported methods for vitamin A and D extraction and their analysis [7]. The commonly used analytical techniques include thin-layer chromatography (TLC), gas-liquid chromatography (GLC), spectrophotometric method and high-performance liquid chromatography (HPLC). Among all the reported method, HPLC based methods with different extraction protocol has been the first choice of researchers. Table 8.2 summarizes various methods used to analyze vitamin A and D in various dairy foods.

8.8 REGULATORY ASPECTS OF VITAMIN FORTIFICATION OF DAIRY FOODS IN INDIA

Across the world, government bodies play an important role for effective implementation of fortification program though food laws and food control systems for population at risk as well as whole population. For international trade, agreement on Technical Barriers to Trade (TBT) covers the measures related to food fortification [83]. At the national level, fortification regulations are formed based on the food supply chain of country, domestically produced staple food and national market for the fortified products. Having this considered, in 2018–19, India contributed 22% in global milk production (i.e., 187.7 million tons) [55].

Availability and daily consumption of milk and milk products across the country makes it efficient and reliable vehicle for micronutrients fortification such as vitamin A and D. In India, Food Safety and Standards Authority (FSSAI) works on the fortification of staple foods like milk, rice, salt, and oil and recently, formulated a comprehensive regulation on fortification of foods, namely, "Food Safety and Standards (Fortification of Foods) Regulations (FSSR), 2018." Under these regulations, standards for fortification of major staple foods like milk, wheat flour, rice, salt, and vegetable oil, etc., has been provided. FSSAI plays an important role not only in production and sale but also for consumption of fortified foods in India. The Government of India is committed to the cause of tackling vitamin A and D deficiency through fortification of food and FSSAI has raised standards for fortification of toned milk, double toned and skimmed milk with vitamin A and D among other nutrients with 270–450 μg RE/L as retinyl acetate or retinyl palmitate and 5–7.5 μg/L as vitamin D_3 or D_2 (only from plant source), respectively [28].

8.9 SUMMARY

Milk, a nutritious food, is enriched with quality protein, calcium, and vitamins. Vitamin A and D play key role in health and perform various metabolic and physiological functions in our body. Processing of milk for low fat products causes depletion of these vitamins. In urban Indian population, especially children and adults, vitamin A and D deficiency is in abundance. The reasons for deficiency are inactive life style, minimum sunlight exposure and changing food habits. Dairy products are part of the daily diet of

TABLE 8.2 Analysis of Vitamins (A and D) in Various Dairy Foods

Vitamin	Dairy Foods	Extraction Method	Detection Method	[a]Recovery or [b]Limit of detection	References
Vitamin A	Fluid milk	Saponification and extraction with hexane	Liquid chromatography (LC)-UV detector	[a]1.5–5.7 % RSD	[64]
	Milk	Saponification followed by hexane-ethyl acetate (8:2) extraction, stream N2	HPLC-photodiode array detector	[a]1.3 ng per injection	[12]
	Milk	One-step extraction by the iEx Mila vial	Portable fluorometer	[a]0.60 mg RE/kg	[52]
	Milk powder	Thermo stable protease based extraction	High pressure liquid chromatography-fluorescence-UV detector	[b]97.2 to 101.9%	[91]
	Infant formula	Solvent extraction using papain enzyme-based matrix disintegration and methanolic protein precipitation	Supercritical fluid chromatography-mass spectrometry	[b]102.1–116.4% (Vitamin D 101.0–97.0%)	[62]
	Milk	Ascorbic acid and ethanol assisted extraction and separation on solid-phase material of tetracycline-grafted polyacrylamide polymer (PAA-T) with acetonitrile	HPLC-Diode array detector (DAD)	[a]5.71 ng/mL	[49]

TABLE 8.2 *(Continued)*

Vitamin	Dairy Foods	Extraction Method	Detection Method	[a]Recovery or [b]Limit of detection	References
Vitamin D	Milk, infant formula	Saponification and extraction with n-heptane	Liquid chromatography-UV detector	[b]93.9 % for infant formula	[63]
	Processed cheese	Saponification followed by petroleum ether/diethyl ether extraction, stream N2	High pressure liquid chromatography-UV detector	[b]100%	[87]
	Cheese, yogurt, ice cream	Methanol/chloroform extraction, stream nitrogen	High pressure liquid chromatography-UV detector	[b]Cheese: 95% to 97% yogurt: 99% to 102% Ice cream 98% to 99%	[45]
	Infant formula, milk, and milk powder	Extraction in one step into isooctane and derivatization using 4-phenyl-1,2,4-triazoline-3,5-dione	Reverse-phase chromatography-triple quadruple mass spectrometer detector	[a]0.008 μg/100 mL	[1]
	Fortified milk	Extraction using hexane and ethanol with nitrogen flushing	HPLC-UV detector	[b]96.46% to 99.05%.	[44]
	Milk and Yogurt	Dispersive liquid-liquid micro-extraction technique	HPLC-UV detector	[b]86.6 to 113.3%	[29]

people. Therefore, fortification of dairy products with these vitamins is helpful in addressing deficiency. New and simple technological intervention is required for improving the stability and bioavailability of fortified vitamins. At the national level, awareness program and policies should also be implemented under food sector related initiatives of government, which should not only educate people about vitamin deficiency but also risk associated with hypervitaminosis.

KEYWORDS

- **bioavailability**
- **methods of estimation**
- **milk fortification**
- **stability**
- **vitamins**

REFERENCES

1. Abernethy, G. A., (2012). A rapid analytical method for cholecalciferol (vitamin D_3) in fortified infant formula, milk, and milk powder using Diels-alder derivatization and liquid chromatography-tandem mass spectrometric detection. *Analytical and Bioanalytical Chemistry, 403*(5), 1433–1440.
2. Allen, L., Benoist, B. D., Dary, O., & Hurrell, R., (2006). *Guidelines on Food Fortification with Micronutrients* (p. 376). World Health Organization and Food and Agriculture Organization of the United Nations (WHO/FAO).
3. Ball, G. F. M., (2005). *Vitamins in Foods: Analysis, Bioavailability, and Stability* (p. 842). CRC Press, Boca Raton; FL.
4. Banville, C., Vuillemard, J. C., & Lacroix, C., (2000). Comparison of different methods for fortifying cheddar cheese with vitamin D. *International Dairy Journal, 10*(5, 6), 375–382.
5. Bector, B. S., & Rani, S. A., (1998). Methodology to fortify low fat milk with vitamin A without using a homogenizer. *Indian Journal of Dairy Science, 51*(1), 69–72.
6. Belatik, A., Hotchandani, S., Bariyanga, J., & Tajmir-Riahi, H. A., (2012). Binding sites of retinol and retinoic acid with serum albumins. *European Journal of Medicinal Chemistry, 48*, 114–123.
7. Blake, C. J., (2007). Status of methodology for the determination of fat-soluble vitamins in foods, dietary supplements, and vitamin premixes. *Journal of AOAC International, 90*(4), 897–910.

8. Blanco, D., Fernandez, M. P., & Gutierrez, M. D., (2000). Simultaneous determination of fat-soluble vitamins and provitamins in dairy products by liquid chromatography with a narrow bore column. *Analyst, 125*(3), 427–431.
9. Borel, P., Caillaud, D., & Cano, N. J., (2015). Vitamin D bioavailability: State of the art. *Critical Reviews in Food Science and Nutrition, 55*(9), 1193–1205.
10. Bourassa, P., Nsoukpoe-Kossi, C. N., & Tajmir-Riahi, H. A., (2013). Binding of vitamin A with milk α- and β-caseins. *Food Chemistry, 138*(1), 444–453.
11. Ceglia, L., (2009). Vitamin D and its role in skeletal muscle. *Current Opinion in Clinical Nutrition & Metabolic Care, 12*(6), 628–633.
12. Chauveau-Duriot, B., Doreau, M., Noziere, P., & Graulet, B., (2010). Simultaneous quantification of carotenoids, retinol, and tocopherols in forages, bovine plasma, and milk: Validation of a novel UPLC method. *Analytical and Bioanalytical Chemistry, 397*(2), 777–790.
13. Chen, T. C., Shao, A., Heath, H. III., Holick, M. F., (1993). An update on the vitamin D content of fortified milk from the United States and Canada. *The New England Journal of Medicine, 329*(20), 1507.
14. FAO, (1996). Food fortification: Technology and quality control: Report of an FAO technical meeting. In: *FAO Food and Nutrition Paper* (p. 102). Food and Agriculture Organization; Rome; Italy.
15. Cohen, Y., Levi, M., Lesmes, U., Margier, M., Reboul, E., & Livney, Y. D., (2017). Re-assembled casein micelles improve *in-vitro* bioavailability of vitamin D in a caco-2 Cell model. *Food & Function, 8*(6), 2133–2141.
16. Cremin, F. M., & Power, P., (1985). Vitamins in bovine and human milks. In: Fox, P. F., (ed.), *Developments in Dairy Chemistry* (p. 346). Essex-England: Elsevier Applied Science Publishers.
17. Crevier, B., Bélanger, G., Vuillemard, J. C., & St-Gelais, D., (2017). Production of cottage cheese fortified with vitamin D. *Journal of Dairy Science, 100*(7), 5212–5216.
18. Dary, O., & Mora, J. O., (2002). Food fortification to reduce vitamin A deficiency: International vitamin A consultative group recommendations. *Journal of Nutrition, 132*(9), 2927S–2933S.
19. Diarrassouba, F., Garrait, G., Remondetto, G., Alvarez, P., Beyssac, E., & Subirade, M., (2014). Increased stability and protease resistance of the β-lactoglobulin/vitamin D_3 complex. *Food Chemistry, 145*, 646–652.
20. Dwyer, J. T., Wiemer, K. L., Dary, O., Keen, C. L., King, J. C., Miller, K. B., Philbert, M. A., et al., (2015). Fortification and health: Challenges and opportunities. *Advances in Nutrition, 6*(1), 124–131.
21. Fairweather-Tait, S. J., Johnson, A., Eagles, J., Ganatra, S., Kennedy, H., & Gurr, M. I., (1989). Studies on calcium absorption from milk using a double-label stable isotope technique. *British Journal of Nutrition, 62*(2), 379–388.
22. Fathi, M., Mozafari, M. R., & Mohebbi, M., (2012). Nanoencapsulation of food ingredients using lipid based delivery systems. *Trends in Food Science & Technology, 23*(1), 13–27.
23. Fortification Basics: OMNI/Roche/USAID. Milk. (2001). http://www.roche.com (accessed on 15 November 2021).
24. Fox, P. F., Uniacke-Lowe, T., McSweeney, P. L. H., & O'Mahony, J. A., (2015). Vitamins in milk and dairy products. In: *Dairy Chemistry and Biochemistry* (pp. 271–297). Springer International Publishing, Cham.

25. Ganesan, B., Brothersen, C., & McMahon, D. J., (2011). Fortification of cheddar cheese with vitamin D does not alter cheese flavor perception. *Journal of Dairy Science, 94*(7), 3708–3714.
26. Gannon, B., Kaliwile, C., Arscott, S. A., Schmaelzle, S., Chileshe, J., & Kalungwana, N., (2014). Biofortified orange maize is as efficacious as a vitamin a supplement in Zambian children even in the presence of high liver reserves of vitamin a: A community-based, randomized placebo-controlled trial. *The American Journal of Clinical Nutrition, 100*(6), 1541–1550.
27. Garcia-Casal, M. N., Layrisse, M., Pena-Rosas, J. P., Ramirez, J., Leets, I., & Matus, P., (2003). Iron absorption from elemental iron-fortified corn flakes in humans. Role of vitamin A and C. *Nutrition Research, 23*(4), 451–463.
28. FSSAI, (2018). *Gazette and Notification.* Food safety and standards (fortification of foods) regulations. Food safety and standards authority of India. Ministry of health and family welfare.
29. Ghalebi, M., Tamizi, E., Ahmadi, S., Sheikhloo, A., & Nemati, M., (2019). A dispersive liquid-liquid micro-extraction technique for the pre-concentration and quantification of vitamin D_3 in milk and yogurt samples using a non-aqueous HPLC method. *Iranian Journal of Pharmaceutical Research, 18*(2), 677–685.
30. Gupta, C., Arora, S., Sharma, A., & Sharma, V., (2019). Evaluation of effective storage conditions and in-vitro bio-accessibility of vitamin a from native and modified sodium caseinate-vitamin A complexes. *LWT, 111*, 284–290.
31. Gupta, C., Arora, S., Syama, M. A., & Sharma, A., (2018). Physicochemical characterization of native and modified sodium caseinate-vitamin A complexes. *Food Research International, 106*, 964–973.
32. Haham, M., Ish-Shalom, S., Nodelman, M., Duek, I., Segal, E., Kustanovich, M., & Livney, Y. D., (2012). Stability and bioavailability of vitamin D nanoencapsulated in casein micelles. *Food & Function, 3*(7), 737–744.
33. Hanson, A. L., & Metzger, L. E., (2010). Evaluation of increased vitamin D fortification in high temperature short-time-processed 2% milk, UHT-processed 2% fat chocolate milk and low-fat strawberry yogurt. *Journal of Dairy Science, 93*(2), 801–807.
34. Hartman, A. M., & Dryden, L. P., (1974). The vitamins in milk and milk products. Chapter 7. In: Webb, B. H., Johnson, A. H., & Alford, J. A., (eds.), *Fundamentals of Dairy Chemistry* (2nd edn., pp. 325–401). AVI Publishing Co., Westport; Connecticut.
35. He, W., Guo, X., Feng, M., & Mao, N., (2013). *In vitro* and *in vivo* studies on ocular vitamin a palmitate cationic liposomal in situ gels. *International Journal of Pharmaceutics, 458*(2), 305–314.
36. Holick, M. F., (1995). Defects in the synthesis and metabolism of vitamin D. *Experimental and Clinical Endocrinology & Diabetes, 103*(4), 219–227.
37. Jafari, T., Askari, G., Mirlohi, M., Javanmard, S. H., Faghihimani, E., & Fallah, A. A., (2016). Stability of vitamin D_3 in fortified yogurt and yogurt drink (Doogh). *Advanced Biomedical Research, 5*, 52.
38. Jamileh, L., & Villota, R., (1992). Effect of acylation on substructural properties of proteins: A study using fluorescence and circular dichroism. *Journal of Agricultural and Food Chemistry, 40*(4), 553–560.
39. Jones, G., (2008). Pharmacokinetics of vitamin D toxicity. *The American Journal of Clinical Nutrition, 88*(2), 582S–586S.

40. Jovani, M., Barbera, R., & Farre, R., (2003). Effect of lactoferrin addition on the dialysability of iron from infant formulas. *Journal of Trace Elements in Medicine and Biology, 17*(2), 139–142.
41. Kansal, V. K., (1998). Milk offers dietary calcium in the best available form. *Indian Dairyman, 50*(8), 23–27.
42. Kaushik, R., Sachdeva, B., & Arora, S., (2014a). Vitamin D_2 stability in fortified milk during processing, packaging, and storage. *LWT-Food Science and Technology, 56*(2), 421–426.
43. Kaushik, R., Sachdeva, B., Arora, S., Kapila, S., & Wadhwa, B. K., (2014b). Bioavailability of vitamin D_2 and calcium from fortified milk. *Food Chemistry, 147,* 317–321.
44. Kaushik, R., Sachdeva, B., Arora, S., & Wadhwa, B. K., (2014c). Development of an analytical protocol for the estimation of vitamin D_2 in fortified toned milk. *Food Chemistry, 154*, 567–573.
45. Kazmi, S. A., Vieth, R., & Rousseau, D., (2007). Vitamin D_3 fortification and quantification in processed dairy products. *International Dairy Journal, 17*(7), 753–759.
46. Kim, Y. S., Strand, E., Dickmann, R., & Warthesen, J., (2000). Degradation of vitamin A palmitate in corn flakes during storage. *Journal of Food Science, 65*(7), 1216–1219.
47. Kiran, R., Amma, M. K. P., & Sareen, K. N., (1977). Milk fortification with a system containing both iron and ascorbic acid. *The Indian Journal of Nutrition and Dietetics, 14*, 260–266.
48. Klassen-Wigger, P., & Barclay, D. V., (2018). Market-driven fortification. Chapter 6. In: Mannar, M. G. V., & Hurrell, R. F., (eds.), *Food Fortification in a Globalized World* (1st edn., pp. 63–67) London-UK: Academic Press.
49. Köseoğlu, K., Ulusoy, H. İ., Yilmaz, E., & Soylak, M., (2020). Simple and sensitive determination of vitamin A and E in the milk and egg yolk samples by using dispersive solid-phase extraction with newly synthesized polymeric material. *Journal of Food Composition and Analysis, 90*, 103482.
50. Kreutler, P. A., (1980). Fat-soluble vitamins. Chapter 6. In: *Nutrition in Perspective* (p. 579). Englewood Cliffs-NJ: Prentice-Hall Inc.
51. Kutsky, R. J., (1981). Vitamin D. Chapter 17. In: *Handbook of Vitamins, Minerals, and Hormones* (p. 191). Van Nostrand Reinhold Co., New York.
52. Laillou, A., Renaud, C., Berger, J., Moench-Pfanner, R., Fontan, L., & Avallone, S., (2014). Assessment of a portable device to quantify vitamin a in fortified foods (flour, sugar, and milk) for quality control. *Food and Nutrition Bulletin, 35*(4), 449–457.
53. Liu, K., Kong, X. L., Li, Q. M., Zhang, H. L., Zha, X. Q., & Luo, J. P., (2020). Stability and Bioavailability of vitamin D_3 encapsulated in composite gels of whey protein isolate and lotus root amylopectin. *Carbohydrate Polymers, 227*, 115337.
54. Loveday, S. M., & Singh, H., (2008). Recent advances in technologies for vitamin A protection in foods. *Trends in Food Science & Technology, 19*(12), 657–668.
55. Ministry of Animal Husbandry and Dairying, Government of India. http://www.dahd.nic.in/ (accessed on 15 November 2021).
56. Moeller, H., Martin, D., Schrader, K., Hoffmann, W., & Lorenzen, P. C., (2018). Spray-or freeze-drying of casein micelles loaded with vitamin D_2: Studies on storage stability and *in vitro* digestibility. *LWT, 97*, 87–93.
57. Mondloch, S., Gannon, B. M., Davis, C. R., Chileshe, J., Kaliwile, C., & Masi, C., (2015). High provitamin a carotenoid serum concentrations, elevated retinyl esters, and

saturated retinol-binding protein in Zambian preschool children are consistent with the presence of high liver vitamin A stores. *The American Journal of Clinical Nutrition, 102*(2), 497504.
58. Mortensen, B. L., & Gotfredsen, P., (1996). Fortification-nutritional improvement of dairy products. *Danish Dairy Food Industry Worldwide*, 64, 65.
59. Mousseron-Cadet, M., (1971). Photochemical transformation of vitamin A. *Methods in Enzymology, 18C*, 591–615.
60. NIN, (2003). *Annual Report 2003–2004* (p. 143). Food and Nutrient Intake. National Institute of Nutrition, ICMR; Hyderabad. https://www.nin.res.in/annualreports/2003-04.pdf (accessed on 15 November 2021).
61. O'Brien, A., & Robertson, D., (1993). Vitamin Fortification of Foods (Specific Applications). In: Ottaway, P. B., (ed.), *The Technology of Vitamins in Food* (pp. 114–142). London: Blackie Academic & Professional.
62. Oberson, J. M., Campos-Giménez, E., Rivière, J., & Martin, F., (2018). Application of supercritical fluid chromatography coupled to mass spectrometry to the determination of fat-soluble vitamins in selected food products. *Journal of Chromatography B, 1086*, 118–129.
63. Official Method of Analysis of AOAC International, (2012). Method 2002.05. Determination of cholecalciferol (vitamin D_3) in selected foods. Chapter 45. In: *Official Methods of Analysis of AOAC International* (pp. 35–37). Gaithersburg-MD: AOAC International.
64. Official Method of Analysis of AOAC International, (2003). Method 2002.06. Determination of retinyl palmitate in fortified fluid milk. Chapter 45. In: *Official Methods of Analysis of AOAC International* (pp. 1, 2). Gaithersburg-MD: AOAC International.
65. Ohelke, K., Adamiuk, M., Behsnilisn, D., Graf, V., Mayor-Miebach, E., & Walz, E., (2014). Potential bioavailability enhancement of bioactive compounds using food-grade engineered nanomaterials: A review of the existing evidence. *Food & Function, 5*(7), 1341–1359.
66. Pike, R. L., & Brown, M. L., (1967). *Nutrition, an Integrated Approach* (p. 542). New York-USA: Wiley.
67. Public Health Service/U.S. Food and Drug Administration. Revision. *Grade "A" Pasteurized Milk Ordinance*. http://www.idfa.org/docs/default-source/d-news/2015-pmo-final.pdf (accessed on 15 November 2021).
68. Ramanathan, V. S., Hensley, G., French, S., Eysselein, V., Chung, D., Reicher, S., & Pham, B., (2010). Hypervitaminosis A inducing intra-hepatic cholestasis: A rare case report. *Experimental and Molecular Pathology, 88*(2), 324–325.
69. Rana, S., Arora, S., Gupta, C., & Kapila, S., (2019). Effect of sodium caseinate and vitamin A complexation on bioaccessibility and bioavailability of vitamin A in caco-2 cells. *Food Research International, 121*, 910–918.
70. Rasmussen, S. E., Andersen, N. L., Dragsted, L. O., & Larsen, J. C., (2006). A safe strategy for addition of vitamins and minerals to foods. *European Journal of Nutrition, 45*(3), 123–135.
71. Riaz, M. N., Asif, M., & Ali, R., (2009). Stability of vitamins during extrusion. *Critical Reviews in Food Science and Nutrition, 49*(4), 361–368.
72. Richardson, D. P., (1997). The addition of nutrients to foods. *Proceedings of the Nutrition Society, 56*(3), 807–825.

73. Ron, N., Zimet, P., Bargarum, J., & Livney, Y. D., (2010). Beta-lactoglobulin-polysaccharide complexes as nanovehicles for hydrophobic nutraceuticals in non-fat foods and clear beverages. *International Dairy Journal, 20*(10), 686–693.
74. Sachdeva, B., Kaushik, R., Arora, S., & Indumathi, K. P., (2015). Impact of fortification with iron salts and vitamin A on the physicochemical properties of laboratory-pasteurized toned milk and bioaccessibility of the added nutrients. *International Journal of Dairy Technology, 68*(2), 253–260.
75. Salgueiro, M. J., Zubillaga, M., Lysionek, A., Caro, R., Weill, R., & Boccio, J., (2002). Fortification strategies to combat zinc and iron deficiency. *Nutrition Reviews, 60*(2), 52–58.
76. Salvia-Trujillo, L., Fumiaki, B., Park, Y., & McClements, D. J., (2017). The influence of lipid droplet size on the oral bioavailability of vitamin d_2 encapsulated in emulsions: An *in vitro* and *in vivo* study. *Food and Function, 8*(2), 767–777.
77. Sauvant, P., Cansell, M., Sassi, A. H., & Atgié, C., (2012). Vitamin A enrichment: Caution with encapsulation strategies used for food applications. *Food Research International, 46*(2), 469–479.
78. Scanlon, K. S., Blank, S., Sinks, T., Lett, S., Mueller, P., & Freedman, D. S., (1995). Subclinical health effects in a population exposed to excess vitamin D in milk. *American Journal of Public Health., 85*(10), 1418–1422.
79. Semo, E., Kesselman, E., Danino, D., & Livney, Y. D., (2007). Casein micelle as a natural nano-capsular vehicle for nutraceuticals. *Food Hydrocolloids, 21*(5, 6), 936–942.
80. Shilpashree, B. G., Arora, S., Chawla, P., & Tomar, S. K., (2015). Effect of succinylation on physicochemical and functional properties of milk protein concentrate. *Food Research International, 72*, 223–230.
81. Syama, M. A., Arora, S., Gupta, C., Sharma, A., & Sharma, V., (2019). Enhancement of vitamin D_2 stability in fortified milk during light exposure and commercial heat treatments by complexation with milk proteins. *Food Bioscience, 29*, 17–23.
82. Tanumihardjo, S. A., Gannon, B. M., Kaliwile, C., & Chileshe, J., (2015). Hypercarotenodermia in Zambia: Which children turned orange during mango season? *European Journal of Clinical Nutrition, 69*(12), 1346–1349.
83. WTO, (1995). *The Results of the Uruguay Round of Multilateral Trade Negotiations: The Legal Texts*. Geneva, World Trade Organization. http://www.wto.org/english/docs_e/legal_e/legal_e.htm (accessed on 15 November 2021).
84. Thompson, A. K., Couchoud, A., & Singh, H., (2009). Comparison of hydrophobic and hydrophilic encapsulation using liposomes prepared from milk fat globule-derived phospholipids and soya phospholipids. *Dairy science & technology, 89*(1), 99–113.
85. Thompson, L. U., & Reyes, E. S., (1980). Modification of heat coagulated whey protein concentrates by succinylation. *Journal of Dairy Science, 63*(5), 715–721.
86. Uchida, T., Oda, T., Sato, K., & Kawakami, H., (2006). Availability of lactoferrin as a natural solubilizer of iron for food products. *International Dairy Journal, 16*(2), 95–101.
87. Upreti, P., Mistry, V. V., & Warthesen, J. J., (2002). Estimation and fortification of vitamin D_3 in pasteurized process cheese. *Journal of Dairy Science, 85*(12), 3173–3181.
88. Wagner, D., Rousseau, D., Sidhom, G., Pouliot, M., Audet, P., & Vieth, R., (2008). Vitamin D_3 fortification, quantification, and long-term stability in cheddar and low-fat cheeses. *Journal of Agricultural and Food Chemistry, 56*(17), 7964–7969.
89. Weaver, C. M., (1998). Calcium in food fortification strategies. *International Dairy Journal, 8*(5, 6), 443–449.

90. Winuprasith, T., Nitithamyong, A., McClements, D. J., Suphantharika, M., Khomein, P., & Mitbumrung, W., (2018). Encapsulation of vitamin D_3 in Pickering emulsions stabilized by nanofibrillated mangosteen cellulose: Impact on *in vitro* digestion and bio-accessibility. *Food Hydrocolloids, 83*, 153–164.
91. Woollard, D. C., Bensch, A., Indyk, H., & McMahon, A., (2016). Determination of vitamin A and vitamin E esters in infant formulae and fortified milk powders by HPLC: Use of internal standardization. *Food Chemistry, 197*, 457–465.
92. Yates, A. A., Schilcker, A. S., & Suitor, C. W., (1998). Dietary reference intake: The new basis for recommendations for calcium and related nutrients, B vitamins, and choline. *Journal of the American Dietetic Association. 98*(6), 699–706.
93. Yeung, A. C., Glahn, R. P., & Miller, D. D., (2002). Effects of iron source on iron availability from casein and casein phosphopeptides. *Journal of Food Science, 67*(3), 1271–1275.
94. Zile, M. H., & Cullum, M. E., (1983). The function of vitamin A: Current Concepts. *Proceedings of the Society for Experimental Biology and Medicine, 172*(2), 139–152.

CHAPTER 9

MINERAL FORTIFICATION OF DAIRY FOODS

RICHA SINGH, SUMIT ARORA, and PRIYANKA S. RAO

ABSTRACT

Malnutrition is truly a universal problem leading to social and economic issues. Nearly everyone suffers from malnutrition in one or the other stage in their life, from infancy to old age. Every geographical location, including every segment of society, is affected by at least one form of malnutrition. According to the report [29], the cost of malnutrition to society was up to US$ 3.5 trillion per year. Micronutrient deficiency is a type of malnutrition that occurs due to the lack of intake and absorption of one or more vitamins and minerals. Micronutrients (vitamins and minerals) are the nutrients which are required essentially in small quantities for maintaining proper health. As per WHO [89], about two billion populace in the globe suffer from dietary deficiency of minerals and vitamins, and about 340 million children suffer from deficiency of minerals and vitamins micronutrients [33]. Minerals perform a variety of functions in the human body, e.g., building blood and strong bones, transmitting nerve impulse, etc. Deficiency of minerals and vitamins play a significant role in the health of human beings, especially children and pregnant women. Mineral deficiency may occur either due to poor availability of variety of foods or inappropriate dietary practices due to poverty. Persistent deficiencies of minerals lead to long term ill effects on health. Therefore, initiatives are needed to address the problems associated with mineral deficiency. Fortification of dairy products allows incorporation

Functional Dairy Ingredients and Nutraceuticals: Physicochemical, Technological, and Therapeutic Aspects. Megh R. Goyal, N. Veena, PhD & Santosh K. Mishra (Eds.)

of essential and beneficial micronutrients that are either lost during processing or are not present in milk. The consumption of fortified dairy products will help in reducing the instances of mineral deficiency among our population and also decrease the rate of associated diseases.

9.1 INTRODUCTION

9.1.1 *MINERALS AND THEIR REQUIREMENTS*

Minerals being part of bone, blood, hormones, and enzymes perform a variety of physiological functions in the human body and thus are called essential minerals. Sodium (Na), potassium (K), calcium (Ca), iodine (I), manganese (Mn), chloride (Cl), selenium (Se), chromium (Cr), cobalt (Co), fluorine (F), arsenic (As), boron (B), molybdenum (Mb), Silicon (Si), and nickel (Ni), are the minerals, which are essential for human nutrition [28]. Essential minerals can also be divided into two categories: major minerals and trace minerals. Ca, P, Mg, Na, K, Cl, and S are the examples of major or macro minerals and Fe, Cu, Zn, I, Mn, F, Co, and Se are the micro or trace minerals. Minerals in both the groups are equally important, however, the only difference lies in the quantity of minerals required. Trace minerals are required in smaller quantities than the macro minerals, but the quantity of minerals required does not indicate its importance. Calcium, phosphorus, and fluoride are important for teeth; calcium, magnesium, manganese, phosphorus, boron, and fluoride are required for bones, whereas, copper, iron, manganese, magnesium, and selenium are structural part of different enzymes; calcium and potassium have an important role in controlling blood pressure; calcium, magnesium, copper, selenium, and zinc are involved in immune function; chromium and manganese are important for brain development.

About 15% of the Indian population was reported to be undernourished during 2014–2016 as per the report submitted by Food and Agricultural Organization, 2017 [23] after conducting a survey on "Food Security and Nutrition in the World." According to WHO micronutrients are essential nutrients which are very important as they help in combating many health-related diseases, especially malnutrition. These are required only in small quantities by humans for growth; therefore, deficiency is referred to as 'hidden hunger' that does not produce hunger (lack of food) but impacts health, learning abilities and productivity. Micronutrient deficiency leads to malnutrition in people, which is related to the physiological phenomenon that occurs because of consumption of wrong content and/or type of food and includes both under and over nutrition. Micronutrient deficiency is considered to be a type of under nutrition where either the intake is low or if intake is sufficient, the absorption of minerals and vitamins is low and is not able to provide sustainable good health with normal physical and mental functions in human body [30]. Malnutrition is a complex problem faced by the world today because of the change in food habits from traditional based diets to processed foods, which are energy dense with macronutrients but lack micronutrients. Minerals are essentially obtained from the diet since the human body cannot produce them, and if diet is deficient in minerals, the role of enrichment and fortification of food comes into the picture.

9.1.2 POPULATION AT RISK FOR MINERAL DEFICIENCY

Micronutrient deficiency can occur in any group of population anywhere in the world; however, children and pregnant women are at utmost risk not for the reason that of lower intake but due to their compromised immune system and higher specialized physiological requirements. During pregnancy; starting from the time of conception wherein the micronutrient requirements are higher and have a long-standing effect on the growth and development of fetus. Development delays during the first 1000 days of life of an infant since conception are irreversible and therefore considered to be the most critical period [77]. During childhood, the deficiency of iodine causes brain damage in newborn along with reduced mental capacity and goiter, zinc deficiency results into weakened immune system that causes frequent infections and stunting, and iron deficiency leads to anemia that enhances the risk of mortality during childbirth as well as premature delivery [28].

In 2013, the National Iron Plus Initiative (NIPI) scheme was launched in India, wherein, iron and folic acid supplementation was implemented from an infant of six months to an adult woman of reproductive age twice a year and on a daily basis during pregnancy followed by supplementation after 100 days of delivery to lactating women for controlling anemia [48].

The 4th National Family Health Survey, 2018 conducted by Ministry of Health and Family Welfare reported that 58.6, 53.1, 50.4, and 22.7% of children (<5 years), women (15–49 years), pregnant women and men (15–49 years) are suffering from anemia in India. About 43.8, 49.4 and 64.6% of children (<5 years), adolescents, and pregnant women are suffering from iron deficiency in India [47]. Although women and children are affected the most due to mineral deficiency, every stage of life is affected by the deficiency of minerals (Figure 9.1). Another risk group for mineral deficiency is the elderly population which will be 22% of the total global population and 20% of the total Indian population by 2050 [3, 19]. The deficiency of minerals is a common problem in older population, however, the requirements for minerals are not excessively high for this group but with age the capacity to absorb and utilization of minerals decreases which causes deficiency. This chapter focuses on mineral fortification of dairy foods.

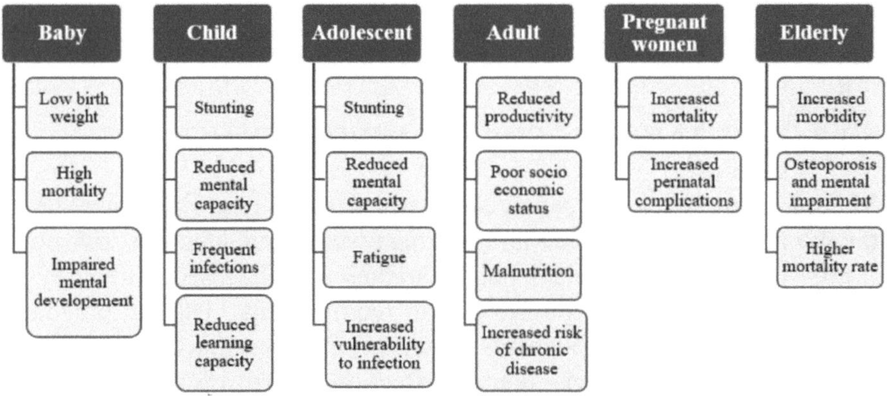

FIGURE 9.1 Effect of deficiency of minerals at different stages of life.

9.2 STRATEGIES TO CONTROL MINERAL DEFICIENCIES

Micronutrient malnutrition is a prominent health issue in both developing and developed nations; different approaches have been recommended to address this issue, e.g., dietary modification, supplementation, and fortification of food [18]. Fortification is a strategy to deliberately improve the nutritional quality of food that further helps in improving health of people. Enrichment is a term synonymously used with fortification; however, enrichment majorly involves the make-up of nutrients that are lost during processing of food and fortification is the addition of nutrients irrespective of the fact whether they were originally present in food or not or lost during processing.

Fortification can be classified as either voluntary or mandatory. When a government legally mandates the food business operators to add micronutrients to a particular food with specified guidelines, it is called mandatory fortification and is generally adopted if a huge population is facing deficiency of certain micronutrients. Mandatory food fortification can be of different types:

1. Mass Fortification: Also called universal fortification wherein micronutrients are added to commonly consumed foods, e.g., rice, flour, milk, oil, etc.;
2. Targeted Fortification: Addition of micronutrient to foods designed for specific population subgroup, e.g., infant formula.
   Voluntary fortification is mainly market driven fortification, where a food business operator initiates the addition of micronutrients within regulatory guidelines, especially for health concerned higher socioeconomic zone of population.

9.3 BIOAVAILABILITY OF DIETARY MINERALS

The absorption, retention, and utilization of minerals are known as bioavailability. Minerals are ingested through food enters into the intestine are absorbed except iron and calcium, which are absorbed in the duodenum. Minerals must be present in a suitable form to be taken up into the absorptive cells of the intestine. The first requirement is the solubility of minerals from the food matrix under gastric and intestinal conditions and further the ability of minerals to bind with the lumen that determines their transportation into mucosal cells. Minerals present in cereals are less bioavailable due to the presence of phytates and polyphenols. The individual binding affinities of inorganic elements with phytate are ranked as follows: Cu > Zn > Mn > Fe > Ca. These are strong chelators of minerals, and these cannot be hydrolyzed in the human intestine; therefore, their presence drastically reduces the mineral absorption through the diet [72].

The bioavailability of zinc depends upon phytate content, i.e., if the molar ratio of phytate and zinc is more than 15, then zinc will be poorly absorbed leading to its deficiency [26]. More than one mineral can also complex with phytates at the same time. These bimetal complexes are more stable than mono metal complexes and exert antagonist effect on the bioavailability of other minerals [88].

The physicochemical form of the dietary minerals also influences its availability for absorption, e.g., iron. Iron from diet is of two forms: heme and non-heme iron. Heme iron is available from non-vegetarian diets (meat, fish, and poultry) and plants are the major source for non-heme iron. Heme iron is more bioavailable than non-heme iron because of its low solubility at intestinal pH. The fact that ferric salts of iron are precipitated as ferric hydroxide at pH above 3.0; thus, these salts are not available for absorption in the duodenum. The presence of fibers may also affect the absorption of minerals. The absorption of calcium was found to decrease (~12%) in the presence of locust bean gum and high methoxyl pectin, whereas supplementation with inulin had a positive effect (~30%) on absorption of calcium [10]. Hence, dietary factors that modulate absorption are of primary interest since they affect the bioavailability of minerals. The effect of some dietary constituents on the absorption of minerals is represented in Table 9.1.

TABLE 9.1 Examples of Dietary Factors That Modify Absorption of Minerals

Dietary Constituent	Modifying Influence	Minerals Affected
Simple sugars	Increase absorption	Cr
Fructose	Increase absorption	Ca, Mg, Cu, Zn, Fe, Mn
Some amino acids	Increase absorption	Mg, Cr
Meat protein	Increase absorption	Zn, Fe, Mg
Phytate	Decrease absorption	Ca, Mg, Zn, Fe, Cr, Mn
Phosphate	Decrease absorption	Mg, Fe, Mn
Oxalate	Decrease absorption	Ca
	Increase absorption	Cr
Polyphenols	Decrease absorption	Fe
Thiols	Increase absorption	Se
Ascorbic acid	Increase absorption	Fe, Cr, Se
	Decrease utilization	Cu
Presence of Other Metal Ions		
Zn	Decrease absorption	Fe
Fe	Decrease absorption	Ca, Zn, Mg
Cu	Decrease absorption	Zn
Mn	Decrease absorption	Ca, Fe

9.4 MILK PRODUCTS AS A MEDIUM FOR MINERAL FORTIFICATION: SUITABILITY AND CHALLENGES

Selection of an appropriate vehicle for delivering the fortificant is most critical for a successful fortification program. The basic foods such as cereals, flour, breads, etc., eaten by the common populace are always the first choice of fortification. However, the presence of phytates in cereal products is major interference in the bioavailability of these minerals. Milk does not contain phytates; therefore, may serve as a suitable vehicle for fortification of minerals. Milk products are also consumed in notable amounts by each age group which makes them an obvious vehicle for dietary mineral fortification. According to a report published by global news wire in 2018 [29], which pointed out that fortified dairy products will generate global revenue of US $150,000 million by 2026 and mineral fortification is expected to register the fastest growth.

Dairy products like milk, flavored milk, cheese, and yogurts are majorly used for fortification with minerals. Milk products are considered as almost complete foods because they possess excellent nutritional properties in terms of high value macro components, e.g., fat, protein, carbohydrate, and minerals like calcium, phosphorus, sodium, potassium. However, other dietary minerals like iron, zinc, and magnesium, etc., are not present in significant amounts in dairy products. Increasing the quantity of some of these minerals could improve dietary balance and health in malnourished individuals. The presence and concentration of minerals in milk and milk products depend upon the presence of minerals in cattle feed and processing of milk in terms of heat treatment and others. Minerals in milk may be present in the inorganic form or may be a part of organic molecules [36] like protein and fat. The concentration of major and trace minerals is shown in Tables 9.2 and 9.3.

The addition of minerals into milk is not easy because the minerals are chemically reactive elements and their fortification into milk poses many technological problems in terms of changes in flavor and color of the product. The kind and extent of reaction in milk system depends upon the type of added mineral (its solubility and reactivity). The following factors are need to be addressed while fortifying milk and milk products with minerals:

1. **Solubility vs. Mineral Content:** Solubility of minerals is a major issue for its fortification in milk products and more specifically in liquid milk. Depending upon the milk products which are to be fortified with minerals, considerable attention should be paid to the different mineral forms available along with their solubility profiles.
2. **Bioavailability:** The added mineral should be highly bioavailable. The presence of milk constituents may affect the bioavailability of the added mineral.

3. **Organoleptic Taste:** When minerals are added in high amounts, the insoluble salts are likely to produce a chalky mouthfeel and even astringent or bitter taste at high concentrations. Addition of some minerals may lead to spoilage of dairy products because of oxidation and also cause loss of other nutrients, e.g., ascorbic acid [11]. This may be detrimental to the acceptability of milk products.

TABLE 9.2 Major and Trace Minerals Concentration in Buffalo and Cow Milk

Minerals	Buffalo Milk	Cow Milk
Major Minerals (mg/Kg)		
Calcium	1740 ±15[1]	1220±15[1]
	1786±7[2]	1202±7[2]
	2042.3±7.9[3]	1254.3–1348.7[3]
	750±172[4]	516±78[4]
Potassium	641±11[1]	1190±18[1]
	1180.5±7.8[3]	1425.7–1448.8[3]
	698±194[4]	1242±279[4]
Phosphorus	1190±14[1]	775±12[1]
	1092±12[2]	881±7[2]
	1174.5±5.26[3]	867.1–900.2[3]
Sulphur	357±15[1]	266±19[1]
Magnesium	183±3[2]	127±3[2]
	235.3±1.33[3]	102.4–108.7[3]
	58.5±11.3[4]	83.2±10.3[4]
Chloride	1100±3[2]	1300±1[2]
Trace Minerals (μg/Kg)		
Iron	301±18[1]	325±13[1]
	1100±20[3]	910–1160[3]
	3010±240[4]	1450±520[4]
Zinc	6488±20[1]	3814±20[1]
	5100 ±10[3]	5100–6200[3]
	4000±840[4]	4360±127[4]
Manganese	2.47±0.24[1]	31.9±0.17[1]
	500±2[3]	400–600[3]

Note: Refs. [1][8], [2][46], [3][84], [4][14].

TABLE 9.3 Major and Trace Minerals Concentration in Different Milk Products

Minerals	Evaporated Milk	Dried Milk	Butter	Cheese	Yogurt	Ice Cream
Major Minerals (mg/100 g)						
Calcium	290	1280	15	720	200	130
Chloride	250	1070	1150	1030	170	110
Magnesium	29	130	2	25	19	13
Phosphorus	260	970	24	490	170	110
Potassium	360	1590	15	77	280	160
Sodium	180	550	750	670	80	69
Trace Minerals (μg/100 g)						
Iron	260	270	200	300	100	100
Zinc	900	4000	100	2300	700	300

4. **Processing Challenge:** Thermal treatment during processing and pH of food affect the solubility of minerals (may be positive or negative) which may further limit the application in specific products. Addition of minerals to milk affects the salt balance which is directly affect heat stability of milk. Therefore, technological interventions are required to overcome the adverse effect on the stability of milk products on addition of minerals into milk.
5. **Interaction:** There can be potential interactions of added minerals with other minerals, naturally present in milk which might affect the metabolic utilization of the added mineral. The addition of calcium to milk does not affect the solubility of simultaneously added zinc and iron, however, the bioavailability of Fe, Zn was found to be hampered in milk fortified with calcium [61].
6. **Cost and Safety:** Milk and milk products are affordable by the general population; however, too many technological interventions for enhancing stability, solubility, and bioavailability of minerals may enhance the cost of the fortified product and make it non affordable for population suffering from malnutrition. Therefore, the cost involved during production must be taken into consideration while fortifying milk products with minerals.

Milk is also a natural source for many of the minerals, and the addition of extra minerals into milk may lead to over consumption of some specific mineral that may cause toxic effects on human health. To avoid the addition of minerals beyond the safety limit, the recommended dietary allowance of minerals (country-specific) must be taken into account.

The Recommended Dietary Allowance (RDA) of essential minerals according to the Food and Nutrition Board of the Institute of Medicine, National Academy of Sciences [25] are presented in Table 9.4. An RDA is the adequate daily intake that is enough to meet the average nutrient requirement of all healthy individuals.

TABLE 9.4 Recommended Dietary Allowances of Major Minerals

Age	Mineral Requirement (mg/day)			
	Calcium	Iron	Magnesium	Zinc
Infants				
0–6 month	200	0.27	30	2
7–12 month	260	11	75	3
Children (Years)				
1–3	700	7	80	3
4–8	1,000	10	130	5
Male Persons (Years)				
9–13	1,300	8	240	8
14–18	1,300	11	410	11
19–30	1,000	8	400	11
31–50	1,000	8	420	11
51–70	1,000	8	420	11
> 70	1,200	8	420	11
Females (Years)				
9–13	1,300	8	240	8
14–18	1,300	15	360	9
19–30	1,000	18	310	8
31–50	1,000	18	320	8
51–70	1,200	8	320	8
> 70	1,200	8	320	8
Pregnant Women (Years)				
19–30	1,000	27	350	11
31–50	1,000	27	360	11
Lactating Mothers (Years)				
19–30	1,000	9	310	12
31–50	1,000	9	320	12

9.5 SELECTION OF MINERAL SOURCES FOR FORTIFICATION IN MILK PRODUCTS

Milk products are often fortified with minerals, such as Ca, Fe, Zn, Mg, P, etc. The selection of appropriate mineral source for addition into milk products is based on the physicochemical properties of products such as mineral content, solubility of minerals, effect on taste, color, and texture of food vehicle as well as absorbability of the mineral. Economic consideration is also an important parameter and depends upon the targeted population and type of fortification. Therefore, while selecting the appropriate fortificants, solubility, bioavailability, and its effect on sensory attributes of the product need to be considered.

Different calcium salts as per their relative bioavailability and organoleptic characteristics are presented in Table 9.5. It shows that organic calcium salts are more soluble but has relatively lower calcium content than inorganic calcium salts [86, 87]. The calcium chloride possesses good solubility but has a bitter and salty taste that restricts its addition. It is reported in literature that inorganic salts contain more calcium and are more soluble, but the bioavailability of these salts is lower than organic mineral salts [83].

TABLE 9.5 Choice of Calcium Salt Available as Fortificant

Calcium Salt	Attributes
Calcium acetate	• Elemental calcium 25 (% w/w).
	• It is colorless and has good phosphate-binding ability.
Calcium carbonate	• Elemental calcium 40 (% w/w).
	• It provides the highest amount of elemental calcium. Least costly calcium source.
	• It is odorless, colorless, and gives soapy, lemony texture in product.
Calcium chloride	• Elemental calcium 36 (% w/w).
	• It is colorless and imparts salty, bitter taste.
Calcium citrate	• Elemental calcium 24 (% w/w).
	• It is odorless, colorless, and gives tart and clean texture.
	• The solubility is pH-dependent being higher at pH below 4.
Calcium citrate malate	• Elemental calcium 21 (% w/w).
	• Calcium citrate malate has better absorption than calcium carbonate.
	• It is colorless.

TABLE 9.5 *(Continued)*

Calcium Salt	Attributes
Calcium gluconate	• Elemental calcium 9 (% w/w). • It is odorless, white, bland, and more soluble than calcium citrate.
Calcium lactate	• Elemental calcium 13 (% w/w). • It is odorless, white, neutral, and similar solubility as calcium gluconate.
Calcium phosphate-monobasic, dibasic, tribasic; and pyrophosphate	• Elemental calcium 31–38 (% w/w). • Low solubility compared to calcium carbonate. • It is odorless, colorless, and gives sandy texture in product.

Table 9.6 represents the choice of iron salts available. Worldwide, the anemia is mainly caused by iron. Milk and milk products lack iron and hence are poor source of iron, and different attempts have been tried to incorporate iron in dairy products. Iron salts, which are soluble in water are more bioavailable but also highly reactive in catalyzing oxidation of lipids, amino acids, and vitamins, which modifies the sensorial acceptability of food as well as reduces its nutritional qualities. This limits its usage in dairy products [28, 38]. Iron salts, which are not water-soluble, are less reactive than the water-soluble forms of iron; however, during storage for longer intervals, they promote rancidity and also, they cannot be used to fortify products having pH values close to neutral pH due to precipitation issues in the product. Iron salts, which are not soluble in water and somewhat soluble in dilute acids are majorly used for fortification of food due to minimum effect on color and taste of food, but these salts are poorly soluble in gastric juice; therefore, their bioavailability (20–75% in comparison to ferrous sulfate) is limited [21, 54, 60].

TABLE 9.6 Choice of Iron Salts Available as Fortificant [37, 38, 58, 85]

Iron Salt	Attributes
Elemental iron H-reduced, atomized, CO-reduced, electrolytic, carbonyl	• It is not soluble in water but slightly soluble in dilute acid solution. • Its iron content is 96–99% and relative bioavailability depends upon the size of elemental iron; and the smaller the size of particle more is bioavailability. • The finer particles adversely affect the color and bigger particles settle out due to its higher density.

TABLE 9.6 *(Continued)*

Iron Salt	Attributes
Ferric orthophosphate	• It is not soluble in water and slightly soluble in dilute acid. • Its relative bioavailability is 25–32%; and iron content is 29%. • Its cost is relatively high, but still due to its light color and chemical inertness, it is used.
Ferric pyrophosphate	• It is not soluble in water but slightly soluble in dilute acid solution. • Its relative bioavailability is 21–74%; and iron content is 25%.
Ferric ammonium citrate	• It is freely soluble in water. • Its relative bioavailability is 51 and iron content is 17%. • It gives intense brown color in the product.
Ferrous fumarate	• It is poorly water-soluble but soluble in dilute acids. • Its relative bioavailability is 100 and iron content is 33%. • It gives undesirable effects on flavor, color, and aroma as ferrous ion is pro-oxidant.
Ferrous gluconate	• It is freely soluble in water. • Its relative bioavailability is 89 and iron content is 12%. • When thermally processed, it maintains the high bioavailability in the product.
Ferrous lactate	• It is freely soluble in water. • Its relative bioavailability is 67 and iron content is 19%. • Similar to ferrous gluconate, when thermally processed, it maintains the high bioavailability in the product.
Ferrous sulfate	• It is freely soluble in water. • Its relative bioavailability is 100 and iron content is 20–33%. • It gives undesirable effects on flavor, color, and aroma as ferrous ion is pro-oxidant.

Zinc is another major mineral that has been incorporated into milk and milk products. Table 9.7 presents several zinc compounds which have been used for fortification of milk. Zinc sulfate is most bioavailable but expensive. Zinc oxide being inexpensive has been used widely along with zinc sulfate.

TABLE 9.7 Choice of Zinc Salt Available as Fortificant [5, 19, 74]

Zinc Salt	Attributes
Zinc acetate	• Bioavailable.
Zinc citrate	• Bioavailable but impart undesirable after taste.
Zinc gluconate	• Bioavailable as zinc sulfate.
	• No effect on sensory attributes.
Zinc oxide	• Lower bioavailability.
	• Insoluble in basic pH.
	• Lower solubility.
	• Colorless and odorless.
Zinc sulfate	• Soluble at all pH range.
	• Better bioavailability than zinc oxide

9.6 TECHNOLOGICAL ADVANCEMENTS FOR ADDITION OF MINERALS IN MILK PRODUCTS

The mineral salts can be added to milk products in different forms, and the easiest approach is the direct addition of mineral salts. However, direct addition of mineral results in interaction with milk components that may cause sedimentation, lower bioavailability, and negative effect on organoleptic characteristics. The intensity of these above-mentioned effects depends on the type and level of mineral salt used for fortification, e.g., addition of soluble Ca salts such as $CaCl_2$ to milk reduces the thermal stability of milk and also affects the flavor of the milk at high levels of calcium [7]. Direct addition of iron also accelerates oxidation of fat due to its pro-oxidant activity and affects the sensory attributes.

Dietary minerals are usually poorly absorbed because of their poor solubility in the gut. Therefore, to address the micronutrient requirements of general population, enhancing the bioavailability of minerals is of utmost importance rather than enhancing the content of minerals in the diet. The mineral fortificants which are allowed to be added are poorly absorbed;

hence, to meet the RDA higher amounts are added to make-up the lower bioavailability of the fortified minerals.

A higher amount of these added minerals may result in competitive reactions which ultimately hampers the absorption of minerals. Addition of minerals in high amounts to overcome lower bioavailability appears as a quick solution to combat mineral deficiency; however, it may adversely affect the absorption of minerals and may not yield desired results. Absorption of iron and calcium in their inorganic forms was found to be only 3–10% [58, 70] and 30–35% [69, 70], respectively. Higher amounts of minerals added also impacts sensory (color and flavor) as well as storage stability of food products (not only in terms of iron but other minerals also). Necessary interventions are thus required for food fortification to address enhanced bioavailability with lower competitive inhibition of the fortified minerals.

Technological advancements have been carried out to fortify minerals in milk products without affecting the sensorial properties of the product. Protection of minerals with either encapsulation or chelation with protein/peptides/amino acid/sugar could overcome the drawbacks of direct addition of mineral salts into food products. This may reduce the interaction and reactivity of minerals within the food matrix, and thus, the negative effect on taste, color, and other sensory aspects can be overcome, leading to greater acceptability and longer shelf life of the fortified product. Microencapsulation technique has been used for iron fortification in milk products (Table 9.8). Encapsulation (i.e., coating material) for iron includes a mixture of phospholipids, polysaccharides, proteins, or hydrogenated vegetable oils which enhance the solubility and bioavailability of minerals. The stability of mineral compounds is ensured when they are present in encapsulated form during processing and storage of food; the trapped iron is released only on breaking of coat that allows controlled and targeted release in gastrointestinal (GI) tract [12].

TABLE 9.8 List of Different Encapsulated Iron Forms for its Application in Milk Products

Core and Coat of Encapsulated Material	Application	References
Core: Ferric ammonium sulfate Coat: Polyglycerol monostearate	Milk	[55]
Core: Ferric ammonium sulfate Coat: Polyglycerol monostearate	Yogurt	[53]
Core: Ferrous sulfate Coat: Whey protein chelated with sodium alginate	Yogurt	[43]

TABLE 9.8 *(Continued)*

Core and Coat of Encapsulated Material	Application	References
Core: Ferrous sulfate	Milk	[1]
Coat: Polyglycerol monostearate and tween 80		
Core: Ferrous sulfate	Milk	[90]
Coat: Egg phosphatidylcholine, cholesterol, and tween 80		
Core: Ferrous sulfate	Cheese	[40]
Coat: Stearine		
Core: Ferrous sulfate heptahydrate	Yogurt	[31]
Coat: Gum Arabic, maltodextrin, and modified starch		
Core: Ferrous sulfate heptahydrate and L-ascorbic acid	Yogurt	[32]
Coat: L-α-phosphatidylcholine (lecithin) and cholesterol		

Proteins are known to bind with minerals. This interaction is dependent on the mineral and available sites of proteins along with the environmental conditions involved. The clustering of phosphoserine residues in proteins influence the metal-binding properties of proteins as these acidic clusters are able to chelate cations (mainly divalent) [35]. Complexation of proteins with calcium, iron, zinc, and other divalent mineral ions prevents its precipitation in the intestine, thereby promoting absorption of these minerals. The mineral bound with protein are absorbed through a specific pathway along with passive absorption (Table 9.9). This absorption of protein bound minerals occurs via a different mechanism of absorption than absorption of a free mineral; hence, the competition for a common metal ion transporter may be prevented. Fortification of minerals in the complex form with proteins may prevent the inhibitive interaction of minerals like calcium on iron or iron on zinc, etc. [17].

TABLE 9.9 Application of Different Mineral-Protein Complex in Milk Products

Mineral-Protein Complex	Application	References
$FeCl_3$-casein complex	Yogurt	[78]
$FeCl_3$-whey protein complex		
$FeCl_3$-casein complex	Mozzarella Cheese	[68]
$FeCl_3$-whey protein complex		
$FeCl_3$-casein complex	Yogurt	[34]
$FeCl_3$-whey protein complex		

TABLE 9.9 *(Continued)*

Mineral-Protein Complex	Application	References
$FeCl_3$-casein complex	Milk	[91]
Ferric phosphate ($FePO_4$)-whey protein complex		
$FeCl_3$-succinylated casein	–	[65]
Ferric chloride ($FeCl_3$)-whey protein complex	Gouda cheese	[39]
Ferrous sulfate ($FeSO_4$)-whey protein hydrolysates	–	[52]

9.7 MINERAL FORTIFICATION IN MILK PRODUCTS

9.7.1 FORTIFICATION WITH CALCIUM

Milk is always considered a suitable vehicle for fortification of calcium; however, addition of calcium alters the mineral balance in milk thus affects the stability of milk products. Hence, most of the research related to fortification of calcium in milk products is related to its effect on stability of the product as well as bioavailability.

Pirkul et al. [63] fortified yogurt with organic calcium salts (calcium lactate and calcium gluconate). The sensory scores of lactate fortified yogurt were higher than gluconate and combined calcium salt fortified yogurt. The calcium content was increased by about 30% in all the yogurt samples. Vyas and Tong [87] reported decreased solubility as well as heat stability in the skim milk (protein content 1.75 and 3%) added with calcium phosphate (0.15%), lactate (0.18%), and citrate (0.24%) salt; whereas, increase in heat stability was observed with carbonate salt added to skim milk (protein content 1.75%). Ranjan et al. [67] reported fortification of calcium (0.05%) lowered heat stability of buffalo milk samples than that of unfortified milk. Among the different fortified milks, calcium gluconate fortified buffalo milk had the least affected physicochemical characteristics. Calcium was more bioavailable (in vivo) from fortified milk than unfortified milk. Among fortified milk, lowest bioavailability was observed in calcium chloride fortified milk as compared to lactate and gluconate fortified milk.

Perales et al. [61] observed that calcium fortification (1.2–1.6 mg/L) of cow milk not only increased the calcium content but also its bioavailability, transport, and uptake as evaluated using caco-2 cell assay. Singh et al. [81]

prepared calcium-fortified (0.05%) milk with calcium in the form of chloride (inorganic), lactate, and gluconate (organic) salts and reported a drop in heat stability in all fortified milks. The bioavailability of calcium from milk fortified with organic salts was more than that of inorganic calcium-fortified milk.

Coskun and Senoglu [16] prepared calcium-fortified milk with different levels of calcium carbonate (200, 400, 600 mg/100 ml). Calcium addition resulted in an increase of pH and viscosity of milk which led to lower separation of serum. The sensory scores were affected less if calcium salts (200 mg/100 ml) are added before pasteurization.

Fleury et al. [24] prepared calcium-fortified yogurt with calcium content ranging from 500 to 1500 mg per 170 g of yogurt. The viscosity of yogurt was 1500 centipoise (at 5°C) using calcium phosphate after fermentation. The calcium phosphate was added in the form of slurry having pH 4–4.6 after adjusting the pH with organic/mineral acid. Singh et al. [82] fortified yogurt (low-fat mango flavored) with calcium lactate (0.05%) prior to fermentation. The viscosity and water holding capacity was higher in fortified yogurt; however, flavor, and microstructure of yogurt was not affected by fortification.

Aportela-Palacios et al. [5] studied the syneresis of yogurt prepared from calcium citrate (0.05%) and dietary fiber from wheat bran fortified milk. The fortified yogurt had lower syneresis which increased from 20 to 48% during 3 weeks of storage. Singh and Muthukumarappan [83] reported that control and calcium-fortified mango yogurt (0.05% calcium lactate) did not show any significant difference in organoleptic properties (color, flavor, and texture). However, water holding capacity was superior in yogurt fortified with calcium. Saremnezhad et al. [76] prepared calcium-fortified low-fat ice cream using tricalcium citrate, tricalcium phosphate and $CaCl_2$. The viscosities were similar for tricalcium citrate (60 and 130 mg/L) and $CaCl_2$ (60 mg/L) and firmness of ice cream decreased on addition of calcium salt in all ice cream samples.

Puspitasari et al. [64] fortified cottage cheese with different calcium salts (Ca chloride, Ca phosphate, Ca lactate, Ca carbonate and Ca citrate) along with different hydrocolloids of gum Arabic (20%), low viscosity guar gum (2.5%), carrageenan (1.5%), and xanthan gum (0.2%) to mask the bitterness of calcium salts. They reported that Cottage cheese containing 7.5 mg/g mixture (1:1) of guar gum and locust bean gum had no bitter taste at 73 mg added Ca/100 g. Perveen et al. [62] fortified buffalo milk with calcium chloride ($CaCl_2$) and trisodium phosphate (Na_3PO_4) at levels of 10, 20 and 30 mM in milk and further manufactured Cheddar cheese. No obvious significant effects were observed in compositional and organoleptic characteristics of Cheddar cheeses during ripening.

Sakr et al. [73] prepared Karish cheese from calcium-fortified milk (200 and 300 mg/100 ml) and reported that fortification positively affected the yield of cheese (about 10%) with an increase in hardness, cohesiveness, gumminess, and chewiness; however, adhesiveness, and springiness of cheese decreased.

Kaushik et al. [51] observed the effect of adding vitamin D_2 on bioavailability of Ca, Fe, and Zn and reported positive effect on bioavailability of Ca. Further, Kaushik et al. [50] used calcium (500 and 600 ppm) and Vit D_2 (600 IU/L) fortified milk for preparation of dahi. All fortified dahi samples were acceptable from a sensory point of view. Fortification did not affect pH and water holding capacity of dahi; however, curd setting time increased. The firmness and viscosity decreased in milk on addition of Ca phosphate and increased on addition of Ca citrate as compared to control. Kaushik et al. [49] also prepared calcium-fortified yogurt using calcium phosphate/calcium citrate and vitamin D_2 fortified milk and obtained similar results.

9.7.2 FORTIFICATION WITH IRON

Fortification of milk products with iron has been carried out widely. Fortified milk products were evaluated in terms of effect on sensory profile and stability. Indumathi et al. [39] reported lower sensory scores but higher lactose retention in Gouda cheese added with ferrous sulfate and whey protein chelated ferrous sulfate in comparison to control Gouda cheese. There was no significant difference in lipolytic profile of control and fortified Gouda cheese during ripening. Sachdeva et al. [71] reported comparable heat stability (type A heat coagulation time (HCT)–pH curve) and higher in vitro bioaccessibility of Fe in pasteurized milk added with ferric pyrophosphate (0.0025%) and vitamin A (2500 IU).

As iron is a pro-oxidant, its fortification in protected form (microencapsulated) is always a better choice for fortification rather than direct addition into milk and milk products as it has less effect on stability and sensory profile of milk products. Jackson and Lee [41] observed lower oxidation of fat after addition of protected Fe salts (stearine-coated) in cheese than its unprotected form. Mehansho [59] reported effective use of iron liposomes prepared using lecithin for fortification of iron in beverages. Kwak et al. [55] fortified microencapsulated ferric ammonium sulfate in milk and reported that all sensory scores were comparable to control for 3 days of storage except for color, flavor (astringency and metallic) and overall scores. Microencapsulated iron was protected in gastric fluid and only released in intestinal fluid after 60 min of incubation. Siddiqui and Park [80] studied sensory and textural properties of Cheddar cheese made from caprine milk fortified with ferrous sulfate and microencapsulated ferrous sulfate. They reported that iron-fortified cheeses had higher gumminess and hardness than the control cheese.

Kwak et al. [56] observed similar lipolytic profile of control and microencapsulated iron-fortified Cheddar cheese; but bitterness, astringency, and sourness were higher in fortified cheese. However, Siddiqui and Park [80] reported difference in lipolytic profile of caprine milk Cheddar cheeses fortified with iron (ferrous sulfate salt and its microencapsulated form) with its control. Results showed that level of all fatty acid were significantly affected, except for C10:0, 14:1, 16:1, 18:1, and 20:0 fatty acids.

Kim et al. [53] reported that microencapsulated ferric ammonium sulfate did not affect the time of fermentation in drinking yogurt; however, regardless of encapsulating iron the fortified yogurt showed significantly high scores of astringency. Doosh and Sadiq [20] also observed the effect of encapsulated and non-encapsulated iron (0.015%) fortification on yogurt and reported no significant differences ($p<0.05$) in the physiochemical, rheological, and sensory properties between control and fortified yogurt.

Jalili et al. [42] reported least effect on sensory properties of Feta cheese (taste, color, flavor) if microencapsulated ferrous sulfate was added at lower concentration (40 mg/kg) in comparison to directly added ferrous sulfate; however, on increasing the concentration (60 and 80 mg/kg) even microencapsulated form had an effect on sensory properties of cheese. Abbasi and Azari [1] reported no significant difference in sensory profile of milk fortified with encapsulated iron than control except metallic taste in fortified milk. Other studies [16, 31, 41, 57] also reported no effect of iron (encapsulated form) fortification on sensory properties of milk and milk products.

Gupta et al. [32] reported higher bioavailability (in vitro) of encapsulated iron when added into milk. Arce and Ustunol [6] observed the effect of different sizes of microencapsulated ferrous sulfate on sensory characteristics of Cheddar cheese and reported that small size microcapsules (220–422 μm) showed better results as compared with large size microcapsules (700–1000 μm) for iron retention and sensory properties of cheese. Jotarkar et al. [44] attempted fortification of paneer with iron. He prepared an edible coating of Fe (ferrous sulfate and ferric ammonium citrate) complexed with whey protein concentrate that was applied to the paneer. The coating of ferrous sulfate had minimum effect on sensory characteristics and also increased the iron content (3 times) in paneer. However, ferric ammonium citrate had negative effects on sensory parameters of the paneer.

9.7.3 FORTIFICATION WITH ZINC

Abd-Rabou et al. [2] prepared Zn (Zn acetate, Zn chloride and Zn sulfate) fortified cheese (0.15%) in cheese curd. They reported that zinc acetate fortified cheese scored highest organoleptic scores among all fortified samples. Seleet et al. [78] prepared fermented drinks using buffalo milk fortified with 10 and 20 mg/L Zn sulfate and Zn acetate, respectively, and reported better sensory profile and growth of probiotic bacteria. Kahraman and Ustunol [45] fortified Cheddar cheese with zinc sulfate (228 mg of zinc/kg) and reported that addition of zinc did not affect cheese making process parameters. The sensory parameters of fortified cheese were comparable to control except hardness which was higher than control. Rana et al. [66] reported comparable pH, acidity, viscosity, rennet coagulation time (RCT), and alcohol stability of milk after fortification with 22.5 ppm zinc acetate with control.

9.7.4 FORTIFICATION WITH OTHER MICRONUTRIENTS

Achanta et al. [4] prepared mineral fortified yogurt by adding Fe, Mg, Zn, Mn, Mb, Cr, and Se individually at 25% level of their RDA and observed no significant difference in viscosity, flavor, and appearance of fortified yogurt than control; however, addition of Fe, Se, and Mg positively affected the water-holding capacity of yogurt. Santillan-Urquiza et al. [75] prepared yogurt fortified with nano-particles of Ca, Fe, and Zn individually. All fortified samples were well accepted in terms of sensory characteristics and also possessed higher firmness, however, Fe-fortified yogurt scored least in comparison to Ca and Zn fortified yogurt.

El-Din et al. [22] fortified buffalo milk individually with iron (ferrous chloride and ferrous sulfate) and zinc (zinc sulfate, and zinc acetate) at different levels (40 to 120 mg /kg) in milk separately and further milk was converted into cheese. The lower level (40 mg/kg) of addition of zinc and iron salt had minimum effect on organoleptic properties of cheese.

9.8 SUMMARY

Mineral fortification of milk products may serve as a desirable approach since milk is commonly consumed by population of all age groups, its consumption is constant and like cereals, it does not contain anti-nutritional components such as phytates that negatively affect the absorption of minerals that makes milk a suitable vehicle for mineral fortification. However, there are always issues associated with mineral fortification such as affordability, overconsumption, and efficacy. Therefore, selection of appropriate mineral at appropriate amount is a challenging task. Although these minerals are essential but their excess intake can be toxic, and moreover, the addition of these mineral fortificants can adversely affect the sensory attributes of dairy products.

KEYWORDS

- **bioavailability**
- **encapsulation**
- **fortification**
- **milk products**
- **mineral-protein complex**
- **minerals**

REFERENCES

1. Abbasi, S., & Azari, S., (2011). Efficiency of novel iron microencapsulation techniques: Fortification of milk. *International Journal of Food Science & Technology, 46*(9), 1927–1933.
2. Abd-Rabou, N. S., Zaghloul, A. H., Seleet, F. L., & El-Hofi, M. A., (2010). Properties of Edam cheese fortified by dietary zinc salts. *Journal of American Science, 6*(10), 441–446.
3. ACC/SCN, (2000). *Fourth Report on the World Nutrition Situation* (pp. 20–34). United Nations, Geneva.
4. Achanta, K., Aryana, K. J., & Boeneke, C. A., (2007). Fat free plain set yogurts fortified with various minerals. *LWT-Food Science and Technology, 40*(3), 424–429.
5. Aportela-Palacios, A., Sosa-Morales, M. E., & Vélez-Ruiz, J. F., (2005). Rheological and physicochemical behavior of fortified yogurt, with fiber and calcium. *Journal of Texture Studies, 36*(3), 333–349.

6. Arce, A., & Ustunol, Z., (2018). Effect of microencapsulated ferrous sulfate particle size on cheddar cheese composition and quality. *Journal of Dairy Science, 101*(8), 6814–6822.
7. Augustin, M. A., & Williams, R. P. W., (2006). *Nutritional Mineral Fortification of Milk*. United States Patent, Patent No. US 6,991,823 B2.
8. Benincasa, C., Lewis, J., Sindona, G., & Tagarelli, A., (2008). The use of multi element profiling to differentiate between cow and buffalo milk. *Food Chemistry, 110*(1), 257–262.
9. Boccio, J., & Monteiro, J. B., (2004). Food fortification with iron and zinc: Pros and cons from a dietary and nutritional viewpoint. *Revista de Nutrição, 17*(1), 71–78.
10. Bosscher, D., Caillie-Bertrand, M. V., Van, C. R., & Deelstra, H., (2003). Availabilities of calcium, iron, and zinc from dairy infant formulas is affected by soluble dietary fibers and modified starch fractions. *Nutrition, 19*(7, 8), 641–645.
11. Cayot, P., Guzun-Cojocaru, T., & Cayot, N., (2013). Iron fortification of milk and dairy products. In: Preedy, V. R., Srirajaskanthan, R., & Patel, V. B., (eds.), *Handbook of Food Fortification and Health: From Concepts to Public Health Applications* (Vol. 1, pp. 75–89). New York: Springer Science and Business Media.
12. Champagne, C. P., & Fustier, P., (2007). Microencapsulation for the improved delivery of bioactive compounds into foods. *Current Opinion in Biotechnology, 18*(2), 184–190.
13. Chang, Y. H., Lee, S. Y., & Kwak, H. S., (2016). Physicochemical and Sensory Properties of milk fortified with iron microcapsules prepared with water-in-oil-in-water emulsion during storage. *International Journal of Dairy Technology, 69*(3), 452–459.
14. Chen, L., Li, X., Li, Z., & Deng, L., (2020). Analysis of 17 elements in cow, goat, buffalo, yak, and camel milk by inductively coupled plasma mass spectrometry (ICP-MS). *RSC Advances, 10*(12), 6736–6742.
15. Čmejlová, K., Panovská, Z., Váchová, A., & Lukešová, D., (2009). Time-intensity studies of sweeteners. *Czech Journal of Food Sciences, 27*(1), S327–S329.
16. Coşkun, F., & Şenoğlu, C., (2011). The effect of using different levels of calcium carbonate on the physical, chemical, and sensory properties of yoghurt. *The Journal of Food, 36*(3), 129–135.
17. Degerud, E. M., Manger, M. S., Strand, T. A., & Dierkes, J., (2015). Bioavailability of iron, vitamin A, zinc, and folic acid when added to condiments and seasonings. *Annals of the New York Academy of Sciences, 1357*(1), 29–42.
18. Derbyshire, E., Brennan, C. S., Li, W., & Bokhari, F., (2010). Iron deficiency-is there a role for the food industry. *International Journal of Food Science + Technology, 45*(12), 2443–2448.
19. Department of Economic and Social Affairs, Population Division, (2015). *World Population Prospects: The 2015 Revision, Key Findings and Advance Tables* (p. 59). United Nations, New York: Department of Economic and Social Affairs.
20. Sadiq, I. H., & Doosh, K. S., (2019). Study of the physicochemical, rheological, and sensory properties of yoghurt fortified with microencapsulation iron. *The Iraqi Journal of Agricultural Science, 50*(4), 1345–1355.
21. Dutra-de-Oliveira, J. E., Freitas, M. L., Ferreira, J. F., Gonçalves, A. L., & Marchini, J. S., (1995). Iron from complex salts and its bioavailability to rats. *International Journal for Vitamin and Nutrition Research, 65*(4), 272–275.

22. El-Din, A. G., Hassan, A. S., El-Behairy, S. A., & Mohamed, E. A., (2012). Impact of zinc and iron salts fortification of buffalo's milk on the dairy product. *World Journal of Dairy & Food Sciences, 7*(1), 21–27.
23. FAO, IFAD, UNICEF, WFP, and WHO, (2017). *The State of Food Security and Nutrition in the World 2017* (p. 117). Building Resilience for Peace and Food Security. Rome, Italy: Food and Agriculture Organization of the United Nations.
24. Fleury, A. R., Funk, D. F., Patel, M. T., & Vala, W. D., (1998). *Calcium Fortified Yogurt and Methods of Preparation.* United States Patent, Patent No. US 5,820,903.
25. Food and Nutrition Board of the Institute of Medicine National Academy of Sciences, (2017). *Dietary Reference Intakes (DRIs): Recommended Dietary Allowances and Adequate Intakes, Elements.* National Academic Press.
26. Fordyce, E. J., Forbes, R. M., & Robbins, K. R., (1987). Phytate x Calcium/ zinc molar ratios: Are they predictive of zinc bioavailability. *Journal of Food Science, 52*(2), 440–444.
27. (n.d.). *Fortified Dairy Products Market Forecast, Trend Analysis and Competition Tracking-Global Market Insights 2019 to 2026.* [WWW Document], Fact.MR. URL https://www.factmr.com/report/4466/fortified-dairy-products-market (accessed on 15 November 2021).
28. Gharibzahedi, S. M. T., & Jafari, S. M., (2017). The importance of minerals in human nutrition: Bioavailability, food fortification, processing effects and nanoencapsulation. *Trends in Food Science & Technology, 62*, 119–132.
29. Global Nutrition Report, (2018). Shining a Light to Spur Action on Nutrition: Executive Summary (p. 19). https://globalnutritionreport.org/reports/global-nutrition-report-2018/ (accessed on 15 November 2021).
30. Gonmei, Z., & Toteja, G. S., (2018). Micronutrient status of Indian population. *The Indian Journal of Medical Research, 148*(5), 511–521.
31. Gupta, C., Chawla, P., & Arora, S., (2015). Development and evaluation of iron microcapsules for milk fortification. *CETA-Journal of Food, 13*(1), 116–123.
32. Gupta, C., Chawla, P., Arora, S., Tomar, S. K., & Singh, A. K., (2015). Iron microencapsulation with blend of gum Arabic, maltodextrin, and modified starch using modified solvent evaporation method-milk fortification. *Food Hydrocolloids, 43*, 622–628.
33. Hayashi, C., Kumapley, R., Wauchope, S., Alnaqshbandi, A. A. I., Perellon, C., Montorzi, G., Keeley, B., et al., (2019). *The State of the World's Children 2019: Children, Food, and Nutrition-Growing Well in a Changing World.* UNICEF.
34. Hekmat, S., & McMahon, D. J., (1997). Manufacture and quality of iron-fortified yogurt. *Journal of Dairy Science, 80*(12), 3114–3122.
35. Hettiarachchy, N. S., Sato, K., Marshall, M. R., & Kannan, A., (2011). *Bioactive food Proteins and Peptides: Applications in Human Health-Application in Human Health* (p. 368). Boca Raton-FL: CRC Press.
36. Holland, B., Brown, J., & Buss, D. H., (1993). *McCance and Widdowson's The Composition of Foods.* Royal Society of Chemistry and Ministry of Agriculture, Fisheries, and Food.
37. Hurrell, R., (2002). How to ensure adequate iron absorption from iron-fortified food. *Nutrition Reviews, 60*(7 Pt 2), S7–15.
38. Hurrell, R. F., (2002). Fortification: Overcoming technical and practical barriers. *The Journal of Nutrition, 132*(4), 806S–812S.

39. Indumathi, K. P., Kaushik, R., Arora, S., & Wadhwa, B. K., (2015). Evaluation of iron-fortified gouda cheese for sensory and physicochemical attributes. *Journal of Food Science and Technology, 52*(1), 493–499.
40. Jackson, L. S., & Lee, K., (1991). Microencapsulated iron for food fortification. *Journal of Food Science, 56*(4), 1047–1050.
41. Jalili, M., (2016). Chemical composition and sensory characteristics of feta cheese fortified with iron and ascorbic acid. *Dairy Science & Technology, 96*(4), 579–589.
42. Jalili, M., Ehsani, M. R., & Mazloumi, M. T., (2017). Evaluation of iron-fortified feta cheese for physicochemical and sensory properties. *International Journal of Dairy Technology, 70*(4), 526–532.
43. Jayalalitha, V., Balasundaram, B., & Kumar, C. N., (2012). Fortification of encapsulated iron in probiotic yoghurt. *International Journal of Agriculture: Research and Review, 2*(2), 80–84.
44. Jotarkar, P. S., Panjagari, N. R., Singh, A. K., & Arora, S., (2018). Effect of whey protein-iron based edible coating on the quality of paneer and process optimization. *International Journal of Dairy Technology, 71*(2), 395–407.
45. Kahraman, O., & Ustunol, Z., (2012). Effect of zinc fortification on cheddar cheese quality. *Journal of Dairy Science, 95*(6), 2840–2847.
46. Kapadiya, D. B., Prajapati, D. B., Jain, A. K., Mehta, B. M., Darji, V. B., & Aparnathi, K. D., (2016). Comparison of surti goat milk with cow and buffalo milk for gross composition, nitrogen distribution, and selected minerals content. *Veterinary World, 9*(7), 710–716.
47. Kapil, U., & Jain, K., (2011). Magnitude of zinc deficiency amongst under-five children in India. *The Indian Journal of Pediatrics, 78*(9), 1069–1072.
48. Kapil, U., Kapil, R., & Gupta, A., (2019). National iron plus initiative: Current status & future strategy. *The Indian Journal of Medical Research, 150*(3), 239–247.
49. Kaushik, R., & Arora, S., (2017). Effect of calcium and vitamin D_2 fortification on physical, microbial, rheological, and sensory characteristics of yoghurt. *International Food Research Journal, 24*(4), 1744–1752.
50. Kaushik, R., Sachdeva, B., & Arora, S., (2017). Effect of calcium and vitamin D_2 fortification on quality characteristics of dahi. *International Journal of Dairy Technology, 70*(2), 269–276.
51. Kaushik, R., Sachdeva, B., Arora, S., Kapila, S., & Wadhwa, B. K., (2014). Bioavailability of vitamin D_2 and calcium from fortified milk. *Food Chemistry, 147*, 307–311.
52. Kim, S. B., Seo, I. S., Khan, M. A., Ki, K. S., Lee, W. S., Lee, H. J., Shin, H. S., & Kim, H. S., (2007). Enzymatic hydrolysis of heated whey: Iron-binding ability of peptides and antigenic protein fractions. *Journal of Dairy Science, 90*(9), 4033–4042.
53. Kim, S. J., Ahn, J., Seok, J. S., & Kwak, H. S., (2003). Microencapsulated iron for drink yogurt fortification. *Asian-Australasian Journal of Animal Sciences, 16*(4), 581–587.
54. Kosonen, T., & Mutanen, M., (1992). Relative bioavailability of iron in carbonyl iron and complex ferric orthophosphate to rat. *International Journal for Vitamin and Nutrition Research, 62*(1), 60–65.
55. Kwak, H. S., Ju, Y. S., Ahn, H. J., Ahn, J., & Lee, S., (2003). Microencapsulated Iron fortification and flavor development in cheddar cheese. *Asian-Australasian Journal of Animal Sciences, 16*(8), 1205–1211.
56. Kwak, H. S., Yang, K. M., & Ahn, J., (2003). Microencapsulated iron for milk fortification. *Journal of Agricultural and Food Chemistry, 51*(26), 7770–7774.

57. Lee, J. B., Ahn, J., Lee, J., & Kwak, H. S., (2004). L-ascorbic acid microencapsulated with polyacylglycerol monostearate for milk fortification. *Bioscience, Biotechnology, and Biochemistry, 68*(3), 495–500.
58. McMillan, J. A., Oski, F. A., Staples, L., & Landaw, S. A., (1977). Iron absorption from human milk, simulated human milk and proprietary formulas. *Pediatric Research, 11*(4), 447.
59. Mehansho, H., (2006). Iron fortification technology development: New approaches. *The Journal of Nutrition, 136*(4), 1059–1063.
60. Pennell, M. D., Wiens, W. D., Varasper, J., Motzok, I., & Ross, H. U., (1975). Factors affecting the relative biological value of food-grade elemental iron powders for rats and humans. *Journal of Food Science, 40*(4), 879–883.
61. Perales, S., Barberá, R., Lagarda, M. J., & Farré, R., (2006). Fortification of milk with calcium: Effect on calcium bioavailability and interactions with iron and zinc. *Journal of Agricultural and Food Chemistry, 54*(13), 4901–4906.
62. Perveen, R., Sadiq, M., & Sikandar, H. A., (2018). Sensory attributes of mineral fortified low fat buffalo milk cheddar cheese. *Progress in Nutrition, 20*(1–S), 298–304.
63. Pirkul, T., Temiz, A., & Erdem, Y. K., (1997). Fortification of yoghurt with calcium salts and its effect on starter microorganisms and yoghurt quality. *International Dairy Journal, 7*(8, 9), 547–552.
64. Puspitasari, N. L., Lee, K., & Greger, J. L., (1991). Calcium fortification of cottage cheese with hydrocolloid control of bitter flavor defects. *Journal of Dairy Science, 74*(1), 1–7.
65. Raja, K. B., Jafri, S. E., Dickson, D., Acebròn, A., Cremonesi, P., Fossati, G., & Simpson, R. J., (2000). Involvement of iron (ferric) reduction in the iron absorption mechanism of a trivalent iron-protein complex (iron protein succinylate). *Pharmacology & Toxicology, 87*(3), 108–115.
66. Rana, B., Kaushik, R., Kaushal, K., Arora, S., Kaushal, A., Gupta, S., Upadhyay, N., et al., (2018). Physicochemical and electrochemical properties of zinc fortified milk. *Food Bioscience, 21*, 117–124.
67. Ranjan, P., Arora, S., Sharma, G. S., Sindhu, J. S., Kansal, V. K., & Sangwan, R. B., (2005). Bioavailability of calcium and physicochemical properties of calcium-fortified buffalo milk. *International Journal of Dairy Technology, 58*(3), 185–189.
68. Rice, W. H., & McMahon, D. J., (1998). Chemical, physical, and sensory characteristics of mozzarella cheese fortified using protein-chelated iron or ferric chloride. *Journal of Dairy Science, 81*(2), 318–326.
69. Saarinen, U. M., & Siimes, M. A., (1979). Iron absorption from breast milk, cow's milk, and iron-supplemented formula: An opportunistic use of changes in total body iron determined by hemoglobin, ferritin, and body weight in 132 infants. *Pediatric Research, 13*(3), 143–147.
70. Saarinen, U. M., Siimes, M. A., & Dallman, P. R., (1977). Iron absorption in infants: High bioavailability of breast milk iron as indicated by the extrinsic tag method of iron absorption and by the concentration of serum ferritin. *The Journal of Pediatrics, 91*(1), 36–39.
71. Sachdeva, B., Kaushik, R., Arora, S., & Indumathi, K. P., (2015). Impact of fortification with iron salts and vitamin a on the physicochemical properties of laboratory-pasteurized toned milk and bioaccessibility of the added nutrients. *International Journal of Dairy Technology, 68*(2), 253–260.

72. Saini, R. K., Nile, S. H., & Keum, Y. S., (2016). Food science and technology for management of iron deficiency in humans: A review. *Trends in Food Science & Technology, 53*, 13–22.
73. Sakr, H. S., El-Nawasany, L. I., & Mehanna, N. M., (2019). Fortification of karish cheese with calcium by means of using calcium chloride or casein co-precipitates containing high calcium. *Journal of Food and Dairy Sciences, 10*(9), 309–313.
74. Salgueiro, M. J., Zubillaga, M., Lysionek, A., Caro, R., Weill, R., & Boccio, J., (2002). Fortification strategies to combat zinc and iron deficiency. *Nutrition Reviews, 60*(2), 52–58.
75. Santillán-Urquiza, E., Méndez-Rojas, M. Á., & Vélez-Ruiz, J. F., (2017). Fortification of yogurt with nano and micro-sized calcium, iron, and zinc, effect on the physicochemical and rheological properties. *LWT-Food Science and Technology, 80*, 462–469.
76. Saremnezhad, S., Zargarchi, S., & Kalantari, Z. N., (2020). Calcium fortification of prebiotic ice-cream. *LWT-Food Science and Technology, 120*, 108890.
77. Schwarzenberg, S. J., & Georgieff, M. K., (2018). Advocacy for improving nutrition in the first 1000 days to support childhood development and adult health. *Pediatrics, 141*(2), e20173716.
78. Seleet, F., El-Kholy, W., & Abd-Rabou, N., (2011). Evaluation of milk drinks fermented by probiotic bacteria and fortified with zinc salts. *Polish Journal of Food and Nutrition Sciences, 61*(1), 55–60.
79. Siddique, A., & Park, Y. W., (2019). Evaluation of correlation between acid degree value and peroxide value in lipolysis of control and iron-fortified caprine milk cheeses during 4 months storage. *Open Journal of Animal Sciences, 9*(1), 1–11.
80. Siddique, A., & Park, Y. W., (2019). Effect of iron fortification on microstructural, textural, and sensory characteristics of caprine milk cheddar cheeses under different storage treatments. *Journal of Dairy Science, 102*(4), 2890–2902.
81. Singh, G., Arora, S., Sharma, G. S., Sindhu, J. S., Kansal, V. K., & Sangwan, R. B., (2007). Heat stability and calcium bioavailability of calcium-fortified milk. *LWT-Food Science and Technology*, *40*(4), 625–631.
82. Singh, G., Biswas, A., & Muthukumarappan, K., (2005). Development of technology for the calcium-fortified mango yoghurt. In: *2005 ASAE Annual Meeting* (p. 1) Tampa-Florida; USA; American Society of Agricultural and Biological Engineers.
83. Singh, G., & Muthukumarappan, K., (2008). Influence of calcium fortification on sensory, physical, and rheological characteristics of fruit yogurt. *LWT-Food Science and Technology, 41*(7), 1145–1152.
84. Singh, M., Rajan, S. S. R., Ranvir, S., Gandhi, K., & Mann, B., (2019). Profiling and distribution of minerals content in cow, buffalo, and goat milk. *Indian Journal of Dairy Science, 72*(5), 480–488.
85. Swain, J. H., Newman, S. M., & Hunt, J. R., (2003). Bioavailability of elemental iron powders to rats is less than bakery-grade ferrous sulfate and predicted by iron solubility and particle surface area. *The Journal of Nutrition, 133*(11), 3546–3552.
86. Vavrusova, M., & Skibsted, L. H., (2014). Calcium nutrition. Bioavailability and fortification. *LWT-Food Science and Technology, 59*(2), 1198–1204.
87. Vyas, H. K., & Tong, P. S., (2004). Impact of source and level of calcium fortification on the heat stability of reconstituted skim milk powder. *Journal of Dairy Science, 87*(5), 1177–1180.

88. Wise, A., (1983). Dietary factors determining the biological activities of phytate. *Nutrition Abstracts and Reviews, 53*, 791–806.
89. WHO, (2006). *The World Health Report 2006: Working Together for Health* (p. 209). Geneva-Switzerland; World Health Organization.
90. Xia, S., & Xu, S., (2005). Ferrous sulfate liposomes: Preparation, stability, and application in fluid milk. *Food Research International, 38*(3), 289–296.
91. Zhang, D., & Mahoney, A. W., (1989). Effect of iron fortification on quality of cheddar cheese. *Journal of Dairy Science, 72*(2), 322–332.

CHAPTER 10

DELIVERY OF PROBIOTICS THROUGH DAIRY FOODS

M. P. RAHILA, ARCHANA CHANDRAN, and R. L. RATHISH

ABSTRACT

Probiotics, the living microorganisms that exert a positive health benefit have become quite popular nowadays. The administration of probiotics will help to maintain a healthy intestinal microbiota by the control of pathogens and immunomodulation. The best way of delivering probiotics to humans is through food. Probiotic foods should contain live-probiotic flora in adequate concentration suspended in a suitable matrix, so that it gives postulated health benefit beyond that of the usual nutrient supply after their ingestion. Nowadays, milk and milk products face growing competition from other food items as a delivery agent for probiotics. The protective effects to probiotics strains are more accomplished in dairy matrix. Fermented milk products such as cheese, yogurt, and delicacies like milk chocolates and ice cream are the most suitable products for delivering probiotics to humans.

10.1 INTRODUCTION

With the rapid advances in the research on gut microbiota, the term 'Probiotics' has emerged as constructive effigy in promoting the human and animal wellbeing. The word 'probiotic' has its origin from Latin words, 'pro' means 'for' and 'bios' means 'life.' World Health Organization (WHO) defines probiotics as "*live microorganisms that, when administered in adequate*

Functional Dairy Ingredients and Nutraceuticals: Physicochemical, Technological, and Therapeutic Aspects. Megh R. Goyal, N. Veena, PhD & Santosh K. Mishra (Eds.)

amounts, confer a health benefit on the host" [33]. The prophylactic and therapeutic value of probiotics have been established by various studies on immunomodulation, antimicrobial activity, anti-inflammatory effects against chronic inflammation, improved gut barrier function and cardiovascular health effects. Whatsoever, the characteristics of probiotics are unique for a particular strain and distinctive of one cannot be extrapolated to others. Hence probiotics should be well defined phenotypically and genotypically. The major probiotic bacterial strains include the genera of *Lactobacilli* and *Bifidobacteria.*

The probiotic yeast *Saccharomyces cerevisiae* var. *boulardii* is also widely used for human use. The strains that are having GRAS status (generally recognized as safe) and qualified presumption of safety (QPS) status from the European Food Safety Authority (EFSA) are generally used as probiotics. The mechanism of action of probiotics on the gut is diverse and critical in maintaining the functions of gut flora. Hence the strain should be host specific with ability to alter the host epithelial and immunological responses.

Traditionally fermented milk products were a part of human culture and civilization. In recent years, foods incorporated with probiotics have received extensive attention among consumers worldwide. The most common genera that have been used commercially with proven health benefits are *Bifidobacterium* and *Lactobacillus.*

This chapter attempts to introduce the concept of probiotics, its beneficial effects, and the scope for delivery of probiotics through dairy products.

10.2 CONCEPT OF PROBIOTICS, PREBIOTICS, SYNBIOTICS, AND POSTBIOTICS

10.2.1 PROBIOTICS

The concept of probiotics was first introduced in 1907 by Elie Metchnikoff. He mentioned in his book "Studies on Optimism," "with various foods undergoing lactic acid fermentation and consumed raw (sour milk, kefir, sauerkraut, and pickles), humans introduced huge amounts of proliferating lactic acid bacteria (LAB) to their alimentary tracts." He proposed that the consumption of fermented milks containing *Lactobacilli* may displace harmful intestinal microorganisms and increase longevity [25, 49].

Our current knowledge on probiotics was actually his foresight and hence he is considered as the 'father of probiotics.' Even though the concept was put forward a century back, the science of probiotics is recent and in constant evolution. The international organizations such as the Codex, International Scientific Association for Probiotics and Prebiotics (ISAPP) and the EFSA have adopted the WHO definition for probiotics. The current definition states that a probiotic must be a viable bacteria and dead bacterial cells or cell components are excluded from this. Also, the potential of a strain to offer a health benefit on the host should be proven scientifically.

10.2.2 PREBIOTICS

Development of probiotic research has led to the exploration of prebiotics. WHO defines prebiotics are a non-viable food component that offers a health benefit to the host by modulation of the gut microbiota. Natural source of prebiotics are vegetables, fruits, cereals/grains and edible plant sources. Onions, tomatoes, asparagus, garlic, artichokes, bananas, berries, chicory, legumes, oats, linseed, barley, and wheat are potential prebiotic sources [15]. Artificially synthesized prebiotics includes lactulose, galactooligosaccharides, fructooligosaccharides, malto oligosaccharides, cyclodextrins, and lacto saccharose. Inulin and oligofructose were found to be highly supportive to the growth of probiotics [59].

10.2.3 SYNBIOTICS

The term "synbiotics," as described by Gibson and Roberfroid [28], are a combination of probiotics and prebiotics that acts synergistically. The synergistic effect is evidenced when both act together in the host environment and thereby improving the survivability of probiotics in the gastrointestinal (GI) tract. The commercial interest in synbiotic food products is increasing steadily with the heightened public interest in healthy foods having therapeutic effects.

10.2.4 POSTBIOTICS

Recent research suggests that metabolic end-products and some non-viable probiotic microorganisms are having biological activity in the host. Such products are termed as postbiotics. Metabolic by-products such as bacteriocins, ethanol, organic acids, diacetyl, hydrogen peroxide and acetaldehyde are generally considered as postbiotics. It was also found that heat killed probiotics can also exert certain biological activity in the host. These are non-toxic, non-pathogenic, and resistant to lysis by mammalian enzymes and many of them have antimicrobial action against potential pathogens.

10.3 PROBIOTIC STRAINS: SELECTION CRITERIA AND REQUIREMENTS

According to the suggestions of the WHO and EFSA, probiotic strains must stipulate to safety, functionality, and technological criteria. The safety criteria are ascertained by its origin, antibiotic resistance profile and absence of virulence. Functional criteria include their survival in the GI tract and various effects on the host. Technological usefulness of a probiotic strain means their ability to survive and express their activity during the processing, packing, and distribution operations (Table 10.1).

TABLE 10.1 Criteria for the Selection Probiotic Strains

Safety Requirements	Functionality Requirements	Technological Requirements
• Source: GI tract, breast milk, raw milk, fermented food	• Acid tolerance (stomach pH)	• Easy production of high biomass
• History of safe use (GRAS/ QPS)	• Bile tolerance	• Acid production and proteolysis
• Phenotypic and genotypic identification	• Growth and survivability in GI tract	• Ability to survive freeze-drying or any preservation techniques
• Non-pathogenic, nontoxic	• Competitiveness (including closely related species)	• Survivability during long term storage
• Absence of antibiotic resistance genes	• Adherence and ability to Antagonistic activity towards pathogens colonize	• Resistance to bacteriophages
• Post-market surveillance of adverse incidents on consumer	• Resistance to bacteriocins and acids	• Genetically stable

10.4 GUIDELINES FOR THE LABELLING OF PROBIOTICS

Few commercially available probiotics are shown in Table 10.2. The International Probiotic Association (IPA) and WHO have stipulated the guidelines regarding the labeling of probiotics [26]. Under these guidelines, the probiotic product should contain the following information [53]:

- Address of manufacturers;
- Genus name, species name and strain name should follow standard international nomenclatures;
- Health claims based on evidence should be mentioned;
- Manufacture lot number and expiry date;
- Nutrition supplements facts;
- Proper storage condition and shelf life should be mentioned;
- Suggested serving size for minimum dose of probiotics should be clearly stated;
- The minimum viable count of each probiotic strain and effective dosage should be clearly mentioned till the date of expiry.

TABLE 10.2 Examples of Commercially Available Probiotic Cultures

Manufacturer	Strain
Arla Dairy, Sweden	*Lb. paracasei* F19
BioGaia, Sweden	*Lb. reuteri* MM53
Chr. Hansen, Denmark	*B. animalis* BB12
Chr. Hansen, USA	*Lb. casei* CRl-431
Danisco, France	*B. lactis* HN019
Danone, France	*Lb. animalis* DN-173010
Nestle, Switzerland	*Lb. acidophilus* NCFM
Urex Biotech Inc., London	*Lb. fermentum* RC-14
Valio, Finland	*Lb. rhamnosus* GG
Yakult, Japan	*Lb. casei Shirota*

10.5 HEALTH BENEFITS OF PROBIOTICS

Trillions of microflora inhabiting the skin, GI tract, and urogenital tract are significantly involved in the routine physiology of the host. Knowledge of beneficial effects of probiotics was classically considered to be limited to the niche filling properties in the GI tract alone. Researches have also proven that the probiotics can produce certain important bioactive molecules like vitamin K. Studies have also shown that changes in the physiology of the host may result in the alteration of density as well as the diversity of the microbiome. This has led to theories that there exists a bidirectional relationship between the probiotic microbiota and host [10]. This bidirectional relationship associated with the host physiology has opened up many new avenues that provide evidence that probiotics can positively influence the health of its host.

Probiotic organisms exert its beneficial effects by virtue of its functionalities at different levels. This can be microbiological, as seen in filing up of the receptor sites in the intestinal or vaginal mucosa, rendering the sites unavailable for invading pathogens or excluding the invaders by competing for available nutrient resources. The probiotics check harmful microflora by producing antimicrobial substances like bacteriocins and short-chain fatty acids (SCFA). Probiotics also aid in regular physiological functions like digestion. Addition of *Bifidobacterium* in milk to supplement lactase for lactose intolerant individual is a typical example. Probiotics are also known to alter the physiology of the host. Altering the gut microbiome is known to reduce gas production in the intestine, improve the ion absorption ability of intestinal epithelial cells or reduce serum cholesterol level with the help of probiotics that have bile salt hydrolase activity [30].

10.5.1 PROBIOTIC AS DETERRENT OF INFECTIONS

All animals possess an ecosystem of microbial organisms on the skin and over many mucosal surfaces within the organ systems. These organisms are called the endogenous organisms. Probiotics are either derived from these endogenous populations or very closely related to these endogenous organisms. Endogenous microbes occupy the mucosal surfaces and skin, inhibiting any receptor sites available for microbial attachment and colonization. Occupation of the available sites by the endogenous population of microbes fills up the niche and utilizes the nutrients and resources, there by denying any resources for incoming pathogenic and opportunistic organisms. This is called competitive exclusion [14, 17].

Other than occupying the niche and preventing competition, a more reactive approach adopted by the gut microbes include production of bioactive chemicals that act as a deterrent to invading pathogens. Microbes produce acids like SCFA that will balance the pH of the gut and inhibit growth of extraneous microbes. Certain strains of probiotics express bacteriocins, small peptides that have bacteriostatic activity to regulate the gut microbial environment [50]. LAB strains PM206 and 229 was found to impede the adhesion and colonization of *Proteus mirabilis* in the urinary tract [41]. Antimicrobial action is also exerted by the production of free radicals and reactive oxygen species like peroxides. Probiotic microorganisms stimulate host defense mechanisms by upregulating expression of the defensins by the host, thereby reducing the intestinal epithelial damage [24].

10.5.2 BENEFITS TO THE GASTROINTESTINAL TRACT

Probiotics are known to neutralize and degrade certain types of toxins that can be acquired through food and feedstuff. For example, heat-killed *Lb. rhamnosus* and *Lb. acidophilus* were able to reduce aflatoxin M1 in ultra-high temperature (UHT) treated milk. Similarly, Probiotic strains *Lb. gallinarum* PDP 10, *Lb. fermentum* PDP 24, *Lb. paracasei* PL 120 and *Lb. gallinarum* PL 149 were able to almost completely eliminate aflatoxin B1 in poultry feed. The probiotic *Pediococcus pentosaceous* L006 is known to hydrolyze fumonisin, another fungal toxin that causes spoilage of food [39]. Another mode of probiotic action on foodborne toxin is when the beneficial microflora produces SCFA that makes toxin like shigatoxin produced by entero-toxigenic *E. coli* ineffective [12].

The championing health benefits of probiotics is the supposed effect of these beneficial organisms in curing and prevention of acute infectious diarrhea. Even though there are certain apparent benefits of using probiotics in managing such cases in children, the effect seems to be more strain dependent. Many of the strains of probiotics that claim to have the said effect were found to have only mild to moderate efficacy in controlling acute-onset diarrhea.

Age of the patient plays an important role on the efficacy of a particular strain. Randomized studies recommended use of *Lb. rhamnossus* GG and *S. boluardii* with a moderate effect [9, 44]. *E. faecium* SF68 was found to be beneficial to dogs and cats housed in animal shelters [6], but was found to be a possible recipient of vancomycin-resistant, questioning its safety to be used as a probiotic. The incidence of antibiotic associated diarrhea was significantly reduced in children by the administration of probiotics.

Bae et al. [3] critically reviewed research works on various probiotics to manage traveler's diarrhea and suggested that there is compelling evidence that probiotics have moderate to significant effect in controlling traveler's diarrhea. Irritable bowel syndrome was amenable to therapy that included probiotic strains like *Bifidobacterium longum* NCC3001 [52]. Probiotics are also found to be beneficial in treating *Helicobacter pylori* infections of the stomach. Probiotic administration in the treatment of infections caused by *Helicobacter pylori* was studied in humans and found that co-administration of probiotics containing *Lb. acidophilus/B. animalis* and a mixture of eight strains along with antibiotics improved the clearing rate of *H. pylori* than using antibiotics alone [48]. Another potential benefit of using probiotics is to prevent necrotic enterocolitis in infants. This is a major cause of mortality among infants. The current consensus is that concomitant breast feeding and oral supplementation of probiotics prevents necrotizing enterocolitis among preterm infants.

10.5.3 ALLERGIC REACTIONS AND AUTOIMMUNITY

The "hygiene hypothesis" due to absence of exposure to microbial stimuli in initial stages of life could contribute to development of allergies in the later stages of life. Studies on mice model have shown that lack of normal gut microbiota results in defective immune status among these animals [42]. Their gut associated lymphoid tissue and spleen lack germinal zones and the normal physiological functions of the T-lymphocytes and dendritic cells are also affected [31]. However, it was also noted that supplementation of probiotic bacteria in the initial stages of life helps in the proper development of these immune cells. [23]. Hence, probiotics and normal gut flora of an organism play a significant role in modulating the immune system of the host. This happens as a result of the interaction between the intestinal flora and the gut associated lymphoid systems across the microfold cells of the intestine that act as the portal between mucosa and the lymphoid tissues. A balanced gut microflora in the early stages of life provides an adequate stimulus to the immune system and help in developing a healthy immune system in the host.

Probiotics and commensal gut flora are known to cause the development and differentiation of regulatory and helper T cells that are better able to differentiate self and non-self antigens (e.g., Th1 and Th2). The dendritic cells that process and deliver the probiotic antigens to the T-lymphocytes of cell-mediated immune system. This will result in differentiation of Interleukin-10 (IL-10) that suppress activities of cytotoxic T (Tc) and Th cells that may poorly differentiate self-antigens (Th1), regulatory T cells (T_{reg}) and transforming growth factor β (TGF-β). Administration of *Lb. casei Shirota* orally has resulted in increased recruitment of Th2 over Th1 cells that released cytokines which resulted in reduced immunoglobulin E (IgE) in mouse models with allergic response to ovalbumin [45]. Similar responses were also observed on feeding *Lb. fermentum* and *S. serveciae* probiotics [5, 22].

Patients with allergic rhinitis and asthma symptoms under supplementation with probiotic strains of *B. longum* and *Lb. acidophilus* were found to have less severe episodes in comparison with a placebo control. The peripheral blood mononuclear cells like monocytes were found to have a better response to interferon-gamma (INF-γ) and could be responsible for alleviating the symptoms [27].

Atopic dermatitis is a chronic inflammatory condition of the skin predominant among young children. The histological feature in atopic dermatitis is the presence of Th2 and Th22 in case of chronic atopy, but Th1 is involved in acute cases. A combination of seven *lactobacillus* strain and *Bifidobacterium* could improve atopic dermatitis in a mouse model, evidenced by relief from skin irritation as well as increase in T_{reg} cell deployment at the sites of lesions. Evidences from randomised control trials involving feeding *Lactococcus lactis* 11/19-B1 strain to children resulted in suppressed pro-inflammatory Th cells, resulting in improved skin condition [58].

10.5.4 PROBIOTICS VERSUS LIFESTYLE DISEASES

The annual death toll from non-communicable diseases is about forty million. This constitutes for about 70% of annual human death rate. Conditions like cardiovascular diseases, stroke, diabetes, and certain forms of diseases contribute heavily to this tally. Since these diseases are heavily linked to the lifestyle choices that one makes, the diseases are called lifestyle diseases. Along with risk factors like sedentary lifestyle and abuse of alcohol and tobacco, unhealthy diet is also incriminated. Lifestyle diseases are not limited to the wealthy and affluent. The disease has a greater impact on low and middle-income countries and poor people compared to other strata of the society.

Even though the relationship between lifestyle and lifestyle diseases are yet to be fully understood, studies have been done on the impact of diet on type-II diabetes and obesity. Dysbiosis of the intestinal microflora is now considered as a factor that contributes to the occurrence of type-II diabetes mellitus. Using probiotics as an alternative therapy for managing type-II diabetes mellitus had resulted in lower glycaemic levels in diabetic mice models as well as human subjects. Subjects receiving probiotics *Lactobacillus* and *Bifidobacterium* strains were found to have reduced rate of beta-cell destruction in islets of Langerhans, improved insulin sensitivity, and improved insulin binding potential in glucose transporter moieties of the muscle cells. The studies by Hu et al. [35] revealed that the treatment period had more impact than the dose administered and sub-strains of the probiotics used had no significant effect on altering the effect.

Nowadays, obesity is a serious health issue affecting all age groups. Its global prevalence has doubled since 1980 and WHO has pledged to control this epidemic. The pathophysiology of obesity involves the net excess of energy input in a person. When a person consumes energy rich food, and all of it is not dissipated, a net excess occurs, which gets deposited as fat. Though the genetic factors of an individual play a role, the current research focus is on the host and environmental factors of energy balance. The role of microflora in managing the weight homeostasis of individuals came to focus with the work of Turnbaugh [59] when they could reproduce obese phenotype in germfree mice when these mice intestines were populated with intestinal microflora of obese mice. One theory regarding the association between obesity and enteric microflora states that fat and sugar rich diet promotes the growth of firmicutes at the expense of Bacteroides. Conversely, intestinal microflora predominantly consists of Bacteroides among individuals with a low fat, low sugar diet. It was also observed that the microfloral balance tips to the favor of Bacteroides when the high fat diet is switched to low fat high fiber diet.

Several theories have been put forward about the importance of gut microflora in prompting the energy balance of the host. SCFAs produced by the intestinal microflora can improve the absorption of glucose and other carbohydrates and thus excess conversion of fat and resulting in increased deposits of adipose tissues in the body. Certain SCFAs are known to retard gut motility and aiding in greater absorption of nutrients. Certain SCFAs like butyrate and propionate are known to reduce food intake among mice models whereas aromatic and branched-chain SCFAs are predominant in obese subjects [21].

VSL#3 (a commercial probiotic mixture) administration in mouse model could suppress gain in body weight as well as the insulin resistance when fed on fat-rich diet. The study found that the probiotic altered the gut microflora composition, increased the gut butyric acid levels and upregulated the glucagon like peptide-1 (GLP-1) hormone, leading to a reduced food intake and improved glucose tolerance [1]. Thus, modification of intestinal microflora can function as an adjuvant in managing obesity and improving metabolic profiles of obese individuals.

10.5.5 GUT-BRAIN AXIS VERSUS PROBIOTICS

WHO defines health as not mere absence of diseases or infirmity, but complete mental, spiritual, and social well-being. The fast-paced life and modern social structure is associated with undesirable emotional traits and associated deterioration of health. There exists a bi-directional interaction between the peripheral nervous system in the intestine and the emotional and cognitive centers of the central nervous system. Current evidences strongly suggest that the intestinal microflora engage in active communication with this gut-brain axis and thereby influence the emotional and mental well-being of the individual [19].

The gut microbiome-brain communication is exerted through neural, endocrine, and immunological signaling. One of the best studied phenomena is the effect of probiotics in the emotional well-being of individuals suffering from irritable bowel syndrome by reducing anxiety and improving the bowel habits. Probiotics were found to improve the attitude towards food among children suffering from autism and regulate mood to modify the attitude towards addictive and rewards behavior.

10.5.6 INTERACTION BETWEEN HOST AND PROBIOTICS

Once the probiotic enters the host intestine, various levels of interaction takes place at the host epithelial and mucus layer [47]. The interaction of gut-associated lymphoid tissue (GALT) and probiotics plays an important role in immunomodulation [29].

The mechanisms of action of probiotics are:

- Competitive exclusion;
- Epithelial barrier enhancement;
- Immunomodulation; and
- Secretion of antimicrobial factors.

Competitive exclusion is highly relevant in case of infections which involve direct interaction of pathogen with the intestinal epithelium [32]. In order to reduce the entry of pathogens to target molecules on the host epithelium, probiotics interact with the cell-surface proteins and produce adhesion factors to mediate cell attachment and subsequent exclusion of the pathogen [51].

Intestinal epithelium plays a significant role in maintaining the transport of ions and water. Integrity of intestinal epithelium is enhanced by the maintenance and upregulation of tight junction proteins. Disruption of barrier function during infection will result in inflammation leading to diarrhea. Probiotic interaction at this level will improve the barrier function by reinforcing the tight junctions and seal the lateral surfaces of enterocytes. The gene expression of E-cadherin and β-catenin, the adherens junction proteins were found to be upregulated on administration of *Lactobacilli* strains in T84 cell line model [56].

The integrity of intestinal epithelium is maintained by probiotics through immunomodulation. Thus, the regulation of adaptive immunity plays a significant role in gut health. Deterioration of intestinal barrier during inflammatory bowel disease is attributed to the production of pro-inflammatory cytokines. Downregulation of pro-inflammatory mediators has been observed during the administration of several probiotic mixtures. The probiotic mixture VSL#3 was found to lessen the epithelial barrier disruption induced by IFN-γ in an *in vitro* (HT29) cell model [40].

In another study *B. infantis* and *Lb. acidophilus* acted against IL-1 beta-induced disruption of intestinal barrier [29]. Modification of cytokine profiles is another method of attenuating adaptive immune response by inhibiting inflammation and maturation of lymphocytes [36]. This may be by the production of immunomodulatory peptides capable of downregulating inflammatory mediators.

Probiotics are found to produce antimicrobial and bioactive factors like organic acids and heat-stable peptides, which enhances their competitive advantage against the pathogens. Lactic or acetic acid produced by probiotic bacteria exerts an antimicrobial effect against gram negative pathogens like *Salmonella typhimurium* [8]. Several strains of probiotics are capable of producing heat-stable peptides known as bacteriocins that act against pathogens by hindering cell wall synthesis or initiate pore formation in cell wall. It was found that *L. curvatus* DN317 produced a bacteriocin curvaticin DN317, having antimicrobial action against *Campylobacter jejuni* [60]. SCFAs synthesized by probiotic bacteria like butyric acid have also got a major role in the upkeep of intestinal epithelial barrier integrity.

10.6 DELIVERY OF PROBIOTICS THROUGH DAIRY PRODUCTS

Beneficial effects of probiotics can be ascertained only through an efficient transit through the gut by a suitable carrier. Milk and milk products serve as an ideal food matrix that protects probiotics from various stresses while production, processing, storage as well as the adverse conditions in the GI tract.

Earlier fermented milk products were usually associated with the delivery of probiotics, but due to the increasing demand, other dairy products like ice cream, infant milk powders and chocolates are also now commercially available. The components like protein, fat, and lactose in dairy products improves the efficiency of food matrix-like ice cream. But the pH values should be in the range of 5.5–6.5 to increase the survivability of probiotics [16].

10.6.1 PROBIOTIC CHEESE

The fermented milks usually have a lower shelf life when compared to cheese and hence nowadays cheese is used as a vehicle of probiotics. The physicochemical characteristics of cheese like high pH, low titratable acidity, high fat, high buffering capacity and increased nutrient availability, more solid consistency and low oxygen content makes it an ideal candidate for delivery and offers protection during transit through gut [37].

Cheese works as a good probiotic delivery vehicle due to its inherent buffering system against the acidic pH of the digestive tract and makes a favorable environment for probiotic persistence in the host's digestive system. Moreover, high total solids especially fat percentage may offer additional protection to probiotic bacteria in the stomach [4]. The presence of the prebiotics such as inulin and oligo-fructose can promote growth rates of probiotic organisms like *Bifidobacteria* and *Lactobacillus* species [7]. Incorporation of probiotic culture does not cause any quality changes in the final product, and the major ripening changes (lipolysis and proteolysis) may be similar or even a little higher in probiotic cheese as compared to the cheese without any added probiotic bacteria [18].

A cottage cheese with *Lactobacillus delbrueckii* UFV H2b20 and inulin (act as prebiotic) was developed and found that this symbiotic relationship does not affect the physicochemical and sensory parameters of cottage cheese [2]. Similarly, *Lb. acidophilus* was added for the development of Turkish white cheese and found that vacuum packed cheese has more consumer acceptance than the brine stored cheese [38].

10.6.2 PROBIOTIC YOGURT

A fermented dairy product consumed throughout the world with great consumer acceptability due to its potential health benefits. It is produced by lactic acid fermentation by *Lb. delbrueckii* ssp. *bulgaricus* and S*trepto-coccus salivarius* ssp. *thermophillus* as starter cultures. Apart from these two starter cultures, probiotic yogurt contains one specific or more strains of probiotic bacteria. Probiotic culture is better added to the yogurt at the end of processing prior to packaging (Figure 10.1). In probiotic yogurt, the probiotic strains showed a slower growth rate because of the inhibitory substances produced by the yogurt culture and because of the reduction in the available nutrients for probiotics due to the rapid growth of starter cultures [54].

Milk

↓

Addition of Sugar (12%)

↓

Pasteurization (90°C)

↓

Holding (30 minutes)

↓

Cooling (42 – 43°C)

↓

Inoculation with starter culture isolates

↓

Incubation (40 – 45°C for 10 hours)

↓

Inoculation with Probiotic isolate

↓

Packing and Storage at 4°C

FIGURE 10.1 Flow diagram for the preparation of probiotic yogurt.

10.6.3 PROBIOTIC ICE CREAM

Here the probiotic cultures may be added in two ways (Figure 10.2): either added to the ice cream mix after pasteurization or added to milk then fermented to produce frozen yogurt ice cream [16]. Probiotic ice creams may be prepared by using *Bifidobacterium lactis* and *Lb. acidophilus* [13]. Similarly, probiotic ice cream with free and microencapsulated *Lb. casei* and *Bifidobacterium lactis* was developed and found that encapsulation can significantly increase the survivability of probiotic bacteria with an extended shelf-life [34].

The influence of altered overrun levels on vanilla flavored probiotics ice cream incorporated with *Lb. acidophilus* has been studied and found that lesser air incorporation should be adopted for probiotic ice cream to conserve its functional status throughout the storage period [20].

Standardization of ice cream mix

↓

Pasteurization (85oC for 15 seconds)

↓

Homogenization (70-75oC, 17-18.5 MPa)

↓

Ripening (3-5oC/ 10-24 h)

↓

Addition of Probiotic strains and stirring (4-6oC/10 minutes)

↓

Freezing (-3 to -40C/ 80-90 % Overrun)

↓

Packaging

↓

Hardening (-25 to -300C/ 10-20 h)

FIGURE 10.2 Production flow chart for ice cream with probiotics.

10.6.4 PROBIOTIC DAIRY DRINKS

10.6.4.1 ACIDOPHILUS MILK

This fermented milk product is prepared by *Lb. acidophilus* culture. *Lb. acidophilus* populate in the gastrointestinal system of the host and inhibit dangerous gas-forming microbes. It is helpful to individuals suffering from diarrhea and gastric gas complications. It has antimicrobial effect because of the occurrence of organic acids, H_2O_2 and antibiotic substances like acidophilin, acidolin, and lactocidin.

10.6.4.2 BIFIDUS MILK

It is considered as the first baby food prepared with *Bifidobacteria. Bifidobacterium bifidum* and *Bifidobacterium longum* isolated from healthy humans feces and added to milk at 10% rate. The final product contains 10^8–10^9 cfu/ ml bacteria.

10.6.4.3 YAKULT

It is a fermented dairy product prepared by using *L. casei Shirota* culture. This probiotic strain is unaffected by GI acids and can able to populate in the small intestine. It has the capability to increase the activity and amounts of macrophages. This strain was identified by Dr. Shirota, a Japanese microbiologist, in 1935. This product is commercialized by Yakult Honsha Co. Ltd.

10.6.4.4 MIL-MIL

This fermented dairy product is originated in Japan (developed by Yakult Honsha Co. Ltd.). *Lb. acidophilus, Bifidobacterium bifidum*, and *Bifidobacterium breve* strains are utilized for the fermentation process. This product is rich in beta-carotene since it is usually enriched with small amounts of carrot juice.

10.6.4.5 KEFIR

Kefir is originated in North Caucasia. The word 'Kefir' means "good food" in Turkish. It is a slightly acidic and naturally carbonated product which is prepared by using kefir grains. Kefir grain contains both bacteria and yeast with a symbiotic relationship in a polysaccharide matrix. During the preparation of kefir, lactic acid and alcohol fermentation takes place. Due to its *Bifidobacterium bifidum* content, it is recommended for infants of above 6 months age.

10.6.4.6 KOUMISS

Koumiss is a dairy beverage prepared from mare milk by using *Lb. acidophilus* and *Torula koumiss*. It is an acidic-alcoholic product with 2% alcohol and 0.5–1.5% lactic acid. Koumiss is useful for people suffering from cardiovascular diseases, tuberculosis, asthma, and gynecological diseases.

10.6.4.7 ACIDOPHILIN

It is prepared primarily by using *Lb. acidophilus* culture. Besides *Lb. acidophilus*, this culture contains *L. lactis* ssp. *lactis* or *Lb. delbrueckii* ssp. *bulgaricus*. The acidity of the final product is around 0.67–1.08%.

10.7 THE DIVERSITY OF PROBIOTIC DAIRY PRODUCTS

Fermented milk products like yogurts, sour milk and cheese can be widely used for delivering probiotics to humans. Delivering probiotics through milk products may support in enduring adverse GI environment than delivering through non-dairy foods. This may be because of the buffering action of milk especially milk fat, that might protect probiotics during exposure to harsh conditions [54]. The bioactive ingredients in food format like prebiotics and other fibers, vitamins, enzymes, minerals, preservatives, and flavors may also affect the efficacy of probiotics. Milk is a rich source of biologically active components like immunoglobulins, lactoferrin, glycosphingolipids, oligosaccharides, and calcium. Some of these components have antimicrobial, immunomodulatory, and prebiotic properties (Table 10.3). The calcium in milk was found to increase resistance of probiotic strains to enterotoxigenic *E. coli* infection [55]. The adhering abilities of probiotic strains to the piglet jejunal epithelial cell line IPEC-J2 was improved significantly by the presence of calcium [43].

TABLE 10.3 Therapeutic Application of Probiotic Strains Through Dairy Foods in Humans

Disease	Probiotic Organism	Dairy Carrier
Antibiotic-associated diarrhea	*Lb. casei* *Lb. bulgaricus* *S. thermophilus*	Yogurt drink Yakult
Cholesterolemia	*Bifidobacterium spp.* *Lb. acidophilus.*	Yogurt Bifidus milk
Colon cancer	*Lb. acidophilus* *B. longum*	Acidophilus milk Acidophilin
Coronary heart diseases and cardiovascular diseases	*Lb. acidophilus*	Fermented milk Acidophilus milk
Crohn's disease	*Lb. acidophilus* La-5, *Bifidobacterium* Bb 12	Fermented milk Acidophilus milk
Gastrointestinal and liver diseases	*B. bifidum* *B. longum*	Bifidus milk Mil-mil
IgE associated eczema	*Lb. rhamnosus,* *B. animalis* spp. *lactis* Bb-12 *Lb. acidophilus* La-5	Fermented milk Acidophilus milk
Irritable bowel syndrome symptoms	*Lb. acidophilus,* *Lb. casei,* *Lb. delbrueckii,* *Lb. plantarum,* *S. salivarius*	Yogurt Fermented milk Acidophilus milk Acidophilin Fermented skimmed-milk
Tuberculosis and cardiovascular diseases	*Lb. acidophilus* *Torula koumiss.*	Koumiss
Type-2 diabetes and obesity	*Lb. paracasei*	Fermented skim milk

Probiotic survivability can be improved by techniques like microencapsulation and induction of cellular stress tolerance pathways [11]. Encapsulation of probiotics often require specific processing steps, which in turn complicates the process and increases the cost of production. Hence, encapsulation is done only if some functionality issue during processing and storage is expected. Different approaches of microencapsulation include spray coating, spray drying, liposomes, extrusion, and emulsions. Due to the convenience and easy availability non-conventional probiotic formulations like cheese, yogurts, ice cream, and chocolates are preferred over the pharmaceutical capsules and tablets by the consumers. But the dosage per serving has to be ensured before marketing. Intestinal flora in the gut may vary from person to person depending on age, race, food habits, etc. Ensuring the composition of gut flora before probiotic administration is detrimental in deciding the effectiveness of probiotics.

10.8 SUMMARY

There is a wide variety of functional diets like probiotic, prebiotic, products supplemented with antioxidants, polyphenols, fat replacers, artificial sweeteners, etc. Out of these products, probiotic functional products gained a leading choice to employ an advanced health benefits on the human body beyond its nutritive role. Dairy and non-dairy food products may be utilized as a vehicle for probiotic delivery to the human body. The dairy probiotic foods comprising yogurt, drinking yogurt, dried yogurt, ice cream, cheese, kefir, and dried milk powder have been suggested as the chief carrier for probiotic bacteria.

Current studies on the outcome of probiotics on human and animal diseases have revealed that apart from modulating the gut microbiota, they can traverse through diverse locations like gut-brain axis. Similarly, certain strains of the most frequently used probiotics failed to populate the gut but are colonized in the intestine. Even though *in vitro* and *in vivo* trials on health benefits of potential strains is going on and there is a dare need for the clinical trials to prove their efficacy. Knowing the molecular mechanism behind the proven benefits of probiotics are also important for the designing of strategies in developing emerging or next-generation probiotics and probiotic functional foods.

KEYWORDS

- **bifidobacterium**
- **dairy products**
- **fermentation**
- **lactobacillus**
- **prebiotics**
- **probiotics**
- **synbiotics**

REFERENCES

1. Amabebe, E., Robert, F. O., Agbalalah, T., & Orubu, E. S., (2020). Microbial dysbiosis-induced obesity: Role of gut microbiota in homoeostasis of energy metabolism. *British Journal of Nutrition, 123*(10), 1127–1137.
2. Araùjo, E. A., Dos, S. P. A. C., Pinto, M. S., Jan, G., & De Carvalho, A. F., (2012). Probiotics in dairy fermented products. Chapter 6. In: Rigobelo, E., (ed.), *Probiotics* (pp. 129–148). Rijeka: IntechOpen.
3. Bae, J. M., (2018). Prophylactic efficacy of probiotics on travelers' diarrhea: An adaptive meta-analysis of randomized controlled trials. *Epidemiology and Health, 40*(1), 1–4.
4. Bergamini, C. V., Hynes, E. R., Quiberoni, A., Suarez, V. B., & Zalazar, C. A., (2005). Probiotic bacteria as adjunct starters: Influence of the addition methodology on their survival in a semi-hard Argentinean cheese. *Food Research International, 38*(5), 597–604.
5. Bhat, M. I., Kumari, A., Kapila, S., & Kapila, R., (2019). Probiotic lactobacilli mediated changes in global epigenetic signatures of human intestinal epithelial cells during *Escherichia coli* challenge. *Annals of Microbiology, 69*(6), 603–612.
6. Bybee, S. N., Scorza, A. V., & Lappin, M. R., (2011). Effect of the probiotic *Enterococcus faecium* SF68 on presence of diarrhea in cats and dogs housed in an animal shelter. *Journal of Veterinary Internal Medicine, 25*(4), 856–860.
7. Cardarelli, H. R., Saad, S. M. I., Gibson, G. R., & Vulevic, J., (2007). Functional petit Suisse cheese: Measure of the prebiotic effect. *Anaerobe, 13*(5), 200–207.
8. Castillo, N. A., Perdigon, G., & De Moreno, D. L. A., (2011). Oral administration of a probiotic *Lactobacillus* modulates cytokine production and TLR expression, improving the immune response against *Salmonella entericasero var Typhimurium* infection in mice. *BMC Microbiology, 11*(1), 177–189.
9. Chen, J., Wan, C. M., Gong, S. T., Fang, F., Sun, M., Qian, Y., & Dong, M., (2018). Chinese clinical practice guidelines for acute infectious diarrhea in children. *World Journal of Pediatrics, 14*(5), 429–436.
10. Cheng, D., Song, J., Xie, M., & Song, D., (2019). The bidirectional relationship between host physiology and microbiota and health benefits of probiotics: A review. *Trends in Food Science & Technology, 91*(1), 426–435.
11. Corcoran, B. M., Stanton, C., Fitzgerald, G., & Ross, R. P., (2008). Life under Stress: The probiotic stress response and how it may be manipulated. *Current Pharmaceutical Design, 14*(14), 1382–1399.
12. Cordonnier, C., Thévenot, J., Etienne-Mesmin, L., Alric, M., Livrelli, V., & Blanquet-Diot, S., (2017). Probiotic and enterohemorrhagic *Escherichia coli*: An effective strategy against a deadly enemy. *Critical Reviews in Microbiology, 43*(1), 116–132.
13. Corrales, A., Henderson, M., & Morales, I., (2007). Survival of probiotic microorganisms *Lactobacillus acidophilus* and *Bifidobacterium lactis* in whipped ice cream. *Revista Chilena de Nutrición, 34*(2), 157–163.
14. Cremon, C., Barbaro, M. R., Ventura, M., & Barbara, G., (2018). Pre- and probiotic overview. *Current Opinion in Pharmacology, 43*(1), 87–92.
15. Crittenden, R., & Playne, M. J., (2008). Prebiotics. In: Lee, Y. K., & Salminen, S., (eds.), *Handbook of Probiotics and Prebiotics* (pp. 535–561). Hoboken-New Jersey: Wiley.
16. Cruz, A., Antunes, A., Sousa, A., Faria, J., & Saad, S., (2009). Ice cream as a probiotic food carrier. *Food Research International, 42*(9), 1233–1239.

17. Cunha, S., Mendes, A., Rego, D., Meireles, D., Fernandes, R., Carvalho, A., & Da Costa, P. M., (2017). Effect of competitive exclusion in rabbits using an autochthonous probiotic. *World Rabbit Science, 25*(2), 123–134.
18. Da Cruz, A. G., Buriti, F. C. A., De Souza, C. H. B., Faria, J. A. F., & Saad, S. M. I., (2009). Probiotic cheese: Health benefits, technological, and stability aspects. *Trends in Food Science & Technology, 20*(8), 344–354.
19. Dinan, T. G., & Cryan, J. F., (2017). The microbiome-gut-brain axis in health and disease. *Gastroenterology Clinics, 46*(1), 77–89.
20. Ferraz, J., Cruz, A., Cadena, R., Freitas, M., Pinto, U., Carvalho, C., Faria, J., & Bolini, H., (2012). Sensory acceptance and survival of probiotic bacteria in ice cream produced with different overrun levels. *Journal of Food Science, 71*(1), 524–528.
21. Fluitman, K. S., Wijdeveld, M., Nieuwdorp, M., & IJzerman, R. G., (2018). Potential of butyrate to influence food intake in mice and men. *Gut, 67*(7), 1203–1204.
22. Fonseca, V. M. B., Milani, T. M., Prado, R., Bonato, V. L. D., Ramos, S. G., Martins, F. S., Vianna, E. O., & Borges, M. D. C., (2017). Oral administration of *Saccharomyces cerevisiae* UFMG A-905 prevents allergic asthma in mice. *Respirology, 22*(5), 905–912.
23. Fu, L., Cherayil, B. J., Shi, H., Wang, Y., & Zhu, Y., (2019). *Food Allergy: From Molecular Mechanisms to Control Strategies* (p. 216). Springer, Singapore.
24. Fusco, A., Savio, V., Cammarota, M., Alfano, A., Schiraldi, C., & Donnarumma, G., (2017). Beta-defensin-2 and beta-defensin-3 reduce intestinal damage caused by modulating the expression of cytokines and enhancing the probiotic activity of *Enterococcus faecium. Journal of Immunology Research, 10*(1), 1–7.
25. Gaggia, F., Mattarelli, P., & Biavati, B., (2010). Probiotics and prebiotics in animal feeding for safe food production. *International Journal of Food Microbiology, 141*(1), 15–28.
26. Ganguly, N. K., Bhattacharya, S. K., Sesikeran, B., Nair, G. B., Ramakrishna, B. S., Sachdev, H. P. S., Batish, V. K., et al., (2011). ICMR-DBT guidelines for evaluation of probiotics in food. *Indian Journal of Medical Research, 134*(1), 22–25.
27. Garcia-Garcia, J. C., Janusz, M. J., Kiely, B., Murphy, E. F., & Omahony, L., (2019*). U.S. Patent No. 10233433.* Washington, DC: U.S. Patent and Trademark Office.
28. Gibson, R. G., & Roberfroid, M. B., (1995). Dietary modulation of the human colonic microbiota: Introducing the concept of prebiotics. *The Journal of Nutrition, 125*(6), 1401–1412.
29. Goto, Y., & Ivanov, I. I., (2013). Intestinal epithelial cells as mediators of the commensal-host immune crosstalk. *Immunology and Cell Biology, 91*(3), 204–214.
30. Guo, S., Gillingham, T., Guo, Y., Meng, D., Zhu, W., & Walker, W. A., (2017). Secretions of *Bifidobacterium infantis* and *Lactobacillus acidophilus* protect intestinal epithelial barrier function. *Journal of Pediatric Gastroenterology and Nutrition, 64*(3), 404–412.
31. Hara, S., Sasaki, T., Satoh-Takayama, N., Kanaya, T., Kato, T., Takikawa, Y., Takahashi, M., et al., (2019). Dietary antigens induce germinal center responses in Peyer's patches and antigen-specific IgA production. *Frontiers in Immunology, 10*, 2432.
32. Hecht, A. L., Casterline, B. W., Earley, Z. M., Goo, Y. A., Goodlett, D. R., & Wardenburg, J. B., (2016). Strain competition restricts colonization of an enteric pathogen and prevents colitis. *EMBO Reports, 17*(9), 1282–1291.
33. Hill, C., Guarner, F., Reid, G., Gibson, G. R., Merenstein, D. J., Pot, B., Morelli, L., et al., (2014). Expert consensus document: The international scientific association for

probiotics and prebiotics consensus statement on the scope and appropriate use of the term probiotic. *Nature Reviews: Gastroenterology & Hepatology, 11*(8), 506–514.
34. Homayouni, A., Azizi, A., Ehsani, M. R., Yarmand, M. S., & Razavi, S. H., (2008). Effect of microencapsulation and resistant starch on the probiotic survival and sensory properties of synbiotic ice cream. *Food Chemistry, 111*(1), 50–55.
35. Hu, Y. M., Zhou, F., Yuan, Y., & Xu, Y. C., (2017). Effects of probiotics supplement in patients with type 2 diabetes mellitus: A meta-analysis of randomized trials. *Medicina Clínica, 148*(8), 362–370.
36. Kamada, N., Seo, S. U., Chen, G. Y., & Núñez, G., (2013). Role of the gut microbiota in immunity and inflammatory disease. *Nature Reviews Immunology, 13*(5), 321–335.
37. Karimi, R., Mortazavian, A. M., & Da Cruz, A. G., (2011). Viability of probiotic microorganisms in cheese during production and storage: A review. *Dairy Science & Technology, 91*(3), 283–308.
38. Kasımoglu, A., Goncuoglu, M., & Akgun, S., (2004). Probiotic white cheese with *Lactobacillus acidophilus*. *International Dairy Journal, 14*(12), 1067–1073.
39. Khandelwal, P., Upendra, R. S., Rani, B. R., Shivani, R., Prasanna, B. K., & More, S. R., (2017). Studies on bacteriocin producing abilities of indigenously isolated lactic acid bacteria strains. *International Journal of Fermented Foods, 6*(1), 35–43.
40. Krishnan, M., Penrose, H. M., Shah, N. N., Marchelletta, R. R., & McCole, D. F., (2016). VSL#3 probiotic stimulates T-cell protein tyrosine phosphatase-mediated recovery of IFN-gamma-induced intestinal epithelial barrier defects. *Inflammatory Bowel Diseases, 22*(12), 2811–2823.
41. Lai, T. M., Lin, P. P., Hsieh, Y. M., & Tsai, C. C., (2020). Evaluation of inhibitory activity of domestic probiotics for against invasion and infection by *Proteus mirabilis* in the urinary tract. *The Journal of Infection in Developing Countries, 14*(4), 366–372.
42. Lambrecht, B. N., & Hammad, H., (2017). The immunology of the allergy epidemic and the hygiene hypothesis. *Nature Immunology, 18*(10), 1076–1083.
43. Larsen, N., Nissen, P., & Willats, W. G. T., (2007). The effect of calcium ions on adhesion and competitive exclusion of *Lactobacillus* ssp. and *E. coli* O138. *International Journal of Food Microbiology, 114*(1), 113–119.
44. Li, Y. T., Xu, H., Ye, J. Z., Wu, W. R., Shi, D., Fang, D. Q., Liu, Y., & Li, L. J., (2019). Efficacy of *Lactobacillus rhamnosus* GG in treatment of acute pediatric diarrhoea: A systematic review with meta-analysis. *World Journal of Gastroenterology, 25*(33), 4999–5016.
45. Liu, M. Y., Yang, Z. Y., Dai, W. K., Huang, J. Q., Li, Y. H., Zhang, J., Qiu, C. Z., et al., (2017). Protective effect of *Bifidobacterium infantis* CGMCC313-2 on ovalbumin-induced airway asthma and β-lactoglobulin-induced intestinal food allergy mouse models. *World Journal of Gastroenterology, 23*(12), 2149–2158.
46. Louis, P., Hold, G. L., & Flint, H. J., (2014). The gut microbiota, bacterial metabolites and colorectal cancer. *Nature Reviews: Microbiology, 12*(10), 661–672.
47. Maynard, C. L., Elson, C. O., Hatton, R. D., & Weaver, C. T., (2012). Reciprocal interactions of the intestinal microbiota and immune system. *Nature, 489*(7415), 231–241.
48. McFarland, L. V., Huang, Y., Wang, L., & Malfertheiner, P., (2016). Systematic review and meta-analysis: Multi-strain probiotics as adjunct therapy for *Helicobacter pylori* eradication and prevention of adverse events. *United European Gastroenterol Journal, 4*(4), 546–556.

49. Metchnikoff, E., (1908). The prolongation of life. *Nature, 77*, 289–290.
50. Nair, M. S., Amalaradjou, M. A., & Venkitanarayanan, K., (2017). Anti-virulence properties of probiotics in combating microbial pathogenesis. *Advances in Applied Microbiology, 98*, 1–29.
51. Nishiyama, K., Nakazato, A., Ueno, S., Seto, Y., Kakuda, T., & Takai, S., (2015). Cell surface-associated aggregation-promoting factor from *Lactobacillus gasseri* SBT2055 facilitates host colonization and competitive exclusion of *Campylobacter jejuni*. *Molecular Microbiology, 98*(4), 712–726.
52. Pinto-Sanchez, M. I., Hall, G. B., Ghajar, K., Nardelli, A., Bolino, C., Lau, J. T., & Traynor, J., (2017). Probiotic *Bifidobacterium longum* NCC3001 reduces depression scores and alters brain activity: A pilot study in patients with irritable bowel syndrome. *Gastroenterology, 153*(2), 448–459.
53. Prado, F. C., Parada, J. L., Pandey, A., & Soccol, C. R., (2008). Trends in non-dairy probiotic beverages. *Food Research International, 41*(2), 111–123.
54. Ranadheera, C. S., Vidanarachchi, J. K., Rocha, R. S., Cruz, A. G., & Ajlouni, S., (2017). Probiotic delivery through fermentation: Dairy vs. non-dairy beverages. *Fermentation, 3*(67), 2–17.
55. Roy, D., (2005). Technological aspects related to the use of bifidobacteria in dairy products. *Lait, 85*(1, 2), 39–56.
56. Sanders, M. E., & Marco, M. L., (2010). Food formats for effective delivery of probiotics. *Annual Reviews in Food Science and Technology, 1*, 65–85.
57. Seth, A., Yan, F., Polk, D. B., & Rao, R. K., (2008). Probiotics ameliorate the hydrogen peroxide-induced epithelial barrier disruption by a PKC- and MAP kinase-dependent mechanism. *American Journal of Physiology: Gastrointestinal and Liver Physiology, 294*(4), G1060–G1069.
58. Suzuki, T., Nishiyama, K., Kawata, K., Sugimoto, K., Isome, M., Suzuki, S., Suzutani, T., (2020). Effect of the *Lactococcus lactis* 11/19-b1 strain on atopic dermatitis in a clinical test and mouse model. *Nutrients, 12*(3), 763.
59. Turnbaugh, P. J., (2017). Microbes and diet-induced obesity: Fast, cheap, and out of control. *Cell Host and Microbe, 21*(3), 278–281.
60. Zommiti, M., Almohammed, H., & Ferchichi, M., (2016). Purification and characterization of a novel anti-campylobacter bacteriocin produced by *Lactobacillus curvatus* DN317. *Probiotics and Antimicrobial Proteins, 8*(4), 191–201.

CHAPTER 11

STRATEGIES TO REDUCE CHOLESTEROL IN DAIRY FOODS

APURBA GIRI, MONALISA ROY, and N. VEENA

ABSTRACT

Cholesterol intake has been shown to have a positive impact on the blood cholesterol level in our body, and it increases the possibility of coronary heart diseases (CHD). Recently, the demand for low cholesterol dairy products has been increasing rapidly because of health-conscious consumers. Therefore, many attempts have been taken to reduce cholesterol from milk and milk products. There are several physical (molecular distillation, steam stripping, supercritical carbon dioxide extraction (SCE)), chemical (solvent extraction, adsorption or complexation with saponin, β-cyclodextrin and reaction with cyclic anhydrides), and biological processes (use of cholesterol degrading organisms and enzymes) to reduce cholesterol from milk products. Among these methods, complexation with β-cyclodextrin, distillation, and SCE are more preferred with more than 90% reduction of cholesterol.

11.1 INTRODUCTION

The milk is a complete food, which is rich in all essential nutrients that are needed for human growth and activities. Along with other nutrients, milk lipids are the most essential nutrient of milk and milk products. They have not only of pleasant sensory characteristics but also attached with some valuable food components. Fats and oils from other food sources take time

Functional Dairy Ingredients and Nutraceuticals: Physicochemical, Technological, and Therapeutic Aspects. Megh R. Goyal, N. Veena, PhD & Santosh K. Mishra (Eds.)

to digest completely, whereas milk fat is digested easily. Milk fat also has some therapeutic properties for its constituents (phospholipids, sphingomyelin, β-carotene and CLA). Therefore, the vegetarians must include milk and dairy products in their daily diet. However, the presence of cholesterol and saturated fatty acid in fat rich milk products enhance the possibility of coronary heart disease (CHD) [63]. Therefore, the present tendency is to avoid these products to reduce CHD.

Many researchers have indicated that there is a well-established link between blood cholesterol and incidence of CHD. Approximately 0.25–0.40% cholesterol is present in milk fat. Intake of ghee, butter, and fatty milk products creates a significant involvement in total cholesterol consumption. Total cholesterol level less than 200 mg/dL is considered as desirable concentration for adults. Besides, several studies have suggested that an oxidized cholesterol product shows mutagenic, carcinogenic, and atherogenic effects [58]. Due to this the tendency of consumption of fat-rich dairy products has been decreasing rapidly. Particularly, the urban and the educated societies are much conscious about the total cholesterol intake through their diet. Dairy products are mainly consumed by this sector of the community. It was suggested that hypercholesterolemic individuals must avoid cholesterol containing foods. However, the recent trend is to develop reduced cholesterol or cholesterol-free dairy products, and as a result, it has increased the market of dairy products that contain vegetable fats or other dairy products from which milk fat is replaced by other non-fat ingredients.

Several methods have been developed to remove the cholesterol from dairy products due to the adverse effects of cholesterol on health of the human beings. This chapter explores several approaches to reduce cholesterol in dairy products. There are several physical (molecular distillation, steam distillation, SCE), chemical (solvent extraction, adsorption or complexation with saponin, β-cyclodextrin, activated carbon, etc., and reaction with cyclic anhydride) and biological methods (use of cholesterol degrading organisms and enzymes) that have been discussed in detail in this chapter.

11.2 PROPERTIES OF CHOLESTEROL

From the two Greek letters 'Chole' and 'Stereos,' the word cholesterol is derived that means biles and solid, respectively. This hydrophobic sterol is formed of a structure of four rings (Figure 11.1) with a molecular formula $C_{27}H_{46}O$ and molecular weight of 386.66. Cholesterol is a fatty material. It is obvious that it is soluble in organic solvents and warm alcohol and not soluble in water. The melting temperature for cholesterol is 148°C.

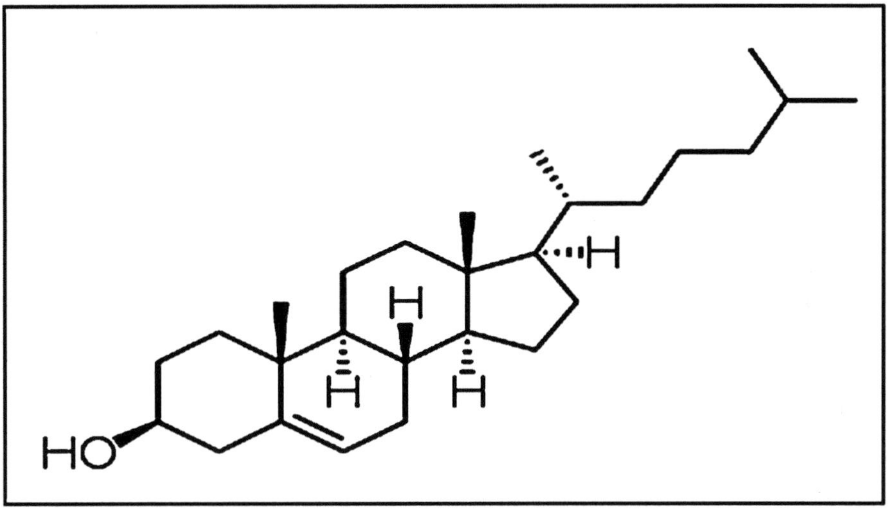

FIGURE 11.1 Structure of cholesterol.

Cholesterol has both hydrophobic and hydrophilic property and that's why it is used for the determination of membrane fluidity. This waxy material is an essential component of the cell membrane and is present in almost all cells of the body. It is principal sterol synthesized by animals. It can be found in some hormones and another part of the body. In the peripheral nerves and brain, cholesterol particularly takes part a crucial role for making of myelin sheaths. Cholesterol also modulates the absorption of fat in the intestine taken from the regular diet and helps in the biosynthesis of steroid hormones and bile acids. It acts as a precursor of vitamin D_3. The requirement for cholesterol in the body can synthesize by the body itself, hence, cholesterol through the diet is not needed very much. Excess amount of cholesterol in the blood can lead to various heart diseases.

11.3 BIOSYNTHESIS OF CHOLESTEROL

Every animal cell produces cholesterol. The rate of production varies with cell difference and organ's function. In liver and intestine, about 80% of cholesterol production occurs of total daily production. Higher rate of synthesis can also be obtained from reproductive organs and adrenal glands. There are 37 complex steps for the synthesis of cholesterol in each cell. The process starts from mevalonate or 3-hydroxy-3-methylglutaryl-CoA (HMG-CoA) reductase pathway carried out the first 18 steps followed by conversion of lanosterol to cholesterol by additional 19 steps [19]:

- Synthesis of cholesterol in the body starts with condensation of two molecules of acetyl Co-A to form an acetoacetyl Co-A followed by 2nd condensation, where acetyl Co-A and acetoacetyl Co-A combined to form HMG-CoA.
- Then, it is reduced by HMG-CoA reductase using NADPH to mevalonate. This step is known as the step for statins, e.g., cholesterol-lowering drug. HMG-CoA reductase is inhibited by statins. Two phosphorylation and one carboxylation steps (require ATP) are involved to convert mevalonate into isopentyl pyrophosphate (IPP).
- In the next step, by the action of geranyl-transferase three molecules of IPP compresses and forms farnesyl pyrophosphate (FPP).
- Then in the endoplasmic reticulum, two fragments of FPP condensed to form squalene. Squalene synthase catalyzes the reaction.
- After squalene formation, oxido-squelene cyclase converts the squalene to lanosterol.
- Further, conversion of lanosterol into cholesterol by additional 19 steps [49].

11.4 TRANSPORTATION OF CHOLESTEROL IN OUR BODY

Cholesterol is a part of lipid molecule and it is least soluble in water. That is why it dissolves in blood at extremely small concentrations. To facilitate the transport through the blood system, cholesterol is packaged within lipoproteins, which are complex particles with external amphiphilic lipids and proteins, whose inward-facing areas are soluble in lipid and outward surfaces are soluble in water and this mechanism permit it to pass through the blood by emulsification [47]. Unbound cholesterol transports in the monolayer area of the lipoprotein molecule containing proteins and phospholipids. The esters of cholesterol and triglycerides are transported within the hydrophobic core.

In the blood, different types of lipoproteins are present. They are chylomicrons, very-low-density lipoprotein (VLDL), low-density lipoprotein (LDL) and high-density lipoprotein (HDL), etc. The density of the lipoprotein depends on its lipid-protein ratio. About 85% triglycerides are present in chylomicron. Besides, it also contains absorbed cholesterol in free or esterified form. VLDL contains approximately 50% of triglycerides and a considerable amount of cholesterol as cholesterol esters. About 15% of the total cholesterol is transported by VLDL in the blood. Approximately 60% of the total blood cholesterol level accounts for LDL and that's why it is known as bad cholesterol.

LDL increases the buildup of plaque by depositing the cholesterol in artery wall, whereas, HDL gathers cholesterol from peripheral cells, and transports to liver and then excretes out. HDL contains only about 20% of the total cholesterol level in blood. There are certain saturated fatty acids like myristic acid and palmitic acid mostly responsible for rising cholesterol level in blood and stearic acid has very less effect on the total blood cholesterol level [57].

11.5 SOURCES OF CHOLESTEROL

In the human body, 80% of total blood cholesterol is synthesized by the liver, whereas 20% comes from the regular diet [3]. Owing to the endogenous synthesis, cholesterol is not considered as indispensable dietary ingredient. Cholesterol synthesis process mainly occurs in the liver and intestine at the rate of 11 mg/ kg body weight/day. Every animal is capable for the synthesis of cholesterol in varying amounts that is why the cholesterol content also varied in the food products made from them.

The important dietary sources of cholesterol are milk and milk products, fish oil, red meat, liver, kidney, egg, etc. Among these, chicken egg contains greater amount of cholesterol (i.e., about 215 mg/egg). Cholesterol accounts for about 0.25 to 0.40% of total fat in milk (Table 11.1), which is mainly associated with fat globules membrane (80%) while small portion in skim milk phase (20%) [64].

TABLE 11.1 Cholesterol Content in Some Milk and Milk Products [64]

Name of Product	Fat (g/100 g)	Cholesterol (mg/100 g of fat)
Butter	82.3	241
Cream (UHT)	34.9	101
Ghee	99.8	310
Pasteurized cow milk	4.0	15
Plain yogurt	3.6	11
Whole milk powder	26.2	79
Cheese		
Camembert	23.7	71
Emmental	33.8	101
Fresh cheese	18.5	54
Gorgonzola	31.2	94
Gruyere	32.0	96
Mascarpone	47.0	141
Parmesan	25.6	77
Quark, cream	15.6	52

11.6 FACTORS AFFECTING THE CONTENT OF CHOLESTEROL

1. **Species or Breed:** Cholesterol content in buffalo ghee (0.267%) is lower than the cow ghee (0.31%). Cow milk (2% fat) contains 0.237% total sterols of fat. Goat ghee contains higher amount of cholesterol (0.236% of fat) than cow ghee (0.230% of fat) and buffalo ghee (0.196% of fat) [59].
2. **Lactation Stage:** During the late phase of lactation cholesterol content increases significantly. Ghee made from old animals contains high level of cholesterol than the young one [71].
3. **Season:** This has an important role on cholesterol level of milk fat. In the spring season, its level in cow milk is 0.24–0.29% of milk lipid and in summer it is 0.18–0.25% of milk lipid. Cholesterol content of Ghee is lower in summer than in winter (0.291 vs 0.301% of fat).

Krzyzewski [38] has showed in their study that cholesterol content significantly decreases during winter that is approximately 16%.

11.7 METHODS TO REDUCE CHOLESTEROL IN DAIRY FOODS

Since milk and milk products contain a considerable amount of cholesterol, several following cholesterol removal approaches (such as: physical, chemical, and biological processes) have been developed to produce cholesterol-reduced milk and milk products.

11.7.1 SUPERCRITICAL CO_2 EXTRACTION (SCE)

SCE under moderate temperature and elevated pressure has been used for removal of cholesterol from milk and milk products. By this extraction method, two fractions of milk fat is obtained at 80°C and 200 bars pressure, in which liquid fraction is rich in cholesterol [32]. Bench-scale with ascending pressure profile [16] and multistage SCE [8] were used to achieve about 90% removal of cholesterol from milk fat. In another study for removal of cholesterol from anhydrous milk fat (AMF), high solubility conditions of AMF (40 MPa at 70°C) have been achieved using supercritical CO_2 for rapid extraction. In addition, the mixture of AMF and CO_2 is further passed through a high-pressure column, which is packed with silica gel and results indicates that 97% of the cholesterol can be removed from extracted AMF [29]. In a study of removal of cholesterol from milk fat by using SCE (320 bars at 50°C) system equipped with a silica gel column and it was found that more than 99% of cholesterol can be removed [31].

Ghosh et al. [23] optimized the process variables for extraction of cholesterol from cream powder using SCE method. It was found that at 204 bars, 75°C and 3.5 hours results in 39% cholesterol and 10.6% fat removal from the cream powder. SCE method for removal of cholesterol from whole milk powder (WMP) was studied by Chitra et al. [18]. In this method, 55.8% cholesterol was removed from powder with the optimal conditions of 207 bar pressure at 68°C for 40 min. In another study, about 46% cholesterol reduction was achieved from WMP using SCE using ethanol as co-solvent at 17 MPa, 48°C and 80 min dynamic time [21].

Supercritical carbon dioxide extraction coupled with adsorption on alumina was applied for 96% removal of cholesterol from butter oil [51]. Supercritical ethane extraction process coupled with adsorption on alumina was used to remove the cholesterol from butter oil [52]. In this study, the level of cholesterol was restricted to about 3% of that in the original product and it also suggested that the solubility of cholesterol and butter oil in ethane was greater than that of supercritical carbon dioxide. It has been reported that because of huge investment and substantial running costs associated with this process avoids the use of this extraction solely for fractionation and no commercial success.

11.7.2 STEAM DISTILLATION

The process for removal of cholesterol by this technique is almost similar to the method of removal of unsaponifiable matter and odorous compounds from vegetable oil. To remove cholesterol by this method, at first deaeration of fat is done under vacuum followed by heating the fat material with steam up to 232°C. Further, it is treated with steam at lower pressure in tall cylindrical chamber. To increase the stripping efficiency, the AMF is passed through a series of plates, in which fat is distributed in many thin layers. When the steam rises, it carries the evaporated sterol which has to be condensed and collected with other volatile. General Mills Inc. (Minneapolis, MN) applied this technique to remove cholesterol from butter oil. The Omega Source Corporation (Burnsville, MN) made a pilot plant, which produced 109.1 kg/d of de-cholesterolized AMF. It reduces 93% cholesterol [14]. This process consumes higher energy, which leads to greater operational cost and possibility of polymerization and oxidation of various components. This process also removes some of the flavoring compounds [62].

11.7.3 MOLECULAR DISTILLATION

Due to the low volatility of sterol compounds (especially cholesterol) than triglycerides of milk lipids is exploited in this process for removal of cholesterol. Lanzani et al. [44] carried out the short-path molecular distillation process under vacuum (10^{-4} torr) for removal of cholesterol from anhydrous butter at a temperature of 185°C. This method can efficiently remove the cholesterol with loss of a small amount of low molecular mass triacylglycerols. Arul et al. [9] applied this process to fractionate the milk lipid into three (liquid, intermediate, and solid) fractions (based on melting point) and measured the cholesterol content of each fraction. In this method, the highest level of cholesterol was reported in the liquid (80%) fraction in comparison to the other fractions. Due to the use of high heat in this process, the resulted product quality is adversely affected. This technology can be applied to reduce cholesterol from milk and milk product.

11.7.4 SOLVENT EXTRACTION

Organic solvents like methanol is used to extract the cholesterol either in a single- or multi-step batch process followed by deaeration with steam under vacuum. In this method, butter oil is combined with ethanol and propane in a mixing chamber. The fatty-solvent mixture is fed to the extraction column. A blend of ethanol, water, and a small amount of propane can be used as an extractant. The solution is divided into two phases in the extraction column. The fat and cholesterol are in the upper phase whereas butterfat and cholesterol mixed with extractant (ethanol, water, and propane) are in the bottom phase. By the next processing step, the upper phase is separated from the column. Approximately 90–95% cholesterol is removed by this method operated at 10 bar pressure and 30°C [20]. These methods are of less commercial interest because of problems linked to the use of organic solvents in the processing of food.

11.7.5 REACTION WITH CYCLIC ANHYDRIDE

Cyclic anhydride (like succinic anhydride) may be used to reduce cholesterol from milk lipid. In this method, succinic anhydride reacted with the hydroxyl group of cholesterol and converted cholesterol into acid derivatives, which can be easily removed from milk fat by aqueous alkali extraction [25]. Acetic acid is added during this reaction because it primarily catalyzes the rate of reaction and on the other hand it decreases the tendency to form distillate of cyclic anhydride from the mixture. Approximately 50% cholesterol is removed from animal fat along with it α-tocopherol and flavor compounds (lactones) also get removed.

11.7.6 ADSORPTION METHOD

Activated carbon, β-cyclodextrin, diatomaceous earth, saponins, etc., are used to remove cholesterol from milk fat by adsorption techniques. Normally adsorbent material is brought in contact with liquid milk fat either in a batch or continuous mode.

11.7.6.1 β-CYCLODEXTRIN

β-Cyclodextrin is a cyclic oligosaccharide. The seven glucose molecules linked together by the bond of α-1,4 linkage to form *β*-cyclodextrin. Its formation is catalyzed by cyclodextrin glycosyltransferase on hydrolysis of starch syrup. It is a chemically stable, non-hygroscopic, edible, non-toxic, non-digestible, and easily separable. Because of these properties, it has greater application in cholesterol reduction in milk and animal products [10].

The structure of *β*-cyclodextrin (i.e., torous like structure) gives the affinity to bind with cholesterol because of central cavity of *β*-cyclodextrin is hydrophobic in nature and it can bind with any non-polar molecule such as cholesterol [19]. In this process, melted fat is mixed with *β*-cyclodextrin solution (aqueous) and when adsorbent forms complex with cholesterol, it is then centrifuged to remove cholesterol from milk fat. The effectiveness of cholesterol adsorption is reliant on concentration of *β*-cyclodextrin, mixing time, speed, temperature, and centrifugation speed [64]. It is 'generally recognized as safe' (GRAS) status in the USA and recommended maximum of 5 mg *β*-cyclodextrin/kg/day in foods by Joint FAO/WHO Expert Committee on Food Additives (JECFA).

Several studies have been carried out to demonstrate the viability of β-cyclodextrin as excellent substrate for reduction of cholesterol from milk [7, 27, 33, 39, 45] and milk products [1, 4–7, 13, 22, 26, 34, 35, 40–43, 55]. The method of manufacturing reduced cholesterol dairy product using *β*-cyclodextrin has been patented by many investigators [24, 30, 40]. Lee et al. [45] reported that the addition of 1.5% of *β*-cyclodextrin in homogenized milk (3.6% fat) results in 95.3% reduction in cholesterol at a mixing temperature of 10°C for 10 min.

Alonso et al. [7] suggested that maximum cholesterol removal was achieved when milk treated with 0.6% *β*-cyclodextrin at 4°C for 20 min. In another study, raw, and homogenized milk with different fat levels (1, 2.5, and 3%) were blended with various amount of *β*-cyclodextrin (0.5, 1 and 1.5%) at two mixing temperatures of 8 and 20°C [48]. Results indicated that in homogenized milk, cholesterol content declined considerably with increased level of *β*-cyclodextrin at 20°C and maximum cholesterol reduction was obtained at 1% concentration of adsorbent. Some milk constituents such as fat, protein, and lactose content also decreased during this process [48].

Crosslinked *β*-cyclodextrin made by adipic acid was used at 1% concentration in homogenized milk for removal of cholesterol over 90% at 5°C for 10 min mixing with speed of 400 rpm [27]. *β*-Cyclodextrin immobilized on glass surface was also used successfully with efficient recycling ability for removal of cholesterol from milk reported by several researchers [39, 68–70].

A process of removal of cholesterol from the butter using *β*-cyclodextrin was patented by Soni [65]. Alonso et al. [5] developed a process for the manufacture of commercial reduced cholesterol butter. For that 0.6% *β*-cyclodextrin was added in pasteurized whole milk. Further cream was separated and churned to make reduced cholesterol butter. The obtained butter contained 95.5% low in cholesterol and contains only 0.04% *β*-cyclodextrin. Alonso et al. [4] also described the scale-up method for the development of low cholesterol butter from milk treated with *β*-cyclodextrin. Results indicated that more than 90% reduction in cholesterol than present in original milk lipids while no effect on compositional and sensory attributes of the prepared butter. Crosslinked *β*-cyclodextrin made from adipic acid was used to remove about 90% cholesterol from butter [34].

Nadeem et al. [55] removed cholesterol from yogurt and ice cream at a rate of 76% and 60%, respectively. It has been reported that overrun of ice cream decreased significantly due to cholesterol removal and fatty acid modification. However, no adverse effect on the sensory attributes of these products. Tahir et al. [68] reduced 78.8% cholesterol from ghee (at 25°C in 2 h and 170 rpm) using *β*-cyclodextrin which is immobilized on glass surface.

β-Cyclodextrin and crosslinked *β*-cyclodextrin have been used for preparing a variety of cheeses from cholesterol reduced milk. A study reported that approximately 97.6% cholesterol reduction was achieved in Manchego cheese prepared from pasteurized ewe's milk while no change in its physicochemical properties [6]. Kim et al. [35] reported that 82.6% and 92% reduction in cholesterol was observed in cream cheese treated with crosslinked *β*-cyclodextrin and powder *β*-cyclodextrin, respectively without significant adverse effects on sensory and chemical properties. Low cholesterol mozzarella cheese was made by treating homogenized milk (70 kg/cm^2 at 70°C) with 1% *β*-cyclodextrin. The result shows that 63.92% reduction in cholesterol and cheese was found to have inferior texture but no effect on flavor and appearance [42].

This process has been commercialized because of less investment, simple, and reduction in flavor of milk fat is comparatively lesser than the distillation method. Nevertheless, huge amount of adsorbent is required because of ineffective recovery from dairy products and consequently high production cost [54].

11.7.6.2 SAPONINS

Cholesterol from dairy products can be removed effectively by using saponins has been reported in many research studies. Saponins (naturally occurring plant derived glycosides) and polymer-supported saponins (digitonin and tomatine) [50] binds with cholesterol to form a complex which is not soluble in an aqueous system and further it can be separated by filtration or centrifugation. Saponins have received GRAS status in Unites States.

Chang et al. [17] treated homogenized milk with saponin (1.5%) and celite (0.95%) for a reaction time of 30 min results in removal of cholesterol by 73.7%. Oh et al. [56] reported that when milk was treated with saponin (1.5%) at 45°C for 30 min, 70.5% cholesterol reduction can be achieved. Further addition of 0.25% celite increased the cholesterol removal to 72%.

Richardson et al. [61] patented the process for the removal of cholesterol from dairy products by using diatomaceous earth and saponin. They recommended that the method can be applicable for milk (raw and pasteurized), cream, and AMF. Saponins from *Quillaja saponaria* [66, 67] and polymer-supported digitonin [50] were used to reduce cholesterol from butter oil has been reported. In one of the studies, 90% and 80% cholesterol reduction can be achieved in AMF and cream, respectively when they were treated with saponin at 65°C for one hour [60].

11.7.7 ENZYMATIC METHOD

In this method, enzymes produced from several bacteria are used to convert cholesterol into some other products [15]. An enzyme such as cholesterol reductase is used for the conversion of cholesterol to coprosterol that is biologically inactive, nonabsorbable, and nontoxic product, which is generally not absorbed or poorly absorbed by the human body [46]. This process seems to be very complex and consequently high production cost [54]. Another enzyme, cholesterol oxidase, which oxidizes cholesterol to non-steroid compounds (cholest-4-en-3-one). However, these oxidized products are toxic and need to demonstrate the safety of reaction products. There is no such application in industrial scale.

11.7.8 USE OF MICROORGANISMS

Cholesterol degrading bacteria are used in foods to reduce cholesterol level. Such bacterial species include Nocardia, Eubacterium, Rhodococcus, Lactobacillus, and Enterococci, which have been isolated from several foods. *Lb. casei* [37], *Lb. plantarum*, *Lb. paracasei* [12], *L. lactis* [36], and *E. faecium* [11, 28] are some of lactic acid bacterial (LAB) strains, which possess cholesterol degrading ability. Of the 58 strains of LAB, seven were checked for their cholesterol-lowering ability in broth and cheese. In broth, 42 to 55% reduction in cholesterol and up to 23% reduction in cholesterol during ripening was observed when *Lb. paracasei* ssp. *paracasei* and *E. lactis* were added during cheese making with no adverse effect on the sensory attributes of the product [2]. Assimilation of cholesterol in milk by kefir grains was reported by Vujičić et al. [72]. Original level of cholesterol (12.8 mg/dL) in milk was decreased to 10.8 and 5.3 mg/dL after 24- and 48-hours incubation of milk by kefir grains, respectively.

11.8 SUMMARY

Many studies have proven the fact that there is well established link between increased blood cholesterol level and incidence of CHD and atherosclerosis. Many animal and clinical trials have revealed that blood cholesterol is elevated by an increased consumption of dietary cholesterol and saturated fat. One of the simplest ways of lowering the bodily cholesterol is reducing the intake of cholesterol in daily diet. Nowadays consumers are also more aware of cholesterol intake and its possible health risk. Due to this there is an increasing market demand for zero-, low-, or reduced-cholesterol products. Several physical, chemical, and biological methods are available for removal of cholesterol from dairy products. Amongst these, distillation, SCE, and adsorption with β-cyclodextrin methods could effectively remove the cholesterol from dairy products while no or negligible effect on sensory and physicochemical properties. The development of such low cholesterol dairy products is not only useful as a food to restrict cholesterol intake but also the consumers will be benefited by its nutritional profile.

KEYWORDS

- **biological process**
- **chemical process**
- **cholesterol**
- **coronary heart disease**
- **dairy products**
- **physical process**

REFERENCES

1. Ahn, J., & Kwak, H. S., (1999). Optimizing cholesterol removal in cream using *β*-cyclodextrin and response surface methodology. *Journal of Food Science, 64*(4), 629–632.
2. Albano, C., Morandi, S., Silvetti, T., Casiraghi, M. C., Manini, F., & Brasca, M., (2018). Lactic acid bacteria with cholesterol-lowering properties for dairy applications: *In vitro* and *in situ* activities. *Journal of Dairy Science, 101*(12), 10807–10818.
3. Allred, J. B., (1993). Lowering serum cholesterol: Who benefits? *The Journal of Nutrition, 123*(8), 1453–1459.
4. Alonso, L., Calvo, M. V., & Fontecha, J., (2019). A scale-up process for the manufacture of reduced-cholesterol butter using beta-cyclodextrin. *Journal of Food Process Engineering, 42*(3), e13009.
5. Alonso, L., Fontecha, J., Cuesta, P., Juárez, M., & Gilliland, S. E., (2010). Industrial application of β-cyclodextrin for manufacturing low cholesterol butter. *Milchwissenschaft, 65*(1), 36–38.
6. Alonso, L., Fox, P. F., Calvo, M. V., & Fontecha, J., (2018). Effect of beta-cyclodextrin on the reduction of cholesterol in ewe's milk Manchego cheese. *Molecules, 23*(7), 1789.
7. Alonso, L., Cuesta, P., Fontecha, J., Juarez, M., & Gilliland, S. E., (2009). Use of beta-cyclodextrin to decrease the level of cholesterol in milk fat. *Journal of Dairy Science, 92*(3), 863–869.
8. Anonymous, (1989). Supercritical extraction holds promise for cholesterol-free butter. *Food Engineering, 61*(2), 83–86.
9. Arul, J., Boudreau, A., Makhlouf, J., Tardif, R., & Grenier, B., (1988). Distribution of cholesterol in milk fat fractions. *The Journal of Dairy Research, 55*(3), 361–371.
10. Astray, G., Gonzalez-Barreiro, C., Mejuto, J. C., Rial-Otero, R., & Simal-Gándara, J., (2009). A review on the use of cyclodextrins in foods. *Food Hydrocolloids, 23*(7), 1631–1640.
11. Ayyash, M., Abushelaibi, A., Al-Mahadin, S., Enan, M., El-Tarabily, K., & Shah, N., (2018). *In vitro* investigation into probiotic characterization of streptococcus and enterococcus isolated from camel milk. *Lebensmittel-Wissenschaft & Technologie (Food Science & Technology), 87*, 478–487.

12. Belviso, S., Giordano, M., Dolci, P., & Zeppa, G., (2009). *In vitro* cholesterol-lowering activity of *Lactobacillus Plantarum* and *Lactobacillus Paracasei* strains isolated from the Italian castelmagno PDO cheese. *Dairy Science & Technology, 89*, 169–176.
13. Bhatia, P., Sharma, V., Arora, S., & Rao, P. S., (2019). Effect of cholesterol removal on compositional and the physicochemical characteristics of anhydrous cow milk fat (cow ghee). *International Journal of Food Properties, 22*(1), 1–8.
14. Boudreau, A., & Arul, J., (1993). Cholesterol reduction and fat fractionation technologies for milk fat: An overview. *Journal of Dairy Science, 76*(6), 1772–1781.
15. Boudreau, A., & Arul, J., (1991). Physical and chemical modification of milk fat: Fractionation. In: *Monograph on Utilizations of Milk Fat* (pp. 7–15). Bulletin, No. 260; Brussels: International Dairy Federation.
16. Bradley, R. L. Jr., (1989). Removal of cholesterol from milk fat using supercritical carbon dioxide. *Journal of Dairy Science, 72*(10), 2834–2840.
17. Chang, E. J., Oh, H. I., & Kwak, H. S., (2001). Optimization of cholesterol removal conditions from homogenized milk by treatment with saponin. *Asian-Australasian Journal of Animal Sciences, 14*(6), 844–849.
18. Chitra, J., Deb, S., & Mishra, H. N., (2015). Selective fractionation of cholesterol from whole milk powder: Optimization of supercritical process conditions. *International Journal of Food Science & Technology, 50*(11), 2467–2474.
19. Courregelongue, J., & Maffrand, J. P., (1989). *Process for Eliminating Cholesterol Contained in a Fatty Substance of Animal Origin and the Fatty Substance with Reduced Cholesterol Obtained.* United States Patent No. 4, 880, 573.
20. Czech, B., Peter, S., & Weidner, E., (1993). Effective removal of cholesterol from butterfat. *Scandinavian Dairy Information, 7*(4), 56–58.
21. Dey, P. I., Jayakumar, C., & Niwas, M. H., (2016). Optimization of process parameters for supercritical fluid extraction of cholesterol from whole milk powder using ethanol as co-solvent. *Journal of the Science of Food and Agriculture, 96*(15), 4885–4895.
22. Dias, H. M. A. M., Berbicz, F., Pedrochi, F., Baesso, M. L., & Matioli, G., (2010). Butter cholesterol removal using different complexation methods with beta-cyclodextrin, and the contribution of photoacoustic spectroscopy to the evaluation of the complex. *Food Research International, 43*(4), 1104–1110.
23. Ghosh, M., Srivastava, C. J., & Mishra, H. N., (2018). Advent of clean and green technology for preparation of low-cholesterol dairy cream powder: Supercritical fluid extraction process. *Food Quality and Safety, 2*(4), 205–211.
24. Graille, J., Pioch, D., Serpelloni, M., & Mentink, L., (1993). *Process for Preparing Dairy Products with a Low Content of Sterols, Particularly of Cholesterol* (p. 14). US Patent No. 5,264,226.; Washington, DC.
25. Gu, Y. F., Chen, Y., & Hammond, E. G., (1994). Use of cyclic anhydrides to remove cholesterol and other hydroxyl compounds from animal fats and oils. *Journal of the American Oil Chemists' Society, 71*, 1205–1209.
26. Ha, H. J., Lee, J. E., Chang, Y. H., & Kwak, H. S., (2010). Entrapment of nutrients during cholesterol removal from cream by cross linked *β*-cyclodextrin. *International Journal of Dairy Technology, 63*(1), 119–126.
27. Han, E. M., Kim, S. H., Ahn, J., & Kwak, H. S., (2005). Cholesterol removal from homogenized milk with crosslinked β-cyclodextrin by adipic Acid. *Asian-Australasian Journal of Animal Sciences, 18*(12), 1794–1799.

28. Hlivak, P., Odraska, J., Ferencik, M., Ebringer, L., Jahnova, E., & Mikes, Z., (2005). One-year application of probiotic strain Enterococcus faecium m-74 decreases serum cholesterol levels. *Bratislavske Lekarske Listy (Slovak), 106*(2), 67–72.
29. Huber, W., Molero, A., Pereyra, C., & Martinez De La, O. E., (1996). Dynamic supercritical carbon dioxide extraction for removal of cholesterol from anhydrous milk fat. *International Journal of Food Science & Technology, 31*(2), 143–151.
30. Hubert, R., Daniel, D., & Robert, B., (1993). *Process for Reducing the Content of Cholesterol and of Free Fatty Acids in an Animal Fat* (p. 17). Washington, DC: US Patent No. 5232725-A.
31. Karkare, V., & Alkio, M., (1993). Removal of cholesterol during milk fat fractionation by supercritical carbon dioxide. *Agricultural Science in Finland, 2*(5), 387–393.
32. Kaufmann, W., Biernoth, G., Frede, E., Merk, W., Precht, D., & Timmens, W., (1982). Fractionation of butter fat by extraction with supercritical carbon dioxide. *Milchwissenschaft (Dairy Science), 37*(2), 92–96.
33. Kim, S. H., Ahn, J., & Kwak, H. S., (2004). Crosslinking of beta-cyclodextrin on cholesterol removal from milk. *Archives of Pharmacal Research, 27*(11), 1183–1187.
34. Kim, J. H., Jung, T. H., Ahn, J., & Kwak, H. S., (2006). Properties of cholesterol-reduced butter made with β-cyclodextrin and added evening primrose oil and phytosterols. *Journal of Dairy Science, 89*(12), 4503–4510.
35. Kim, S. H., Han, E. M., Ahn, J., & Kwak, H. S., (2005). Effect of crosslinked *β*-cyclodextrin on quality of cholesterol-reduced cream cheese. *Asian-Australasian Journal of Animal Sciences, 18*(4), 584–589.
36. Kimoto, H., Ohmomo, S., & Okamoto, T., (2002). Cholesterol removal from media by lactococci. *Journal of Dairy Science, 85*(12), 3182–3188.
37. Klaver, F. A., & Van, D. M. R., (1993). The assumed assimilation of cholesterol by *Lactobacilli* and *Bifidobacterium Bifidum* is due to their bile salt-deconjugating activity. *Applied and Environmental Microbiology, 59*(4), 1120–1124.
38. Krzyzewski, J., Strzakowska, N., Jozwik, A., Bagnicka, E., & Ryniewicz, Z., (2003). Effect of nutrition and season on cholesterol level in milk of Holstein-Friesian cows. *Annals Animal Science, 3*(2), 45–49.
39. Kwak, H. S., Kim, S. H., Kim, J. H., Choi, H. J., & Kang, J., (2004). Immobilized *β*-cyclodextrin as a simple and recyclable method for cholesterol removal in milk. *Archives of Pharmacal Research, 27*(8), 873–877.
40. Kwak, H. S., Ahn, J. J., & Lee, D. K., (2000). *Method for Removing Cholesterol from Milk and Cream* (p. 10). Washington, DC: US Patent No. 6110517.
41. Kwak, H. S., Jung, C. S., Shim, S. Y., & Ahn, J., (2002). Removal of cholesterol from cheddar cheese by β-cyclodextrin. *Journal of Agricultural and Food Chemistry, 50*(25), 7293–7298.
42. Kwak, H. S., Nam, C. G., & Ahn, J., (2001). Low cholesterol mozzarella cheese obtained from homogenized and β-cyclodextrin-treated milk. *Asian-Australasian Journal of Animal Sciences, 14*(2), 268–275.
43. Kwak, H. S., Sun, H. M., Ahn, J., & Kwon, H. J., (2001). Optimization of β-CD recycling process for cholesterol removal in cream. *Asian-Australasian Journal of Animal Sciences, 14*(4), 548–552.
44. Lanzani, A., Bondioli, P., Mariani, C., Folegatti, L., Venturini, S., Fedeli, E., & Barreteau, P., (1994). A new short-path distillation system applied to the reduction of cholesterol in butter and lard. *Journal of the American Oil Chemists' Society, 71*(6), 609–614.

45. Lee, D. K., Ahn, J., & Kwak, H. S., (1999). Cholesterol removal from homogenized milk with β-cyclodextrin. *Journal of Dairy Science, 82*(11), 2327–2330.
46. Macdonald, I. A., Bokkenheuser, V. D., Winter, J., McLernon, A. M., & Mosbach, E. H., (1983). Degradation of steroids in human gut. *Journal of Lipid Research, 24*(6), 675–700.
47. Mahley, R. W., & Innerarity, T. L., (1983). Lipoprotein receptors and cholesterol homeostasis. *Biochimica et Biophysica Acta (BBA), 737*(2), 197–222.
48. Maskooki, A. M., Beheshti, S. H. R., Valibeigi, S., & Feizi, J., (2013). Effect of cholesterol removal processing using β-cyclodextrin on main components of milk. *International Journal of Food Science, 2013*, 215305.
49. Mehta, S., (2013). *Biosynthesis and Regulation of Cholesterol (with Animation).* https://pharmaxchange.info/2013/09/biosynthesis-and-regulation-of-cholesterol-animation (accessed on 15 November 2021).
50. Micich, T. J., (1990). Behaviors of polymer-supported digitonin with cholesterol in the absence and presence of butter oil. *Journal of Agricultural and Food Chemistry, 38*(9), 1839–1843.
51. Mohamed, R. S., Neves, G. B. M., & Kieckbusch, T. G., (1998). Reduction in cholesterol and fractionation of butter oil using supercritical CO_2 with adsorption on alumina. *International Journal of Food Science & Technology, 33*(5), 445–454.
52. Mohamed, R. S., Saldaña, M. D. A., Socantaype, F. H., & Kieckbusch, T. G., (2000). Reduction in the cholesterol content of butter oil using supercritical ethane extraction and adsorption on alumina. *The Journal of Supercritical Fluids, 16*(3), 225–233.
53. Mohamed, R. S., & Mansoori, G. A., (2020). *The Use of Supercritical Fluid Extraction Technology in Food Processing* (Vol. 20, pp. 134–139). Food Technology Magazine.
54. Mortensen, B. K., (2011). Butter and other milk fat products: Modified butters. In: Fuquay, J. W., Fox, P. F., & McSweeney, P. L. H., (eds.), *Encyclopedia of Dairy Sciences* (2nd edn., Vol. 1, pp. 500–505). London-UK: Elsevier Ltd.
55. Nadeem, M., Ullah, R., Imran, M., & Arif, A. M., (2015). Effects on physicochemical characteristics of yogurt and ice cream with fatty acid modification and cholesterol removal. *Biological Sciences-PJSIR, 58*(1), 40–46.
56. Oh, H. I., Chang, E. J., & Kwak, H. S., (1998). Conditions of the removal of cholesterol from milk by treatment with saponin. *Korean Journal of Dairy Science, 20*(4), 253–260.
57. Ohlsson, L., (2010). Dairy products and plasma cholesterol levels. *Food & Nutrition Research, 54*(1), 5124–5132.
58. Orczewska-Dudek, S., Bederska-Lojewska, D., Pieszka, M., & Pietras, M. P., (2012). Cholesterol and lipid peroxides in animal products and health implications: A review. *Annals Animal Science, 12*(1), 25–52.
59. Pandya, A. J., & Ghodke, K. M., (2007). Goat and sheep milk products other than cheeses and yogurt. *Small Ruminant Research, 68*(1, 2), 193–206.
60. Riccomini, M. A., Wick, C., Peterson, A., Jimenez-Flores, R., & Richardson, T., (1990). Cholesterol removal from cream and anhydrous milk fat using saponins. *Journal of Dairy Science, 73*(1), 107–115.
61. Richardson, T., Calif, D., Jimenez, R., & Champaign, F., (1994). *Process to Remove Cholesterol from Dairy Products* (p. 11). Washington, DC: US Patent No. 5,326,579.
62. Schlimme, E., & Kiel, D., (1989). *Removal of Cholesterol from Milk Fat* (pp. 12–21). European Dairy Magazine.

63. Sieber, R., (1993). Cholesterol removal from animal food-can it be justified? *Lebensmittel-Wissenschaft und Technologie (Food Science & Technology), 26*(5), 375–387.
64. Sieber, R., Rehberger, B. S., & Walther, B., (2011). Removal of cholesterol from dairy products. In: Fuquay, J. W., Fox, P. F., & McSweeney, P. L. H., (eds.), *Encyclopedia of Dairy Sciences* (2nd edn., Vol. 3, pp. 734–740). London-UK: Elsevier Ltd.
65. Soni, K. N., (2012). *Low Cholesterol Butter and Process of Preparation* (p. 12). International Patent WO 2012/025931-A1.
66. Sundfeld, E., Yun, S., Krochta, J. M., & Richardson, T., (1993). Separation of cholesterol from butteroil using quillaja saponins, I: Effects of PH, contact time and adsorbent. *Journal of Food Process Engineering, 16*(3), 191–205.
67. Sundfeld, E., Krochta, J. M., & Richardson, T., (1993). Separation of cholesterol from butteroil using quillaja saponins, II: Effects of temperature, agitation, and concentration of quillaja solution. *Journal of Food Process Engineering, 16*(3), 207–227.
68. Tahir, M. N., Bokhari, S. A., & Adnan, A., (2015). Cholesterol extraction from ghee using glass beads functionalized with beta-cyclodextrin. *Journal of Food Science and Technology, 52*(2), 1040–1046.
69. Tahir, M. N., Kwon, C., Jeong, D., Cho, E., Paik, S. R., & Jung, S., (2013). Cholesterol Reduction from milk using β-cyclodextrin immobilized on glass. *Journal of Dairy Science, 96*(7), 4191–4196.
70. Tahir, M. N., & Lee, Y., (2013). Immobilization of β-cyclodextrin on glass: Characterization and application for cholesterol reduction from milk. *Food Chemistry, 139*(1–4), 475–481.
71. Tripathi, P. M., Ingole, S. D., Deshmukh, B. T., Nagvekar, A. S., & Bharucha, S. V., (2010). Serum lipid profile during lactation in buffalo. *Indian Journal of Animal Sciences, 80*(3), 217–219.
72. Vujičić, I. F., Vulić, M., & Könyves, T., (1992). Assimilation of cholesterol in milk by kefir cultures. *Biotechnology Letters, 14*, 847–850.

INDEX

G

I

J

K

L

M

O

P